AF900705

Natural Science in Archaeology

Series editors
Günther A. Wagner
Christopher E. Miller
Holger Schutkowski

More information about this series at http://www.springer.com/series/3703

Christoph Siart • Markus Forbriger •
Olaf Bubenzer
Editors

Digital Geoarchaeology

New Techniques for Interdisciplinary Human-Environmental Research

Springer

Editors
Christoph Siart
Institute of Geography
Physical Geography
Heidelberg University
Heidelberg, Germany

Markus Forbriger
Institute of Geography
Physical Geography
Heidelberg University
Heidelberg, Germany

Olaf Bubenzer
Institute of Geography
Physical Geography
Heidelberg University
Heidelberg, Germany

ISSN 1613-9712
Natural Science in Archaeology
ISBN 978-3-319-25314-5 ISBN 978-3-319-25316-9 (eBook)
DOI 10.1007/978-3-319-25316-9

Library of Congress Control Number: 2017956917

© Springer International Publishing AG 2018
Chapters 1, 2, 6, 8 and 14 are licensed under the terms of the Creative Commons Attribution 4.0 International License (http://creativecommons.org/licenses/by/4.0/). For further details see license information in the chapters.
This work is subject to copyright. All rights are reserved by the Publisher, whether the whole or part of the material is concerned, specifically the rights of translation, reprinting, reuse of illustrations, recitation, broadcasting, reproduction on microfilms or in any other physical way, and transmission or information storage and retrieval, electronic adaptation, computer software, or by similar or dissimilar methodology now known or hereafter developed.
The use of general descriptive names, registered names, trademarks, service marks, etc. in this publication does not imply, even in the absence of a specific statement, that such names are exempt from the relevant protective laws and regulations and therefore free for general use.
The publisher, the authors and the editors are safe to assume that the advice and information in this book are believed to be true and accurate at the date of publication. Neither the publisher nor the authors or the editors give a warranty, express or implied, with respect to the material contained herein or for any errors or omissions that may have been made. The publisher remains neutral with regard to jurisdictional claims in published maps and institutional affiliations.

Printed on acid-free paper

This Springer imprint is published by Springer Nature
The registered company is Springer International Publishing AG
The registered company address is: Gewerbestrasse 11, 6330 Cham, Switzerland

Preface

In this book, the term Digital Geoarchaeology is to be established as a novel approach of interdisciplinary collaboration between classical studies, geosciences and computer sciences. It can be regarded as an intersection of different scholarly concepts, which are combined for common benefits to unravel cultural-historical questions at the human–environmental interface.

In retrospect, the idea of this volume began with several rather spontaneous collaborations between archaeologists, historians, geoscientists and computer experts at Heidelberg University. Given this fruitful cooperation, it was our wish to bring together a broader scientific audience to promote the multilateral exchange, which ultimately led to an eponymous conference at the Heidelberg Academy of Sciences in the end of 2013. One of the symposium's major outcomes was the joint intention of compiling a book that presents an overview of some of the most illustrative, promising applications in the field of Digital Geoarchaeology and in turn encourages to seek a dialogue across disciplinary boundaries.

Accordingly, Digital Geoarchaeology helps bridging the gap between different subject areas by leveraging synergies in the study of past man–nature relationships. It should not be considered as another distinct discipline but rather seen as an auspicious way of common research that allows to better understand palaeoenvironmental interactions and helps to interpret them holistically from different viewpoints.

The different case studies in this volume presented by scholars from all over Europe clearly highlight the interdependencies and interrelationships between archaeology, geosciences and computer sciences, which, when considering the common scientific discourse, have not been fully explored yet. Above all, they underline that technologies developed and used by one particular discipline may frequently be combined with tools and methodologies from others for mutual benefit. No matter from which academic field a specific research question may be brought up, it is always the complementing subjects embedded in the framework of Digital Geoarchaeology that improve insight and assist in solving the overarching topic. As demonstrated by the examples in our book, this strategy easily outperforms mere monodisciplinary approaches.

Taken as a whole, the case studies at hand are to be considered as best practice and kind of idea generators, as the development of Digital

Geoarchaeology can indeed still be seen as a new phenomenon. Apart from that, this book is particularly designed as a teaching volume intended for arts scholars, geoscientists and computer scientists, who are interested in interdisciplinary research and seek insight into other disciplinary concepts and new methodologies. We particularly hope to attract the interest of young researchers and to encourage them to enter into a cross-disciplinary dialogue between subjects like archaeology, computer sciences, ecology, geography, geoinformatics, geology, history, etc. Besides offering incentives for collaboration, common fields of work are to be identified and discussed from different scientific perspectives in order to learn from each other, better understand past relationships between man and environment, protect cultural heritage, develop new ideas for common projects, increase knowledge transfer and, thus, benefit from synergy effects.

As a general introduction to this volume, the theoretical concept and framework of Digital Geoarchaeology are described in the first chapter, taking special account of the developments and shortcomings that can be observed from the current scientific discourse. Subsequently, the book is structured into four parts, based upon a division into major categories of the most promising applications and key technologies, respectively. Each part is introduced by a general thematic overview on the particular methodologies, followed by several case studies that vividly illustrate the broad spectrum of potential techniques and interdisciplinary research designs. Part I deals with topics whose focus is on information systems and spatial analysis, while Part II concentrates on research in the context of remote sensing and image digital image analysis. Part III gives an insight into laser scanning applications used for geoarchaeological questions. Last but not least, geophysical prospecting techniques used in multi-method approaches and combined with the aforementioned applications are presented in Part IV.

Needless to say, the selection of contributions is just exemplary and does not raise any claim to completeness. Certainly, there may be further innovative approaches, which could have been considered just as well. Hence, one of the main intentions of this book is to promote and disseminate the concept of Digital Geoarchaeology and thereby bring together more scholars from different disciplines interested in past human–environmental interactions. As can been seen from all papers included in this volume, bridging disciplines demands scholarly strength and, depending on the specific research question to be unravelled, a profound comprehension of the counterparties. Digital geoarchaeological research may therefore be a great challenge but offers promising future prospects as it contributes to a better understanding of ancient landscapes along with their forming processes.

In closing, it should not go unmentioned that this volume would not have been possible without the assistance and support of many colleagues and institutions. Most of all, we would like to express our sincere gratitude to Günther A. Wagner, who significantly encouraged us to pursue our idea and provided constant advice. Sincere thanks are due to the Heidelberg Academy of Sciences for funding the Digital Geoarchaeology conference in 2013, which substantially paved the way for this book. Special thanks go to all authors for their contributions and willingness to make this endeavour a success. We are also indebted to Max Kanig, who provided profound organizational assistance during the editing process. Finally, a special thank you is dedicated to Annett Büttner and the Springer team for letting us to take our idea to the next level and giving us the opportunity to publish it in the series Natural Science in Archaeology.

Heidelberg
November 2017

Christoph Siart
Markus Forbriger
Olaf Bubenzer

Contents

List of Contributors

Constantin Athanassas Centre de Recherche et d'Enseignement de Géosciences de l'Environnement (C.E.R.E.G.E.), Aix-en-Provence, France

Felix Bachofer Institute of Geography, University of Tuebingen, Tuebingen, Germany

Yannis Bassiakos Laboratory of Archaeometry, N.C.S.R. "Demokritos", Athens, Greece

Andreas Bolten Institute of Geography, University of Cologne, Cologne, Germany

Olaf Bubenzer Institute of Geography, Physical Geography, Heidelberg University, Heidelberg, Germany

Markus Forbriger Institute of Geography, Physical Geography, Heidelberg University, Heidelberg, Germany

Liane Giemsch Archäologisches Museum Frankfurt, Frankfurt am Main, Germany

Martin Hämmerle GIScience & 3D Spatial Data Processing Research Group, Institute of Geography, Heidelberg University, Heidelberg, Germany

Bernhard Höfle GIScience & 3D Spatial Data Processing Research Group, Institute of Geography, Heidelberg University, Heidelberg, Germany

Heidelberg Center for the Environment, Heidelberg, Germany

Stefan Hecht Institute of Geography, Heidelberg University, Heidelberg, Germany

Christine Hertler Heidelberg Academy of Sciences and Humanities, Tuebingen, Germany

Senckenberg Forschungsinstitut, Frankfurt am Main, Germany

Volker Hochschild Institute of Geography, University of Tuebingen, Tuebingen, Germany

Dirk Hoffmeister GIS & Remote Sensing, Institute of Geography, University of Cologne, Cologne, Germany

Ingmar Holzhauer Institute of Geography, Heidelberg University, Heidelberg, Germany

Tuna Kalayci GeoSat ReSeArch Lab, Foundation for Research and Technology, Hellas (F.O.R.T.H.), Rethymno, Crete, Greece

Daniel Knitter Department of Geography, Physical Geography—Landscape Ecology and Geoinformation, Christian-Albrechts-Universität zu Kiel, Kiel, Germany

Karsten Lambers Leiden University, Leiden, the Netherlands

Bertil Mächtle Institute of Geography, Heidelberg University, Heidelberg, Germany

Michael Maerker Heidelberg Academy of Sciences and Humanities, Tuebingen, Germany

Meropi Manataki GeoSat ReSeArch Lab, Foundation for Research and Technology, Hellas (F.O.R.T.H.), Rethymno, Crete, Greece

Ian Moffat GeoSat ReSeArch Lab, Foundation for Research and Technology, Hellas (F.O.R.T.H.), Rethymno, Crete, Greece

Oliver Nakoinz Institut für Ur- und Frühgeschichte, Christian-Albrechts-Universität, Kiel, Germany

Eleftheria Paliou Archäoinformatik/Computational Archaeology, Institute of Archaeology, University of Cologne, Cologne, Germany

Michelle Pfeiffer Junior Research Group Digital Humanities and Digital Cultural Heritage at the Interdisciplinary Center for Scientific Computing and Cluster of Excellence Asia & Europe in Global Context, Heidelberg University, Heidelberg, Germany

Geraldine Quénéhervé Institute of Geography, University of Tuebingen, Tuebingen, Germany

Karl Hjalte Maack Raun Junior Research Group Digital Humanities and Digital Cultural Heritage at the Interdisciplinary Center for Scientific Computing and Cluster of Excellence Asia & Europe in Global Context, Heidelberg University, Heidelberg, Germany

Heiko Riemer Institute of Prehistoric Archaeology, African Archaeology, University of Cologne, Cologne, Germany

Apostolos Sarris GeoSat ReSeArch Lab, Foundation for Research and Technology, Hellas (F.O.R.T.H.), Rethymno, Crete, Greece

Gerd Schukraft Institute of Geography, Heidelberg University, Heidelberg, Germany

Christian Seitz Research Group "Optimization, Robotics, and Biomechanics", Institute for Computer Engineering, Heidelberg University, Heidelberg, Germany

Christoph Siart Institute of Geography, Physical Geography, Heidelberg University, Heidelberg, Germany

Katerina Theodorakopoulou Department of History, Archaeology and Social Anthropology, University of Thessaly, Volos, Greece

Philip Verhagen Faculty of Humanities, Vrije Universiteit Amsterdam, Amsterdam, The Netherlands

Benjamin N. Vis Classical and Archaeological Studies (CLAS) and Kent Interdisciplinary Centre for Spatial Studies (KISS), School of European Culture and Languages, University of Kent, Canterbury, UK

Armin Volkmann Heidelberg University, Junior Research Group Digital Humanities and Cultural Heritage at the Interdisciplinary Center for Scientific Computing and Cluster of Excellence Asia & Europe in Global Context, Heidelberg, Germany

Günther A. Wagner Institute of Geography, Heidelberg University, Heidelberg, Germany

Digital Geoarchaeology: Bridging the Gap Between Archaeology, Geosciences and Computer Sciences

1

Christoph Siart, Markus Forbriger, and Olaf Bubenzer

Abstract

Modern archaeology increasingly crosses academic boundaries by combining different new methodologies in order to answer research questions about ancient cultures and their remains. Above all, the geosciences became an indispensable counterpart of archaeology and cultural heritage management. As to the investigation of past archaeological landscapes and palaeoenvironments, the term Geoarchaeology is commonly used, representing the utilization of traditional and the development of new geoscientific applications for archaeological purposes. In addition, computationally engaged research became absolute state of the art in modern archaeology, in geoscientific landscape reconstructions and in the deciphering of spatio-temporal interactions between man and nature. Despite this multidisciplinary constellation, the thematic and methodological overlap of humanities, natural sciences and informatics is frequently disregarded. It is beyond debate that multidisciplinary approaches, which especially emerge at the interface of adjacent subjects, substantially contribute to a better understanding of ancient landscapes, their forming processes and the resulting cultural heritage. They allow fusing complementary perspectives for the first time and therefore go far beyond unilateral research designs. Digital Geoarchaeology, which is to be established in this chapter as a new concept for the first time, can therefore be regarded as an intersection of disciplines that contributes to the consolidation of different academic perspectives. It represents a novel approach in terms of computer scientific methods combined with geoscientific know-how and archaeological expertise to multi-methodically investigate past human-environmental relationships. Accessing this multidisciplinary interface helps overcome potentially restricted, monodisciplinary perceptions and

C. Siart (✉) • M. Forbriger • O. Bubenzer
Institute of Geography, Physical Geography, Heidelberg University, Heidelberg, Germany
e-mail: qb120@uni-heidelberg.de; m.forbriger@gmail.com; olaf.bubenzer@uni-heidelberg.de

© The Author(s) 2018

C. Siart et al. (eds.), *Digital Geoarchaeology*, Natural Science in Archaeology,
DOI 10.1007/978-3-319-25316-9_1

provides new forms of unbiased approaches for investigating the interplay of man and nature. Thus, closer collaboration and dialogue across disciplinary boundaries will offer promising prospects for future research at the human-environmental interface.

Keywords
Digital geoarchaeology • Geoarchaeology • Geoinformatics • Digital humanities • Human-environmental interactions • Digital methods • Cross-disciplinarity

1.1 Introducing the Concept of Digital Geoarchaeology

Modern archaeology increasingly crosses academic boundaries by combining different new methodologies in order to answer research questions about ancient cultures and their remains. On the one hand, this development is attributed to the fact that human impact and environmental conditions are more and more considered coherent within the context of integral landscape reconstructions. On the other hand, it is the substantial technological progress made during the last decades that fostered this progression. Above all, the geosciences became an indispensable counterpart of archaeology and cultural heritage management. As to the investigation of past archaeological landscapes and palaeoenvironments, the term *Geoarchaeology* (or *Archaeometry*, if with a broader natural scientific focus; see Reindel and Wagner 2009) is commonly used, representing the utilization of traditional and the development of new geoscientific applications for archaeological purposes. In general, this pertains to geographical investigations and field methods (analysis and dating of geoarchives, e.g. sediments, soils, landforms) or geophysical prospecting like earth resistivity tomography, ground penetrating radar and geomagnetics (Sarris et al. 2018; Theodorakopoulou et al. 2018). During the last 50 years, geoarchaeological research was strongly influenced by earth scientists, namely, by geomorphologists (Butzer 1964, 1982, 2008; Renfrew 1976; Gladfelter 1977; Hassan 1979; Rapp and Hill 2006; for the history of the discipline, see Hill 2017), who focused on deciphering the stratigraphy, the formation or the preservation of an archaeological site, the development of the surrounding landscape and the potential influence of environmental conditions like climate on human behaviour and vice versa. However, an increasing number of disciplines became involved (e.g. biology, anthropology, history), so that Engel and Brückner (2014) define *Geoarchaeology* more comprehensively as "*... the science that studies geo-bio-archives in an archaeological context by also considering historical and archaeological data sources in its synthesis. It mainly applies geoscience tools in order to reconstruct the evolution and use of former landscapes and ecosystems, with regard to the interactions between humans and their environment*". Hence, when considering the prevailing view, *Geoarchaeology sensu originali* is generally regarded as an applied and primarily physically oriented subject in terms of methodology and technologies used. In analogy to the definition of *Archaeometry* by Wagner (2007), it represents the interface between geosciences and archaeology and aims at contributing to the solution of cultural-historical questions.

In contrast to this conventional perception, we intend to go one step further by also taking account of the digital dimension of human-environmental studies and thereby aim at highlighting the frequently disregarded thematic overlap of humanities, natural sciences and informatics (Fig. 1.1). As shown by practice, computationally engaged research became absolute state of the art in modern archaeology, in geoscientific landscape reconstructions and in the deciphering of spatio-temporal interactions between man and nature. For instance, area-wide remote sensing based on satellite imagery proves to be of great value for identifying surface findings (Lambers 2018), laser scanning allows for capturing archaeological sites and findings in 3D (Hämmerle and Höfle 2018; Raun et al. 2018) and digital elevation models (DEM) ideally qualify for detecting zones of archaeological interest and for spatially analysing and predicting the relationships between different sites (Siart et al. 2013; Bubenzer et al. 2018; Knitter and Nakoinz 2018). GIS, geoinformatics and computer sciences, have also served as essential tools in cultural heritage management for many years, particularly with regard to protecting archaeological sites (e.g. Ioannides et al. 2014). Moreover, the permanent increase in spatial resolution of available datasets also helps to overcome the traditional problems of scale (e.g. Stein 1993; Schlummer et al. 2014) between archaeologists, who mainly work on specific and spatially restricted sites over distinct human-related time slices, and geoscientists, who rather focus on specific environments and/or landscape-forming processes. Given this broad spectrum of applications and new research designs, the rising impact of computer scientific, digital techniques on archaeological and palaeoenvironmental studies becomes evident.

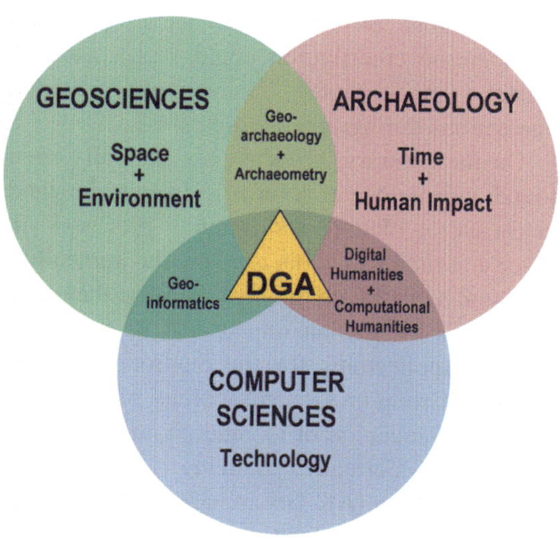

Fig. 1.1 The interdisciplinary intersection that defines the concept of *Digital Geoarchaeology* (DGA). While archaeology generally determines the historical context and the corresponding time slice, geosciences refer to the spatial and environmental dimension of studied objects. No less important is the use of computer-aided tools, particularly as to post-processing and fusion of archaeological and geoscientific datasets. This is why computer science functions as the technological backbone in DGA. Yet, all three domains are to be understood as equal and co-depending, even though each one may have a different share. Their methodological inventories can be combined in due respect of the scientific question to be unravelled (see Figs. 1.2 and 1.3 for details on workflow and applications of DGA)

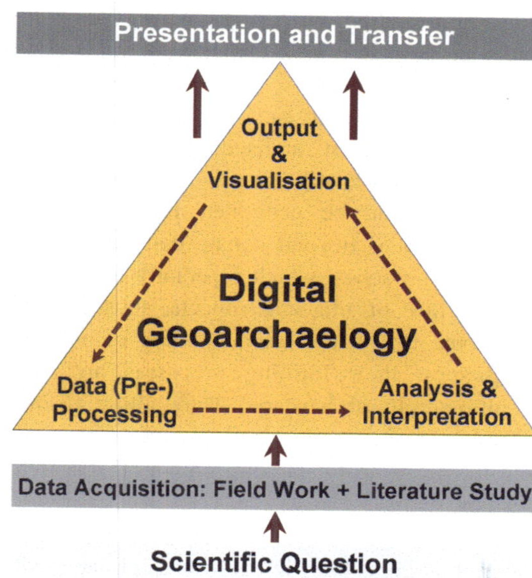

Fig. 1.2 Schematic workflow of Digital Geoarchaeology (modified according to Siart et al. 2013). Based on the topic to be investigated, which is mostly brought up by archaeology itself, acquisition of different datasets is carried out. Geosciences, computer sciences and archaeology come into play then, depending on the respective type of information required for solving specific scientific questions. The interdisciplinary interface (*triangle*) is characterized by data (pre-)processing in terms of desktop studies, analysis and subsequent interpretation of results, as well as final visualization (see Fig. 1.3 for details). This entire workflow is iterative and significantly relies on scholarly exchange. It leads to presentation and transfer of outcomes to both academia and the public—an aspect of special importance, e.g. for cultural heritage management and protection of archaeological sites, in which many different stakeholders are to be addressed with different levels of detail as to scientific outcomes

Data (Pre-) Processing	Analysis & Interpretation	Output & Visualisation
aerial (stereoscopic) photographs	DEM terrain analysis	spatial mobility (road systems & communication networks)
satellite data (orthorectification, mosaicing)	least-cost analysis	candidate site (land use; settlement locations)
LiDAR (point cloud registration, organisation)	predictive modelling	buried archaeological remains (walls, irrigation systems)
generation of digital elevation models	GIS-based spatial analysis	perspective views of the subsurface
maps & vectorisation (historical, topographical, geological, environmental data)	data integration (raster-vector fusion)	landscape visualisation in different time periods & evolutionary stages
	analysis of geophysical data (spatial relationships, 2D-3D data conversion)	

Fig. 1.3 Major work packages and examples of common procedures at the interdisciplinary interface of *Digital Geoarchaeology*

The multi-method overlap associated therewith could be described best using the term *Digital Geoarchaeology* (DGA, Fig. 1.1). Yet, when considering the status quo, comprehensive collaboration between archaeology, geosciences and informatics is still rare, even though useful synergies could be generated for all parties concerned. It is beyond debate that multidisciplinary approaches, which especially emerge at the interface of adjacent subjects, substantially contribute to a better understanding of ancient landscapes, their forming processes and the resulting cultural heritage. They allow fusing complementary perspectives for the first time and therefore go far beyond unilateral research designs.

Considering the common scientific discourse, the mutual benefits of integral research have not been fully explored yet. In fact, several similar trends and developments can be observed amongst different subjects, e.g. an increase of both the quantity and the quality of geoinformatic applications, but almost all of them happen separately within narrow disciplinary boundaries. In archaeology, for example, integration and consultation of geoscientific know-how is still the exception rather than the rule. This fact is mainly caused by the great progress archaeologists made themselves in using and developing further their own digital techniques over the last decades (Forte and Campana 2016). *Digital Archaeology*, a term that directly originates from this evolution (Zubrow 2006), describes these concepts best, but highlights at once that the initiatives are primarily driven by archaeologists and computer scientists without any input from earth sciences (Fig. 1.1). As almost every archaeological dataset is or can be georeferenced, and therefore owns a quintessential spatial attribute, allowance of profound geoscientific contribution could be an asset to better understand human-environmental interactions during specific time slices.

On the contrary, geoscientists generally need sound archaeological records to fully understand landscape evolution and natural processes in space and time. Now this is where the humanities represent a crucial component, because they help to identify the human impact and its historico-cultural implications for the environment. The Working Group on Geoarchaeology of the International Association of Geomorphologists (IAG) and the German Working Group on Geoarchaeology (AK Geoarchäologie) as part of the German Society for Geography are only two examples, which illustrate the increasing significance of archaeological know-how in human-environmental research. However, both are predominantly driven by geomorphologists and only to a minor degree by archaeologists. Here, too, collaboration mostly corresponds to *Geoarchaeology* sensu *originali*, in which human impact on the environment is considered key, but joint research on the basis of digital applications is rare to non-existing. Vice versa, there are also several archaeological working groups with explicit foci on computationally engaged research, such as Computer Applications and Quantitative Methods in Archaeology (CAA) (AG CAA in Germany). Most of them are successfully promoted by archaeologists, mathematicians and computer scientists, whereas it is the geosciences that remain underrepresented despite similarities in research interests and applications.

Apart from the above-mentioned domains governed by either archaeology or the geosciences, it is essential to consider the viewpoint of scientific computing, which has also advanced into interdisciplinary spheres over the years. Computer scientific research is often linked to explicit archaeological and geospatial aspects, e.g. in the context of documentation of archaeological findings and the reconstruction of historical sites (Schäfer et al. 2012; Var et al. 2013; Bogacz et al. 2015) or photogrammetric image processing (Sauerbier 2013; Kersten et al. 2014). Referring to this, Bock et al. (2013) present a broad range of interdisciplinary studies and therefore introduce the term *Computational Humanities*. Unfortunately, the great potential and benefits of scientific computing still have not been fully explored as to geoarchaeological investigations, even though lots of innovative approaches could actually be transferred or adapted to other research questions. In common practice, initial project ideas for studies at the human-environmental interface do not originate from informatics. As the latter represents a very versatile, multipurpose discipline, overarching scientific questions must be developed by humanities or natural sciences, if human-environmental interactions are to be investigated (Fig. 1.2).

By analogy, this predominant unilateral view, in which geoscientific input is rather neglected, also applies to many topics that fall within the scope of *Digital Humanities*. They can be defined as transdisciplinary, computationally engaged research that brings digital tools and methods to the study of the humanities (Burdick et al. 2012), but following Bock et al. (2013), they differ from *Computational Humanities* as to their closer spectrum of concepts and methods applied as well as their stronger focus on information sciences.

To sum up and return to the starting point of this paper, *Geoarchaeology* sensu *originali* generally remains under the aegis of geoscientists, while *Digital Archaeology* or *Computational Humanities* are mainly pursued by archaeologists and computer scientists, respectively. All of the above-mentioned disciplinary interfaces (see also Fig. 1.1) have significantly evolved and developed their own spheres of activity during the last decades, but exchange with disciplinary counterparts was and still is uncommon. The major reason might be the different scholarly attitudes of natural sciences, computer sciences and humanities. Their prima facie distinct perceptions sometimes just originate from ignorance of actual parallels (e.g. with regard to research topics). Besides, attention must be paid to the fact that it is the methodological expertise that governs but also limits the execution of specific research. For instance, archaeologists and geoscientists gained profound IT experience over the years, but their knowledge is mostly very specific and does not necessarily compare to the actual scope of expertise of computer scientists (discrepancy between end users and developers).

Nevertheless, it is important to note, that within the framework of Digital Geoarchaeology, disciplines are meant to support each other and significantly interdepend. Geoscientific know-how, for example, shall not substitute but promote archaeological studies as for methodologies and integration of spatial aspects (e.g. acquisition of appropriate geodata, use of specific techniques that help assess new archaeological archives like sediments). Geosciences may therefore improve the investigation and understanding of geoarchaeological landscapes and allow unravelling the palaeoenvironmental history comprehensively. In contrast, the historico-cultural implications of measured values and geodata are not self-evident. They depend on thorough interpretation, which only evolves if close exchange between disciplines is ensured. In this regard, Maran (2007) points out the different levels of progress achieved amongst those humanities, which focus on the relationships of man and environment. While pre- and early history and earth sciences constantly approach each other and thereby integrate their viewpoints, archaeology is still prone to considering historical findings and geoscientific data in a juxtaposed and unlinked way.

In order to demonstrate the valuable synergies, which arise from interdisciplinary collaboration between humanities, natural sciences and informatics, the concept of *Digital Geoarchaeology* is to be promoted (Figs. 1.2 and 1.3). As shown by the current scholarly discourse, research interest is indeed often the same and similar scientific questions are to be answered. That is why each discipline involved can ultimately benefit from mutual knowledge transfer. *Digital Geoarchaeology* can therefore be regarded as an intersection of disciplines that contributes to the consolidation of different academic perspectives (Fig. 1.1). It represents a novel approach in terms of computer scientific methods combined with geoscientific know-how and archaeological expertise to multi-methodically investigate past human-environmental relationships. Accessing this multidisciplinary interface helps overcome potentially restricted, monodisciplinary perceptions and provides new forms of unbiased approaches for investigating the interplay of man and nature. Thus, closer collaboration and dialogue across disciplinary boundaries will offer promising prospects for future research at the human-environmental interface.

References

Bock H, Jäger W, Winckler M (eds) (2013) Scientific computing and cultural heritage. Contributions in computational humanities. Springer, Heidelberg

Bogacz B, Massa J, Mara H (2015) Homogenization of 2D & 3D document formats for cuneiform script analysis. In: HIP 15 proceedings of the 3rd international workshop on historical document imaging and processing Gammarth, Tunisia – August 22–22, 2015. ACM, New York, pp 115–122

Bubenzer O, Bolten A, Riemer H (2018) In search of the optimal path to cross the desert. Geoarchaeology traces old trans-Saharan routes. In: Siart C, Forbriger M, Bubenzer O (eds) Digital geoarchaeology – new techniques for interdisciplinary human-environmental research. Springer, Heidelberg, pp 139–148

Burdick A, Drucker J, Lunenfeld P, Presner T, Schnapp J (2012) Digital humanities. Open access eBook. MIT Press, Cambridge, MA

Butzer KW (1964) Environment and archaeology: an introduction to Pleistocene geography. Aldine, Chicago

Butzer KW (1982) Archaeology as human ecology: method and theory for a contextual approach. Cambridge University Press, New York

Butzer KW (2008) Challenges for cross-disciplinary geoarchaeology: the intersection between environmental history and geomorphology. Geomorphology 101:402–411

Engel M, Brückner H (2014) Late Quaternary environments and society: progress in geoarchaeology. Z Geomorph N.F. 58(Suppl. 2):1–6

Forte M, Campana S (2016) Digital methods and remote sensing in archaeology. Archaeology in the age of sensing. Springer, Cham

Gladfelter BG (1977) Geoarchaeology: the geomorphologist and archaeology. Am Antiq 42(4):519–538

Hämmerle M, Höfle B (2018) Introduction to LiDAR in Geoarchaeology from a technological perspective. In: Siart C, Forbriger M, Bubenzer O (eds) Digital geoarchaeology – new techniques for interdisciplinary human-environmental research. Springer, Heidelberg, pp 167–182

Hassan FA (1979) Geoarchaeology: the geologist and archaeology. Am Antiq 44(2):267–270

Hill CL (2017) Geoarchaeology, history. In: Gilbert A (ed) Encyclopedia of geoarchaeology. Springer, Dordrecht

Ioannides M, Magnenat-Thalmann N, Fink E, Žarnić R, Yen A, Quak E (2014) Digital heritage. Progress in cultural heritage: documentation, preservation, and protection. In: Proceedings of 5th international conference, EuroMed 2014, Limassol, Cyprus, 3–8 Nov 2014. Springer, Cham

Kersten T, Hinrichsen N, Lindstaedt M, Weber C, Schreyer K, Tschirschwitz F (2014) Architectural historical 4D documentation of the old-segeberg town house by photogrammetry, terrestrial laser scanning and historical analysis. In: Ioannides M, Magnenat-Thalmann N, Fink E, Žarnić R, Yen A, Quak E (eds) Digital heritage. Progress in cultural heritage: documentation, preservation, and protection. Proceedings of 5th international conference, EuroMed 2014, Limassol, Cyprus, 3–8, Nov 2014. Springer, Cham, pp 35–47

Knitter D, Nakoinz O (2018) Point pattern analysis as tool for digital Geoarchaeology – a case study of megalithic graves in Schleswig-Holstein, Germany. In: Siart C, Forbriger M, Bubenzer O (eds) Digital geoarchaeology – new techniques for interdisciplinary human-environmental research. Springer, Heidelberg, pp 45–64

Lambers K (2018) Airborne and spaceborne remote sensing and digital image analysis in archaeology. In: Siart C, Forbriger M, Bubenzer O (eds) Digital geoarchaeology – new techniques for interdisciplinary human-environmental research. Springer, Heidelberg, pp 109–122

Maran J (2007) Mit den Methoden der Gegenwart in die Vergangenheit – Archäologie und

Naturwissenschaften. In: Wagner GA (ed) Einführung in die Archäometrie. Springer, Heidelberg, pp 342–350

Rapp GR, Hill CL (2006) Geoarchaeology: the earth-science approach to archaeological interpretation, 2nd edn. Yale University Press, New Haven

Raun K, Pfeiffer M, Höfle B (2018) Visual detection and interpretation of cultural remnants on the Königstuhl hillside in Heidelberg using airborne and terrestrial LiDAR data. In: Siart C, Forbriger M, Bubenzer O (eds) Digital geoarchaeology – new techniques for interdisciplinary human-environmental research. Springer, Heidelberg, pp 201–214

Reindel M, Wagner GA (2009) Introduction – new methods and technologies of natural sciences for archaeological investigations in Nazca and Palpa, Peru. In: Reindel M, Wagner GA (eds) New technologies for archaeology. Multidisciplinary investigations in Palpa and Nazca, Peru. Springer, Heidelberg

Renfrew C (1976) Archaeology and the earth sciences. In: Davidson DA, Shackley ML (eds) Geoarchaeology: earth science and the past. Westview Press, Boulder, pp 1–5

Sarris A, Kalayci T, Moffat I, Manataki M (2018) An introduction to geophysical and geochemical methods in digital geoarchaeology. In: Siart C, Forbriger M, Bubenzer O (eds) Digital geoarchaeology – new techniques for interdisciplinary human-environmental research. Springer, Heidelberg, pp 215–236

Sauerbier M (2013) Image-based techniques in cultural heritage modeling. In: Bock H, Jäger W, Winckler M (eds) Scientific computing and cultural heritage. Contributions in computational humanities. Springer, Heidelberg, pp 61–69

Schäfer A, Mara H, Freudenreich J, Breuckmann B, Düffort C, Bock G (2012) Large scale Angkor style reliefs: high definition 3D acquisition and improved visualization using local feature estimation. In: Zhou M, Romanowska I, Wu Z, Xu P, Verhagen P (eds) Revive the past. Proceedings of the 39th conference on computer applications and quantitative methods in archaeology, Beijing, 12–16 April. Pallas, Amsterdam, pp 70–80

Schlummer M, Hoffmann T, Dikau R, Eickmeier M, Fischer P, Gerlach R, Holzkämper J, Kalis AJ, Kretschmer I, Lauer F, Maier A, Meesenburg J, Meurers-Balke J, Münch U, Pätzold S, Steininger F, Stobbe A, Zimmermann A (2014) From point to area: upscaling approaches for late quaternary archaeological and environmental data. Earth Sci Rev 131:22–48

Siart C, Brilmayer Bakti B, Eitel B (2013) Digital Geoarchaeology – an approach to reconstructing ancient landscapes at the human-environmental interface. In: Bock H, Jäger W, Winckler M (eds) Scientific computing and cultural heritage. Contributions in computational humanities. Springer, Heidelberg, pp 71–85

Stein JK (1993) Scale in archaeology, geosciences, and geoarchaeology. In: Stein JK, Linse AR (eds) Effects of scale in archaeological and geoscientific perspectives. Geological Society of America, Boulder, Colorado, Special paper, vol 283, pp 1–10

Theodorakopoulou K, Bassiakos Y, Athanassas C, Schukraft G, Holzhauer I, Hecht S, Mächtle B, Wagner GA (2018) A geoarchaeological approach for the localization of the prehistoric harbor of Akrotiri, Thera. In: Siart C, Forbriger M, Bubenzer O (eds) Digital geoarchaeology – new techniques for interdisciplinary human-environmental research. Springer, Heidelberg, pp 237–252

Var P, Phal D, Nguonphan P, Winckler MJ (2013) 3D reconstruction of Banteay Chhmar temple for google earth. In: Bock H, Jäger W, Winckler M (eds) Scientific computing and cultural heritage. Contributions in computational humanities. Springer, Heidelberg, pp 261–272

Wagner GA (2007) Archäometrie in Heidelberg – ein Beispiel für das Werden eines neuen Faches. In: Wagner GA (ed) Einführung in die Archäometrie. Springer, Heidelberg, pp 351–366

Zubrow E (2006) Digital archaeology: a historical context. In: Evans T, Daly P (eds) Digital archaeology: bridging method and theory. Routledge, London, pp 10–31

Open Access This chapter is licensed under the terms of the Creative Commons Attribution 4.0 International License (http://creativecommons.org/licenses/by/4.0/), which permits use, sharing, adaptation, distribution and reproduction in any medium or format, as long as you give appropriate credit to the original author(s) and the source, provide a link to the Creative Commons license and indicate if changes were made.

The images or other third party material in this chapter are included in the chapter's Creative Commons license, unless indicated otherwise in a credit line to the material. If material is not included in the chapter's Creative Commons license and your intended use is not permitted by statutory regulation or exceeds the permitted use, you will need to obtain permission directly from the copyright holder.

Part I

Spatial Analysis and Geographical Information Systems

Spatial Analysis in Archaeology: Moving into New Territories

2

Philip Verhagen

Abstract

GIS has become an indispensable tool for archaeologists to organize, explore and analyse spatial data. In this introductory chapter, an historical overview of the development of GIS use in archaeology is given. It focuses on three major fields of application: site location analysis, modelling movement and transport and visibility analysis. This state of the art is illustrated by discussing three different case studies. Finally, some thoughts on the future of GIS in archaeology are presented.

Keywords

GIS • Archaeological theory • Site location analysis • Least-cost paths • Viewsheds

2.1 Introduction

It is now over 30 years ago that the term GIS was introduced in archaeology (Hasenstab 1983), and it is hard to imagine how archaeologists have ever done research without it. GIS and spatial analysis are now seen by most archaeologists as essential tools to explore, analyse and interpret spatial data and have become standard ingredients in many archaeological research projects. GIS and spatial analysis are extremely convenient techniques for more efficiently carrying out 'traditional' archaeological research. However, there are also those who maintain that the 'spatial turn', boosted by GIS technology, points the way to applying fundamentally different theoretical perspectives in archaeology.

In this chapter, I will give a condensed overview of the current state of GIS use in archaeology and attempt to sketch the current role of GIS and spatial analysis for archaeological interpretation and show its potential for changing theoretical perspectives and research traditions, drawing on examples from recent research. And lastly, I will try to look into the crystal ball and set a tentative agenda for future research.

P. Verhagen (✉)
Faculty of Humanities, Vrije Universiteit Amsterdam, Amsterdam, The Netherlands
e-mail: j.w.h.p.verhagen@vu.nl

© The Author(s) 2018

C. Siart et al. (eds.), *Digital Geoarchaeology*, Natural Science in Archaeology,
DOI 10.1007/978-3-319-25316-9_2

2.2 The Position of GIS in Archaeological Research

If we would have to describe the history of the use of GIS in archaeology in a nutshell, it could be summarized as a cycle of initial enthusiasm and proliferation in the 1980s and early 1990s, followed by severe criticism and (partial) disillusionment in the late 1990s, only to be reappraised again and rapidly gaining momentum in the late 2000s, leading to its current status as an almost indispensable research tool—or rather methodology—for dealing with spatial archaeological data. The main trends in this development have been described by, e.g. Kvamme (1999), Verhagen (2007: 13–25), McCoy and Ladefoged (2009), Wagtendonk et al. (2009) and Verhagen (2012) and need not be repeated here. However, when reading the academic literature on the subject (which has the tendency of being a rather slow detector of longer-term trends), we could be under the impression that archaeologists are still reluctant and hesitant in their appreciation and adoption of GIS-based spatial analysis. This is because of its association with the theoretical school of processual archaeology, with its underlying, naive support of scientism, and with its emphasis on environmental determinism (see Hacıgüzeller 2012). Theoretically oriented archaeologists were seriously concerned about these issues in the 1990s and early 2000s when thinking about how to deal with digital technologies in general. However, archaeological practice has certainly moved on since then, and currently archaeologists have generally embraced geographical database management, digital cartography and spatial analysis, if only for reasons of efficiency. To a lesser extent, they have also gradually adopted computer-based modelling as a research tool, although acceptance here has been a lot slower, due to the fact that it has stood in the middle of the processual versus post-processual controversy (see also Verhagen and Whitley 2012). This is part of a larger debate about computing applications in archaeology that has been described as an 'anxiety discourse' by Huggett (2013) and which is a general characteristic of emerging fields trying to establish their scientific identity.

One of the reasons why the debate on GIS has been quite tense is highlighted by Hacıgüzeller (2012). She distinguishes between two views of understanding the past, the representational and non-representational. In the representational view, the past is supposed to have an objective reality. This is a reality, however, that we cannot touch. For this reason we can only use representations to understand the past. This leads to a dualistic approach to research, separating, e.g. past and present and material and meaning. It also implies that there is a constant search for the right medium to construct representations that are as faithful to 'reality' as possible—and this is exactly where GIS filled a huge gap when it came to the scene in the 1980s. Digital cartography suddenly allowed researchers (not just in archaeology) to take mapping to a much higher level and to collect and manipulate geographical data in a much more sophisticated way.

The critique of this representational viewpoint is very prominent in the post-processual rejection of 'processualist' methods such as GIS (Thomas 1993, 2001, 2004; Tilley 2004, 2008). The pre-occupation of post-processual researchers with bodily understanding as the preferred way to study the past, and in this way to come closer to the mindset of human beings long dead, shows that they were looking for new ways of representation, albeit in a different form than what cartography and other techniques of data complexity reduction could achieve (see, e.g. Tilley 2004). It has however been noted before (Fleming 2006; Verhagen and Whitley 2012) that the rejection of the 'scientific method' by post-processualism contradicts one of its own tenets, i.e. the exploration of multiple and equivalent views of the past. As such, 'scientific' approaches can and should have their place in archaeological research practice, and the predominance given to narrative by post-processualists is not necessarily the best way to represent the past either.

What the early practitioners of GIS and their critics did not perceive is that GIS and other computer-based methods enable pluralism, rather than enforce reductionism. Using these tools, a multitude of representations can be created with little effort, in which there is no longer an easy way to distinguish between right and wrong and between more and less plausible. Because of this, cartography has been effectively democratized, and mapping these days is, more than ever, an exercise in (scientific) rhetoric.

Following Hacıgüzeller's (2012) view, we can gain much more by adopting a non-representational approach to the study of the past, and thus to GIS. In this view, the past is not something that can be understood in a static and definitive way, but rather something that continually changes and is repeatedly reconstructed in the present. It is therefore a plea for eclecticism in using GIS and to consider it more as a constantly changing research practice than as a technology-driven instant solution that can be applied to all forms of spatial data and all archaeological research questions. It also follows that GIS-based spatial analysis and modelling can never be a stand-alone approach, but should be an integral part of what we might call 'hybrid' archaeological research—which of course echoes the strong call for interdisciplinarity in modern science.

We might even go one step further and ask ourselves whether spatial analysis and modelling could not be just one of many approaches, but perhaps constitute a *leitmotiv* for doing archaeological research in the twenty-first century. An important characteristic of computer-based techniques that sets them apart from all other approaches is their ability to deal with data sets that are too big and complex to handle by human minds. Therefore, they can be applied to all situations where we have 'big data'. GIS can deal with big data that also have a spatial dimension and in this way help to discern patterns and to simulate theories of human behaviour over large areas. It is therefore, in all probability, the next frontier for spatial technology in archaeology: to move beyond the boundaries of individual, site-based or micro-regional projects and to have a look at the 'big picture'.

2.3 Spatial Analysis in Action

In the following sections, I will introduce examples of the use of GIS-based spatial analysis that I believe illustrate current research trends, as well as its utility and limitations in practice. The main applications of GIS in archaeology can be classified into *site location analysis*, *modelling movement and transport* and *visibility analysis*, and I will provide examples of applications of all of them. In many cases these approaches are also used in conjunction—although we can suspect that this is partly because they are all available in the same toolbox and are therefore relatively easy to combine. Over the last few years, however, we can see that researchers are becoming more and more interested in coupling GIS to other techniques, such as social network analysis, advanced statistical methods and agent-based modelling.

2.3.1 Site Location Analysis

Without site location analysis, GIS might not have caught on as quickly as it did in archaeology. The earliest examples of GIS-based site location analysis can already be found in the early 1980s in the USA, and it is has never left the scene since then. At the time, it met a strongly felt need for more efficiently analysing site location preferences, a field of study which had received an enormous boost in the mid-1970s through site catchment analysis (Higgs and Vita-Finzi 1972; Findlow and Ericson 1980). The closely related practice of predictive modelling followed in its wake, responding to a demand from Cultural Resources Management to predict the distribution of archaeological remains in areas under threat of destruction (Kvamme 1983, 1984; Judge and Sebastian 1988). However, both methods quickly came under severe attack when post-processual theory made its way into archaeology in the late 1980s and early 1990s. To the post-processualists, site location analysis and predictive modelling were prime examples of how the processualists had chosen to ignore the human dimension in the study of the past.

The first and foremost of these criticisms was the accusation of environmental determinism. The comparison of site locations to various environmental characteristics (such as soil type, slope or distance to natural resources) will inevitably favour environmental-deterministic explanations of site location patterns (Wheatley 1993, 1996, 2004; Gaffney and van Leusen 1995). In fact, it was argued that the use of GIS even forced the interpretation of site location patterns towards this point of view, since it could not handle 'soft' social and cognitive parameters and non-environmental data sets were scarce or unavailable. And even when the method can hardly be blamed for lack of data or for flawed archaeological-theoretical perspectives on site location, in practice it has proved to be very difficult to deal with non-environmental variables in site location analysis, although some real progress has been made in this respect over the last decade (Whitley et al. 2010; Verhagen et al. 2013a, 2016), also using, e.g. point pattern analysis (Bevan and Conolly 2006) and network analysis techniques (Bevan and Wilson 2013).

A second criticism of site location analysis is directed towards its quantitative nature: the results of site location analysis are typically presented as statistical tables and diagrams and offer the perspective of extrapolating the analysis results to other areas in the form of predictive maps—and they are therefore potentially misleading if the data and/or theories used are flawed (Wheatley 2004). In the early days, many GIS practitioners were well aware of the pitfalls of using and interpreting statistical analyses (see, e.g. Judge and Sebastian 1988). However, the backlash against quantitative methods in the early 1990s led to a general distrust in the use of statistics and a loss of proficiency amongst archaeologists that is still evident in university curricula these days. More importantly, dealing with field survey data for site location analysis has proved to be one of the trickier statistical issues in archaeology, since we usually have little control over the representativeness of sampling, and there are no established procedures for dealing with uneven representation (see, e.g. Orton 2000; Verhagen 2007: 115–168).

Thirdly, site location analysis and predictive modelling have been criticized for being static and not taking into account the temporal dimension of human behaviour. Again, this is much more a question of the availability of suitable data rather than of flawed methodology—temporal dynamics of site location patterns can only be studied fruitfully if we have sufficiently reliable dating of archaeological sites and if we have sufficiently detailed palaeogeographical reconstructions.

At a deeper level, this debate shows the everlasting struggle between the application and development of scientific theory. We cannot expect a method or technique to operate in a theory-neutral environment; our choice of research questions, study regions, methods and data sets is governed by what we think we know about the past and by what we think we need to do to expand our knowledge. So if we really did not think that the environment influences site location choice, then we would never choose to analyse it. And on the other hand, patterns that suggest themselves to us, for example, while performing site location analysis, will find their way back into existing theoretical frameworks and either reinforce or challenge established opinions.

Basically, we are still looking for answers to the question of what made people settle where at a particular point in time. For this, site location analysis is not the only possible tool, but it remains a versatile, powerful and relatively straightforward way to explore site location preferences over large areas, to detect patterns and anomalies in settlement distribution, to compare these between areas and time periods and to place these in perspective together with other sources of information.

2.3.2 Example: Long-Term Settlement Pattern Dynamics in the South of France

Over the past 25 years, French scholars have worked extensively on collecting and analysing a large set of archaeological and environmental data in the southeast of France, in various study areas located along the river Rhône south of Valence and in the Provence and Languedoc regions on the Mediterranean coast. In the mid-1990s, considerable efforts were made within the Archaeomedes Project (Durand-Dastès et al. 1998; Van der Leeuw et al. 2003) to develop a protocol for site location analysis that could be applied to all study regions and that would allow to make long-term and cross-regional comparisons of location preferences of rural settlement in the period from 800 BC to 800 AD. This protocol was based on three important principles:

- The establishment of a standardized hierarchical classification of archaeological settlements
- The selection of a reduced number of relevant environmental variables for analysis that could be standardized for all regions
- The analysis of not just the location of the site itself but also its immediate surroundings

The results of the site location analysis clearly showed how the rhythms of occupation and abandonment of settlement changed between study areas and over the southeast of France as a whole. This pointed to a different story for in particular the Late Roman 'agricultural crisis' than was previously assumed. Instead of a political and economic crisis coupled to land abandonment and environmental degradation, it reflects a process of restructuration and stabilization of the settlement pattern, in which settlement is contracting into fewer locations, but not necessarily exploiting smaller areas (Favory et al. 1999; Fovet 2005).

At the time, performing the necessary calculations was a considerable task. The method however proved successful enough to be repeated in various other areas in France during the Archaedyn Project (Gandini et al. 2012), with similar outcomes. Invariably, the site location analyses showed a different story of occupation pattern dynamics than was previously assumed on the basis of local studies. However, the analysis was limited to a comparison of environmental preferences. Verhagen et al. (2013a, 2016) therefore decided to extend the analysis protocol of site location preferences to include non-environmental factors. For this, a distinction was made between what might be called 'socio-environmental' variables, in particular accessibility and visibility, and 'true' socio-cultural variables. These include the influence of long-term occupation ('memory of landscape') on site location preference and the position of settlements in networks.

Modelling of accessibility through so-called cumulative cost paths (Verhagen et al. 2013a) was somewhat inconclusive, with little discernible direct influence on site location patterns. The importance of 'memory of landscape' for new occupation, however, was clearly demonstrated in two study regions in the south of France (Verhagen et al. 2016; Fig. 2.1). These results show that it is possible and desirable to include more sophisticated and archaeologically informed parameters into site location analysis, especially when environmental factors are not very good predictors.

2.3.3 Modelling Movement and Transport

One of the techniques offered by GIS that has attracted much interest in archaeology is least-cost path analysis (see, e.g. Bell and Lock 2000; Llobera 2000; Bell et al. 2002; Fábrega-Álvarez and Parcero-Oubiña 2007; Zakšek et al. 2008; Murrieta-Flores 2014). It is a method to find the optimal path between two or more locations. In a landscape that is characterized by differential accessibility, finding the most efficient route is a non-trivial task that requires determining the trade-off between distance travelled and obstacles on the route. These obstacles can be of a physical nature, like steep slopes, water bodies or vegetation hindering free movement, thus making detours potentially more attractive than taking the straight line. But we can also think of obstacles of a social nature, such as the presence of enemies, toll roads or even taboos on entering certain areas. And in some cases, intermediate locations may need to be visited in order

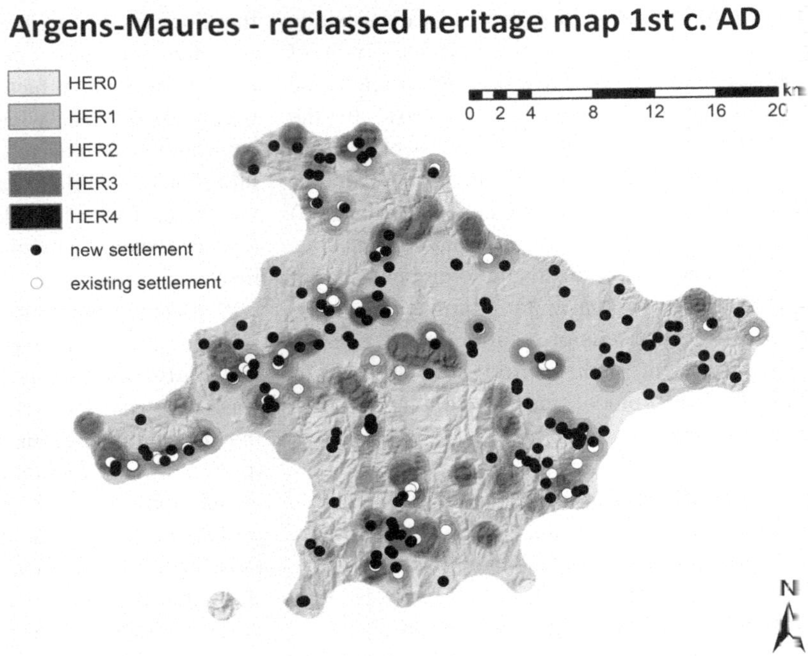

Fig. 2.1 Land use heritage map of the Argens-Maures region (Var, Provence, France) for the first century AD, with archaeological sites. 'Existing settlements' are locations already occupied in the first century BC, with continuing occupation in the first century AD. Source: Verhagen et al. (2016)

to go from A to B, for example, because there is water, food, firewood and/or shelter available.

Tools for finding the most efficient route are implemented in all GIS software. First, a cost surface is defined that determines the costs of crossing one grid cell, usually specified in time or energy units spent. Slope is amongst the most common cost factors considered, and a number of methods have been developed to determine movement costs associated with slope (Herzog 2013). Different types of costs (monetary, social) could be included as well, but this will make things more complicated since we then cannot use the same cost units anymore. An *accumulated cost surface* is then created from a starting point, using Dijkstra's (1959) algorithm or one of its variants, which will provide the cost distance from the starting point to every grid cell in the study area. Finally, we can then determine the *least-cost path* between the starting location and any other location, giving us some idea on where transport and movement may have taken place (see, e.g. Van Leusen 2002: 308–329; Howey 2007; Zakšek et al. 2008). The accumulated cost surface can also be used to find the area that can be reached within a certain amount of time (or by spending a maximum amount of energy). This is often used to analyse the size, location and environmental characteristics of site territories (see, e.g. Robb and Van Hove 2003; Ducke and Kroefges 2008; Whitley et al. 2010). Alternatively, we can also determine cost surfaces calculated from all locations on a raster map and combine these into what has been called a *total accessibility map* or *potential path field* (Llobera 2000; Mlekuž 2014). These can be used to analyse the relative accessibility of a certain location.

The definition of the cost surfaces is seen as the main obstacle in least-cost path modelling. Even in the simplest case, when we only take into account the effect of slope on movement speed, there are a number of complex issues to be dealt with, including the accuracy and level of detail of the digital elevation models used (see Llobera 2000; Ejstrud 2005; Zakšek et al. 2008; Gietl et al. 2008; Herzog 2013), potential changes in topography since prehistory and the ability of humans to overcome topographical constraints by using technological solutions, such as paving or building bridges. Furthermore, different modes of transport can lead to different optimal paths: carts carrying heavy loads will have more difficulty negotiating steep slopes (see, e.g. Bevan 2013). And lastly, route networks will develop dynamically, depending on the already existing route and settlement networks in the region and period of study (Fovet and Zakšek 2014). The optimal path in such a situation may not be the one that takes least time or energy, but one that (also) uses the existing infrastructure.

In practice, therefore, least-cost path analyses have only been moderately successful in reconstructing past routes and movement (see, e.g. Bell and Lock 2000; Becker and Altschul 2008; Fiz and Orengo 2008; Verhagen and Jeneson 2012). Furthermore, validation of the models will be problematic in most cases since most prehistoric roads are difficult to detect—if they have survived at all. Mixed results have been reported as well by researchers who modelled 'natural' travel corridors as variables for site location analysis (Whitley and Hicks 2003; Whitley and Burns 2008; Murrieta-Flores 2012; Standley 2015; Van Lanen et al. 2015). This would lead us to suspect that 'natural' accessibility is a more important factor for locating new routes, rather than for locating new settlements per se. The application of least-cost path modelling is therefore more fruitfully seen in the light of hypothesis development and testing of where people might have moved in the past and what factors may have been influencing movement and transport.

More recently, it is acknowledged that social network analysis (SNA) techniques can be helpful to better understand the interaction between settlements and other transport nodes and to analyse patterns of communication at the regional scale (Verhagen et al. 2013b; Fovet and Zakšek 2014). Thus far, however, combining SNA and LCP modelling is still a complex task, since existing SNA software solutions are not integrated with GIS.

2.3.4 Example: Modelling Transport and Movement in the Dutch Roman *Limes*

In a recent paper, Groenhuijzen and Verhagen (2015) present a method to combine LCP and SNA in an attempt to model local transport and movement in the Dutch part of the Roman *limes* (the border of the Roman Empire). The Kromme Rijn study area is located on the south bank of the river Rhine, where differential accessibility is mainly due to differences in wetness: the low-lying floodplains are more prone to flooding than the levees, and thus movement is spatially constrained by the location of palaeo-channels in the area. By combining data from physiological experiments and specifying different costs for different modes of transport (on foot—unburdened and burdened—and by mule cart), it was possible to create LCP models connecting all Roman settlements in the area for different modes of transport and for different time slices, departing from a detailed palaeo-geographical reconstruction and a comprehensive site database. Since the focus was on local transport, the connections between settlements were only taken into account if they were less than 20 min travelling (approximately 1.5 km) apart.

The modelled networks indicate a strong preference for movement on the levees (as expected) and show different patterns for different modes of transport. This is especially clear when SNA is used to analyse the importance of the connections. *Betweenness centrality*, which measures the number of times a node has to be passed in order to go from one place to another, is a good indicator of the importance of a node in a transport network (Fig. 2.2).

The analysis results indicate that the sites showing signs of higher status (as derived from the presence of stone building remains) are relatively well connected, which might indicate that their position in the network may have contributed to them becoming more important during the Roman period. It is also noteworthy that the network configuration, as well as the importance of settlements, changes with different

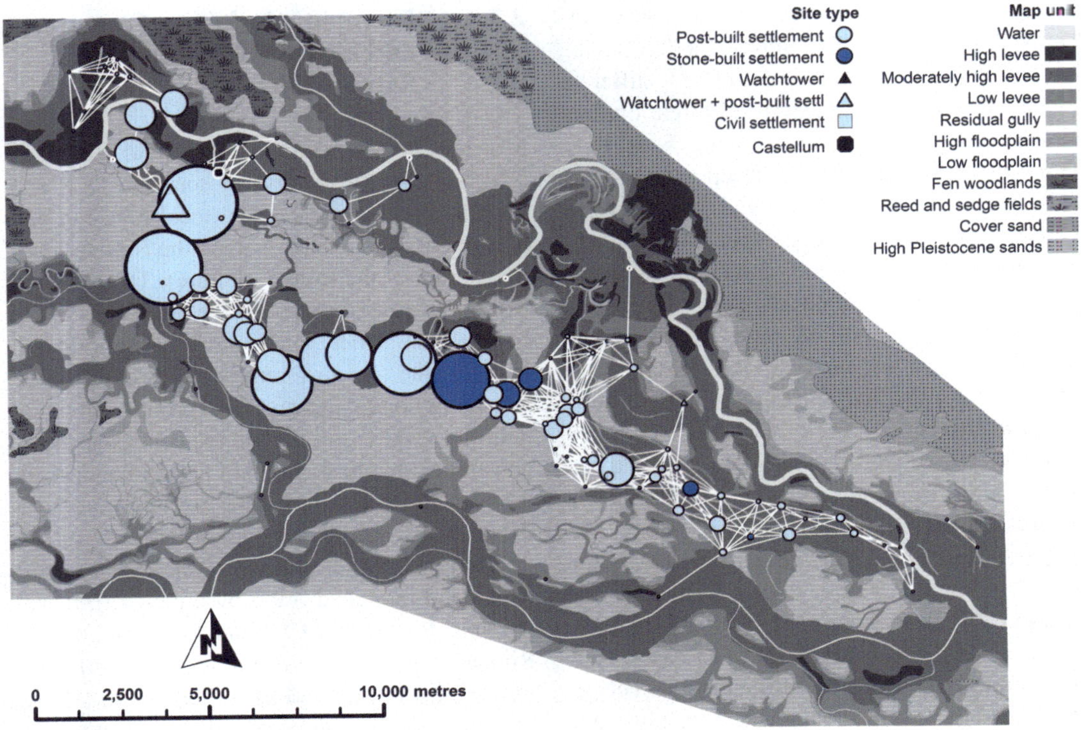

Fig. 2.2 Betweenness centrality measurements of Early Roman sites in the Kromme Rijn area, the Netherlands, based on least-cost paths calculated for mule cart transport. Source: Groenhuijzen and Verhagen (2015)

modes of transport, since the options for mule cart transport are much more limited than for movement on foot.

Furthermore, the military *limes* road, located close to the Rhine, and which has been the subject of most research on transport in the Dutch *limes*, is not very important as a connection between settlement sites. Obviously, this road did serve to connect the *castella* on the Rhine, but the lack of settlements on the Rhine banks means that it may not have been used frequently for local transport; the *castella* and adjoining *vici* were therefore peripheral to the local transport network.

2.3.5 Visibility Analysis

The third major branch of GIS methods that made its way into archaeology is the calculation of lines of sight and viewsheds. It is a technique that originated outside archaeology, where it is used in particular for siting military and telecommunications facilities. Archaeology however is quite unique in how it has used visibility analysis—and it is probably the nearest that GIS can come to representing bodily experience, by determining what places and objects can be seen from a particular vantage point (Tschan et al. 2000; Llobera 2003; Fitzjohn 2007).

In essence, visibility analysis starts by determining the line of sight between two locations, by comparing the elevation of location A to the elevations encountered on a straight line to location B. If there is no higher elevation obstructing the view, then B can be seen from A. In this way, it is possible to calculate, for each and every grid cell in a region, which cells within a certain neighbourhood can be seen: this is the cell's *viewshed*. These viewsheds can then be combined to obtain *cumulative* (Wheatley 1995) or even *total viewsheds* (see Llobera 2003), which show the number of cells from which a location can be seen. Importantly, these viewsheds not only provide information on which locations are most visible but also on those which are hidden from view. Obviously, viewsheds can be calculated for different ranges of view (Wheatley and Gillings 2000; Llobera 2007a), and in this way multiple measures of visibility can be obtained to characterize landscapes and site locations.

Llobera (2003) introduced the concept of visualscape as '*a spatial representation of a visual property generated by or associated with a spatial configuration*'. Using this concept, Llobera explicitly linked visual prominence and exposure to movement, which both are strongly connected to sensory perception. In practice, however, the application of GIS-based visibility analysis to questions of human perception of the landscape has not become very popular, despite several attempts in this direction (e.g. Llobera 1996; Trifković 2006; Gillings 2009; Lock et al. 2014). 3D modelling and virtual reality approaches would now seem to be more effective tools for this, although these generally lack the analytical capabilities offered by GIS.

Viewshed analysis has been applied more regularly and successfully in conjunction with site location analysis (e.g. Sevenant and Antrop 2007; De Montis and Caschili 2012) as well as with least-cost path modelling (e.g. Murrieta-Flores 2014; Lock et al. 2014), not just to test whether visibility may have been a factor influencing site location but more importantly to understand how archaeological sites are visually related. This has been especially of interest for analysing the placement of megalithic monuments, burial mounds, hillforts, castles and other monumental and defensive sites (e.g. Gaffney and Stančič 1991; Wheatley 1995; Ruggles and Medyckyj-Scott 1996; Loots et al. 1999; Lake and Woodman 2003; Bourgeois 2013: 105–158; Čučković 2015).

However, it is also a technique that is fraught with difficulties, since its results highly depend on the quality of the digital elevation models used, both in terms of vertical accuracy (Fisher 1992; Loots et al. 1999; Ruestes Bitrià 2007) and in terms of how well a DEM, which is stripped of vegetation, represents a prehistoric landscape (Llobera 2007b). Viewshed analysis is also highly sensitive to edge effects and can therefore only be applied to large areas, which even today might lead to problems with computing power. Furthermore, the question of what specific elements in the landscape would be important to see is not always addressed, resulting in maps of cumulative viewsheds that only provide information on the proportion of the landscape that is visible from a vantage point. Even though there has been some research done on the level of detail of objects that can be discerned at various distances (e.g. Ogburn 2006), GIS would not seem the best suited tool for this, and many studies interested in understanding visibility of objects, especially in built-up spaces, now tend to use 3D modelling instead.

2.3.6 Example: Studying Visibility and Movement in the Sierra Morena

An exemplary case study that combined visibility analysis, path modelling and site location analysis was presented by Murrieta-Flores (2012, 2014). It is a good example of how to use all three analysis techniques together in an attempt to better understand the development of settlement patterns and the placement of megalithic monuments in the western Sierra Morena (Southwestern Spain). It was long suspected that megalithic monuments in Iberia are linked to potential patterns of movement and might have been used as navigational markers, associated with a long tradition of transhumance. In order to test this hypothesis, Murrieta-Flores first modelled potential pathways of movement in the study region and then analysed the visual characteristics of the megalithic monuments in order to see whether they were located in places where they could be seen and would be dominating the view.

Least-cost paths were created originating from mountain passes at the edge of the study area and crossing the whole area (Murrieta-Flores 2012)—a technique similar to the one used by Whitley and Hicks (2003). The density

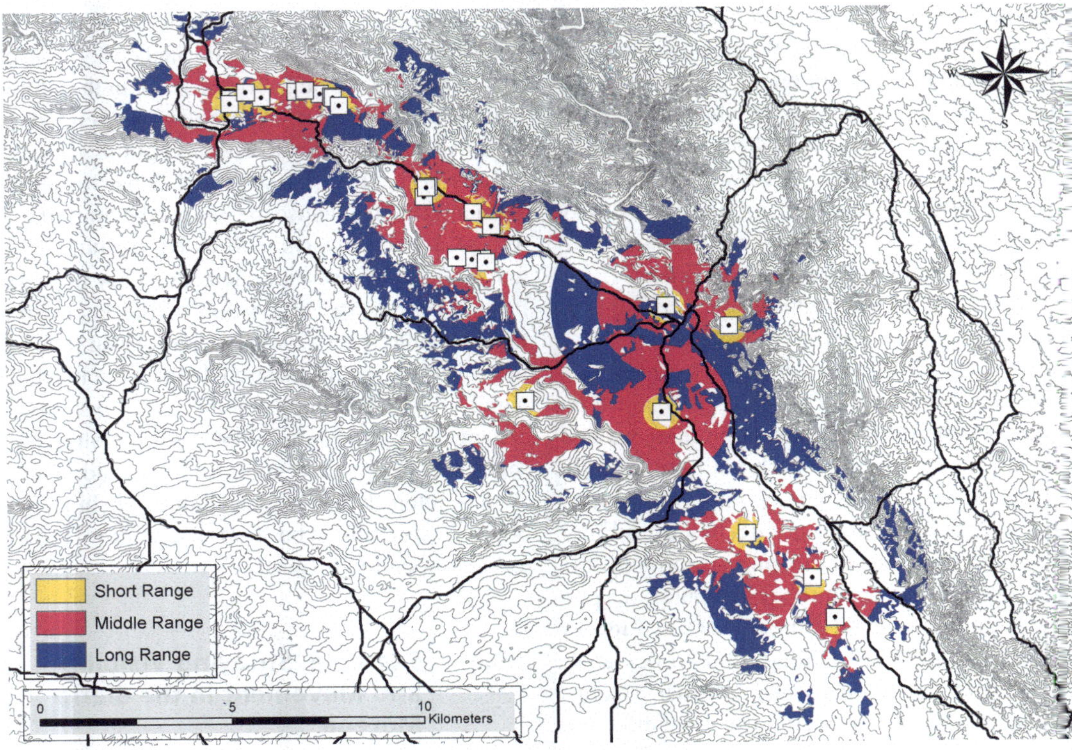

Fig. 2.3 Visual ranges of megaliths along the Viar Valley (Sierra Morena, Andalusia, Spain) and their correspondence with natural corridors. Source: Murrieta-Flores (2014)

of least-cost paths was then used to generate 'natural movement corridors', that were interpreted as potential routes used for the movement of people and cattle in the annual transhumance orbit. These corridors were shown to have a statistically significant relationship with the placement of settlements in the Copper and Bronze Age. It also became clear that most of the megalithic monuments were located quite close to the modelled corridors (Fig. 2.3).

By then creating viewsheds for each of the megalithic monuments, and looking at the direction of view, it proved possible to show that the monuments were in places where they could oversee relatively long stretches of the corridors, considerably longer than from other points in the landscape (Murrieta-Flores 2014). However, it seems less likely that they have acted as navigational markers—intermediate waypoints would have been necessary for effective wayfinding. This would point to a more symbolic role for the monuments, for example, in maintaining collective memory.

2.4 Moving into New Territories

In the preceding sections, I have described existing trends in GIS-based spatial analysis with GIS in archaeology. The methods and approaches outlined are well developed, even when in some cases there are still some technical issues to be solved, for example, where it concerns the best ways to calculate least-cost paths. The more important questions to be tackled, however, are related to the archaeological-theoretical perspective on spatial analysis and to the increasing cross-fertilization between GIS and other digital techniques.

Despite the widespread use of GIS in archaeological practice nowadays, archaeological-theoretical thinking still has to come to terms with the idea of using formalized spatial analysis as a primary line of attack to better understand the regional and diachronic dynamics of settlement and land use. The emphasis placed by post-processual theoreticians on narrative, at the expense of scientific methods, has led to a rift between 'science-based' and mainstream archaeology that is still very evident today. It has also led to an attitude amongst archaeologists of seeing 'hard science' methods and techniques as auxiliary tools that provide helpful data to be used in the construction of an historical narrative, rather than as a possible focus of archaeological research. For this reason, 'digital archaeology' is still mainly the realm of specialists, and developing and diffusing digital tools and models for specific archaeological purposes is often a low priority. It is, in my view, a form of short-sightedness that will be detrimental to the discipline in the long run, since, as pointed out earlier, computer techniques allow us to analyse data and construct models that go far beyond the capacity of the human brain and are therefore essential to push the boundaries of our knowledge of the past. They are also extremely well suited for formalizing hypotheses and testing these, thereby fulfilling an important role as heuristic devices, as tools that help us to think more deeply about our assumptions. But if we don't consider the tools themselves as an object of inquiry, it will only lead us to be dependent on what experts—often from different disciplines—develop, instead of setting our own agenda for digital archaeology.

Users of GIS in archaeology should be aware of this and keep an open eye for developing new tools and for combining spatial analysis with other approaches. As mentioned, network analysis is gaining in importance and will become much better integrated with GIS in the near future. A similar development can be observed where it concerns the combination of GIS and dynamical simulation modelling, which is a rapidly growing field of archaeological research as well (Kohler and Van der Leeuw 2007; Wurzer et al. 2015). The amalgamation of GIS-based analysis techniques into agent-based modelling and advanced statistical software is already on its way and will lead to a further integration and extension of spatial analysis techniques. Perhaps the term GIS will not be that relevant anymore in a decade or so, and the acronym will change meaning to Geographical Information Science.

Furthermore, it would be useless to develop tools and models without taking into account the realities of archaeological data, especially when we are thinking of tackling questions at the supra-regional level. The assessment of the reliability of archaeological and to a lesser extent environmental data sets is probably the most neglected factor in current studies. Archaeology does not qualify as a field with really big data, but it has lots of 'messy' data coming from a multitude of sources with distinct regional and historical peculiarities—one might say that archaeological data has 'character' (Cooper and Green 2015). Obviously, this makes setting up standardized analysis protocols challenging and forces us to be very critical not only of the original data but also of the modelling results produced, since errors in the data will be propagated into our analyses. Tools to perform effective data mining on digital archaeological data are still in their infancy, but will inevitably start to play a more important role in future research aiming at the analysis of data from multiple sources across institutional and even national boundaries (Wilcke 2015; Chiarcos et al. 2016).

GIS will certainly continue to contribute to the debate on the utility of digital and quantitative methods in archaeology—and we need to be aware that it has clear limitations in what it can do for us. In the end, it is the eclectic attitude advocated by Hacıgüzeller (2012) that will bring us further: concepts from GIS need to be combined with other theoretical and methodological approaches to create a new practice of archaeology—one in which doing spatial analysis is as natural as analysing pottery, interpreting stratigraphy or theorizing about identity and in which all these aspects of doing archaeology will be mutually reinforcing.

References

Becker KM, Altschul JH (2008) Path finding: the archaeology of trails and trail systems. In: Altschul JH, Rankin AG (eds) Fragile patterns: the archaeology of the western Papaguería. SRI Press, Tucson, pp 419–446

Bell T, Lock G (2000) Topographic and cultural influences on walking the Ridgeway in later prehistoric times. In: Lock G (ed) Beyond the map. Archaeology and spatial technologies. IOS Press, Amsterdam, pp 85–100

Bell T, Wilson A, Wickham A (2002) Tracking the Samnites: landscape and communications routes in the Sangro Valley, Italy. Am J Archaeol 106:169–186

Bevan A (2013) Travel and interaction in the Greek and Roman world. A review of some computational modelling approaches. In: Dunn S, Mahoney S (eds) The digital classicist 2013. Wiley-Blackwell, London

Bevan A, Conolly J (2006) Multiscalar approaches to settlement analysis. In: Lock G, Molyneaux BL (eds) Confronting scale in archaeology. Issues of theory and practice. Springer, New York, pp 217–324

Bevan A, Wilson A (2013) Models of settlement hierarchy based on partial evidence. J Archaeol Sci 40:2415–2427

Bourgeois Q (2013) Monuments on the horizon. The formation of the barrow landscape throughout the 3rd and 2nd millennium BC. Sidestone Press, Leiden

Chiarcos C, Lang M, Verhagen P (2016) IT-assisted exploration of excavation reports using natural language processing in the archaeological research process. In: Campana S, Scopigno R, Carpentiero G, Cirillo M (eds) CAA 2015. Keep the revolution going. Proceedings of the 43rd annual conference on computer applications and quantitative methods in Archaeology, Siena. Archaeopress, Oxford, pp 87–94

Cooper A, Green C (2015) Embracing the complexities of 'big data' in archaeology: the case of the English Landscape and Identities project. J Archaeol Method Theory 23(1):271–304. https://doi.org/10.1007/s10816-015-9240-4

Čučković Z (2015) Exploring intervisibility networks: a case study from Bronze and Iron Age Istria (Croatia and Slovenia). In: Giligny F, Djindjian F, Costa L, Moscati P, Robert S (eds) CAA2014. 21st century archaeology. Concepts, methods and tools. Proceedings of the 42nd annual conference on computer applications and quantitative methods in archaeology. Archaeopress, Oxford, pp 469–478

De Montis A, Caschili S (2012) Nuraghes and landscape planning: coupling viewshed with complex network analysis. Landsc Urban Plan 105:315–324

Dijkstra E (1959) A note on two problems in connexion with graphs. Numer Math 1:269–271

Ducke B, Kroefges PC (2008) From points to areas: constructing territories from archaeological site patterns using an enhanced Xtent model. In: Posluschny A, Lambers K, Herzog I (eds) Layers of perception. 35th International conference on computer applications and quantitative methods in archaeology (CAA), April 2007, Berlin, Germany. Dr. Rudolf Habelt GmbH, Bonn, pp 245–251

Durand-Dastès F, Favory F, Fiches JL, Mathian H, Pumain D, Raynaud C, Sanders L, van der Leeuw S (1998) Des oppida aux métropoles. Archéologues et géographes en vallée du Rhône. Anthropos, Paris

Ejstrud B (2005) Cost surface analysis and ancient roads: a comparison. In: Berger JF, Bertoncello F, Braemer F, Davtian G, Gazenbeek M (eds) Temps et espaces de l'Homme en société, analyses et modèles spatiaux en archéologie, Actes des XXVe Rencontres internationales d'archéologie et d'histoire d'Antibes, 21–23 Octobre 2004. Éditions APDCA, Sophia-Antipolis, pp 135–139

Fábrega-Álvarez P, Parcero-Oubiña C (2007) Proposals for an archaeological analysis of pathways and movement. Archeologia e Calcolatori 18:121–140

Favory F, Girardot JJ, Nuninger L, Tourneux FP (1999) Archaeomedes II: une étude de la dynamique de l'habitat rural en France méridionale, dans la longue durée (800 av. J.-C. – 1600 ap. J.-C.) AGER 9:15–35

Findlow FJ, Ericson JE (eds) (1980) Catchment analysis. Essays on prehistoric resource space. University of California, Los Angeles

Fisher PF (1992) First experiments in viewshed uncertainty: simulating the fuzzy viewshed. Photogramm Eng Remote Sens 58:345–352

Fitzjohn M (2007) Viewing places: GIS applications for examining the perception of space in the mountains of Sicily. World Archaeol 39:36–50

Fiz I, Orengo H (2008) Simulating communication routes in mediterranean alluvial plains. In: Posluschny A, Lambers K, Herzog I (eds) Layers of perception. 35th International conference on computer applications and quantitative methods in archaeology (CAA), April 2007, Berlin, Germany. Dr. Rudolf Habelt GmbH, Bonn, pp 316–321

Fleming A (2006) Post-processual landscape archaeology: a critique. Camb Archaeol J 16:267–280

Fovet E (2005) Dynamique spatiale du peuplement et analyse des ressources agro-pastorales dans le bassin de Combas et le vallon de l'Aigalade (Gard, France), du Bronze final au bas Moyen Age. In: Berger JF, Bertoncello F, Braemer F, Davtian G, Gazenbeek M (eds) Temps et espaces de l'Homme en société, analyses et modèles spatiaux en archéologie. Actes des XXVe Rencontres internationales d'archéologie et d'histoire d'Antibes, 21–23 Octobre 2004. Éditions APDCA, Sophia-Antipolis, pp 499–503

Fovet E, Zakšek K (2014) Path network modelling and network of aggregated settlements: a case study in Languedoc (southeastern France). In: Polla S, Verhagen P (eds) Computational approaches to movement in archaeology. Theory, practice and interpretation of factors and effects of long term landscape formation and transformation. De Gruyter, Berlin pp 43–72

Gaffney V, Stančič Z (1991) GIS approaches to regional analysis: a case study of the island of Hvar. Znanstveni institut Filozofske fakultete, Ljubljana
Gaffney V, van Leusen M (1995) Postscript-GIS, environmental determinism and archaeology: a parallel text. In: Lock G, Stančič Z (eds) Archaeology and geographical information systems: a European perspective. Taylor and Francis, London, pp 367–382
Gandini C, Favory F, Nuninger L (eds) (2012) Settlement patterns, production and trades from the Neolithic to the Middle Ages, ARCHAEDYN, Seven millennia of territorial dynamics, final conference, Dijon, 23–25 June 2008. Archaeopress, Oxford
Gietl R, Doneus M, Fera M (2008) Cost distance analysis in an alpine environment: comparison of different cost surface modules. In: Posluschny A, Lambers K, Herzog I (eds) Layers of perception. 35th International conference on computer applications and quantitative methods in archaeology (CAA), April 2007, Berlin, Germany. Dr. Rudolf Habelt GmbH, Bonn, p 342 & CD-ROM
Gillings M (2009) Visual affordance, landscapes and the megaliths of Alderney. Oxf J Archaeol 28:335–356
Groenhuijzen MR, Verhagen P (2015) Exploring the dynamics of transport in the Dutch limes. eTopoi J Anc Stud Spec 4:25–47
Hacıgüzeller P (2012) GIS, critique, representation and beyond. J Soc Archaeol 12:245–263
Hasenstab RJ (1983) A preliminary cultural resource sensitivity analysis for flood control facilities construction in the Passaic River basin of New Jersey. US Army Corps of Engineers, Marietta, GA
Herzog I (2013) Theory and practice of cost functions. In: Contreras F, Farjas M, Melero FJ (eds) Fusion of cultures, Proceedings of the 38th annual conference on computer applications and quantitative methods in archaeology, Granada, Spain, April 2010. Archaeopress, Oxford, pp 375–382
Higgs ES, Vita-Finzi C (1972) Prehistoric economy: a territorial approach. In: Higgs ES (ed) Papers in economic prehistory. Studies by members and associates of the British Academy major research project in the early history of agriculture. Cambridge University Press, Cambridge, pp 27–36
Howey MCL (2007) Using multi-criteria cost surface analysis to explore past regional landscapes: a case study of ritual activity and social interaction in Michigan, AD 1200–1600. J Archaeol Sci 34:1830–1846
Huggett J (2013) Disciplinary issues: challenging the research and practice of computer applications in archaeology. In: Earl G, Sly T, Chrysanthi A, Murrieta-Flores P, Papadopoulos C, Romanowka I, Wheatley D (eds) Archaeology in the digital era. Papers from the 40th annual conference of computer applications and quantitative methods in archaeology (CAA), Southampton, 26–29 March, vol 2012. Amsterdam University Press, Amsterdam, pp 13–24
Judge JW, Sebastian L (eds) (1988) Quantifying the present and predicting the past: theory, method and application of archaeological predictive modelling. U.S. Department of the Interior, Bureau of Land Management, Denver
Kohler TA, van der Leeuw S (2007) Model-based archaeology of socionatural systems. School of Advanced Research, Santa Fe
Kvamme KL (1983) A manual for predictive site location models: examples from the Grand Junction District, Colorado. Bureau of Land Management, Grand Junction District
Kvamme KL (1984) Models of prehistoric site location near Pinyon Canyon, Colorado. In: Condie CJ (ed) Papers of the Philmont conference on the archaeology of northeastern New Mexico. New Mexico Archaeological Council, Albuquerque, pp 349–370
Kvamme KL (1999) Recent directions and developments in geographic information systems. J Archaeol Res 7:153–201
Lake M, Woodman PE (2003) Visibility studies in archaeology: a review and case study. Environ Plann B Plann Des 30:689–707
Llobera M (1996) Exploring the topography of mind: GIS, social space and archaeology. Antiquity 70:612–622
Llobera M (2000) Understanding movement: a pilot model towards the sociology of movement. In: Lock G (ed) Beyond the map. Archaeology and spatial technologies. IOS Press, Amsterdam, pp 65–84
Llobera M (2003) Extending GIS-based visual analysis: the concept of visualscapes. Int J Geogr Inf Sci 17:25–48
Llobera M (2007a) Reconstructing visual landscapes. World Archaeol 39:51–69
Llobera M (2007b) Modeling visibility through vegetation. Int J Geogr Inf Sci 21:799–810
Lock G, Kormann M, Pouncett J (2014) Visibility and movement: towards a GIS-based integrated approach. In: Polla S, Verhagen P (eds) Computational approaches to movement in archaeology. Theory, practice and interpretation of factors and effects of long term landscape formation and transformation. De Gruyter, Berlin, pp 23–42
Loots L, Nackaerts K, Waelkens M (1999) Fuzzy viewshed analysis of the Hellenistic city defence system at Sagalassos, Turkey. In: Dingwall L, Exon S, Gaffney V, Laflin S, van Leusen M (eds) Archaeology in the age of the internet. CAA97. Computer applications and quantitative methods in archaeology. Proceedings of the 25th anniversary conference, University of Birmingham, April 1997. Archaeopress, Oxford, p 82 & CD-ROM
McCoy MD, Ladefoged TN (2009) New developments in the use of spatial technology in archaeology. J Archaeol Res 17:263–295
Mlekuž D (2014) Exploring the topography of movement. In: Polla S, Verhagen P (eds) Computational approaches to movement in archaeology. Theory,

practice and interpretation of factors and effects of long term landscape formation and transformation. De Gruyter, Berlin, pp 5–22
Murrieta-Flores P (2012) Understanding movement during late prehistory through spatial technologies: the role of natural areas of transit in south-western Iberia. Trab Prehist 69:114–133
Murrieta-Flores P (2014) Developing computational approaches for the study of movement: assessing the role of visibility and landscape markers in terrestrial navigation during Iberian late prehistory. In: Polla S, Verhagen P (eds) Computational approaches to movement in archaeology. Theory, practice and interpretation of factors and effects of long term landscape formation and transformation. De Gruyter, Berlin, pp 99–132
Ogburn DE (2006) Assessing the level of visibility of cultural objects in past landscapes. J Archaeol Sci 33:405–413
Orton C (2000) Sampling in archaeology. Cambridge University Press, Cambridge
Robb J, van Hove D (2003) Gardening, foraging and herding: neolithic land use and social territories in Southern Italy. Antiquity 77:241–254
Ruestes Bitrià C (2007) A multi-technique GIS visibility analysis for studying visual control of an Iron Age landscape. Internet Archaeology 23. 10.11141/ia.23.4
Ruggles CLN, Medyckyj-Scot DJ (1996) Site location, landscape visibility, and symbolic astronomy: a Scottish case study. In: Maschner HDG (ed) New methods, old problems: geographic information systems in modern archaeological research. Southern Illinois University, Carbondale, IL, pp 127–146
Sevenant M, Antrop M (2007) Settlement models, land use and visibility in rural landscapes: two case studies in Greece. Landsc Urban Plan 80:362–374
Standley A (2015) Making tracks: simulating prehistoric human travel networks. Northland, New Zealand. Vrije Universiteit Amsterdam, Amsterdam
Thomas J (1993) The politics of vision and the archaeologies of landscape. In: Bender B (ed) Landscape, politics and perspectives. Berg, Oxford, pp 19–48
Thomas J (2001) Archaeologies of place and landscape. In: Hodder I (ed) Archaeological theory today. Polity Press, Cambridge, pp 165–186
Thomas J (2004) Archaeology and modernity. Routledge, London
Tilley C (2004) The materiality of stone: explorations in landscape phenomenology 1. Berg, Oxford
Tilley C (2008) Body and image: explorations in landscape phenomenology 2. Left Coast Press, Walnut Creek, CA
Trifković V (2006) Persons and landscapes: shifting scales of landscape archaeology. In: Lock G, Molyneaux BL (eds) Confronting scale in archaeology. Issues of theory and practice. Springer, New York, pp 217–324
Tschan AP, Raczkowski W, Latalowa M (2000) Perception and viewshed: are they mutually inclusive? In: Lock G (ed) Beyond the map. Archaeology and spatial technologies. IOS Press, Amsterdam, pp 28–48
van der Leeuw S, Favory F, Fiches JL (eds) (2003) Archéologie et systèmes socio-environnementaux. Études multiscalaires sur la vallée du Rhône dans le programme Archaeomedes. CNRS Éditions, Paris
van Lanen R, Kosian MC, Groenewoudt BJ, Spek T, Jansma E (2015) Best travel options: modelling Roman and early-medieval routes in the Netherlands using a multi-proxy approach. J Archaeol Sci Rep 3:144–159
van Leusen PM (2002) Pattern to process. Methodological investigations into the formation and interpretation of spatial patterns in archaeological landscapes. Rijksuniversiteit Groningen, Groningen
Verhagen P (2007) Case studies in archaeological predictive modelling. Leiden University Press, Leiden
Verhagen P (2012) Biting off more than we can chew? The current and future role of digital techniques in landscape archaeology. In: Kluiving SJ, Guttmann-Bond E (eds) Landscape archaeology between art and science. From a multi- to an interdisciplinary approach. Amsterdam University Press, Amsterdam, pp 309–320
Verhagen P, Jeneson K (2012) A Roman puzzle. Trying to find the Via Belgica with GIS. In: Chrysanthi A, Murrieta-Flores P, Papadopoulos C (eds) Thinking beyond the tool. Archaeological computing and the interpretive process. Archaeopress, Oxford, pp 123–130
Verhagen P, Whitley TG (2012) Integrating archaeological theory and predictive modeling. A live report from the scene. J Archaeol Method Theory 19:49–100
Verhagen P, Nuninger L, Tourneux FP, Bertoncello F, Jeneson K (2013a) Introducing the human factor in predictive modelling: a work in progress. In: Earl G, Sly T, Chrysanthi A, Murrieta-Flores P, Papadopoulos C, Romanowka I, Wheatley D (eds) Archaeology in the digital era. Papers from the 40th annual conference of computer applications and quantitative methods in archaeology (CAA), Southampton, 26–29 March 2012. Amsterdam University Press, Amsterdam, pp 379–388
Verhagen P, Brughmans T, Nuninger L, Bertoncello F (2013b) The long and winding road. Combining least cost paths and network analysis techniques for settlement location analysis and predictive modelling. In: Earl G, Sly T, Chrysanthi A, Murrieta-Flores P, Papadopoulos C, Romanowka I, Wheatley D (eds) Archaeology in the digital era. Papers from the 40th annual conference of computer applications and quantitative methods in archaeology (CAA), Southampton, 26–29 March 2012. Amsterdam University Press, Amsterdam, pp 357–366
Verhagen P, Nuninger L, Bertoncello F, Castrorao Barba A (2016) Estimating the "memory of landscape" to predict changes in archaeological settlement patterns. In: Campana S, Scopigno R, Carpentiero G, Cirillo M (eds) CAA 2015. Keep the revolution going. Proceedings of

the 43rd annual conference on computer applications and quantitative methods in Archaeology, Siena. Archaeopress, Oxford, pp 623–636

Wagtendonk A, Verhagen P, Jeneson K, Soetens S, de Kleijn M (2009) Past in place: the role of geo-ICT in present-day archaeology. In: Scholten HJ, van de Velde R, van Manen N (eds) Geospatial technology and the role of location within science. Springer, Dordrecht, pp 59–86

Wheatley D (1993) Going over old ground: GIS, archaeological theory and the act of perception. In: Andresen J, Madsen T, Scollar I (eds) Computing the past. Computer applications and quantitative methods in archaeology. CAA92. Aarhus University Press, Aarhus, pp 133–138

Wheatley D (1995) Cumulative viewshed analysis: a GIS-based method for investigating intervisibility, and its archaeological application. In: Lock G, Stančič Z (eds) Archaeology and geographical information systems: a European perspective. Taylor and Francis, London, pp 171–185

Wheatley D (1996) Between the lines: the role of GIS-based predictive modelling in the interpretation of extensive survey data. Analecta Praehistorica Leidensia 28:275–292

Wheatley D (2004) Making space for an archaeology of place. Internet Archaeology 15. 10.11141/ia.15.10

Wheatley D, Gillings M (2000) Vision, perception and GIS: developing enriched approaches to the study of archaeological visibility. In: Lock G (ed) Beyond the map. Archaeology and spatial technologies. IOS Press, Amsterdam, pp 1–27

Whitley TG, Burns G (2008) Conditional GIS surfaces and their potential for archaeological predictive modelling. In: Posluschny A, Lambers K, Herzog I (eds) Layers of perception. 35th International conference on computer applications and quantitative methods in archaeology (CAA), April 2007, Berlin, Germany. Dr. Rudolf Habelt GmbH, Bonn, pp 292–298

Whitley TG, Hicks LM (2003) A geographic information systems approach to understanding potential prehistoric and historic travel corridors. Southeast Archaeol 22:77–91

Whitley TG, Moore G, Goel G, Jackson D (2010) Beyond the marsh: settlement choice, perception and spatial decision-making on the Georgia coastal plain. In: Frischer B, Crawford J, Kollers D (eds) CAA 2009. Making history interactive. Computer applications and quantitative methods in archaeology. Proceedings of the 37th conference, Williamsburg, VA, USA. Archaeopress, Oxford, pp 380–390

Wilcke WX (2015) ARIADNE D16.1: first report on data mining. VU University, Amsterdam

Wurzer G, Kowarik K, Reschreiter H (2015) Agent-based modeling and simulation in archaeology. Springer, Cham

Zakšek K, Fovet E, Nuninger L, Pobodnikar T (2008) Path modelling and settlement pattern. In: Posluschny A, Lambers K, Herzog I (eds) Layers of perception. 35th International conference on computer applications and quantitative methods in archaeology (CAA), April 2007, Berlin, Germany. Dr. Rudolf Habelt GmbH, Bonn, pp 309–315

Open Access This chapter is licensed under the terms of the Creative Commons Attribution 4.0 International License (http://creativecommons.org/licenses/by/4.0/), which permits use, sharing, adaptation, distribution and reproduction in any medium or format, as long as you give appropriate credit to the original author(s) and the source, provide a link to the Creative Commons license and indicate if changes were made.

The images or other third party material in this chapter are included in the chapter's Creative Commons license, unless indicated otherwise in a credit line to the material. If material is not included in the chapter's Creative Commons license and your intended use is not permitted by statutory regulation or exceeds the permitted use, you will need to obtain permission directly from the copyright holder.

3 Methods and Perspectives of Geoarchaelogical Site Catchment Analysis: Identification of Palaeoclimate Indicators in the Oder Region from the Iron to Middle Ages

Armin Volkmann

Abstract

In the last years, site catchment analyses came under criticism and were only used rarely in the geoarchaeological studies. It was assumed something like an eco-determinism, which most often did not consider a lack of free decision-making when choosing a new location for a new settlement. But it cannot be stated that prehistoric settlers were not affected by the geoecological potential given by their surrounding environment and that they chose settlement sites according to it. The more a prehistoric culture was based on agriculture and stock farming, the stronger was their dependency on geoecological site factors in the settlement region. New studies with an extended application of the site catchment analyses illustrate those relations very impressively. Therefore, it is essential to identify the prehistoric records, like the settlements belonging to an economic area, very accurately in order to evaluate the geoecological environment data. Different models can be applied in the geographic information systems (GIS) for the reconstruction of prehistoric surroundings that one wants to compare. In this way, it is possible to understand the human-environment interactions better in prehistoric settlement surroundings, and plausible thesis based upon a well-founded data basis can be verified or falsified as well.

As an example, Iron Age to early Middle Ages settlements in the Oder river region illustrate the high significance of site catchment analysis. Due to the acquisition and evaluation of environment data, statistically significant conclusions can be made by a degree of representation. In this way, a method was developed in which indirect signals of the palaeoclimate can be

A. Volkmann (✉)
Heidelberg University, Junior Research Group Digital Humanities and Cultural Heritage at the Interdisciplinary Center for Scientific Computing and Cluster of Excellence Asia & Europe in Global Context, Heidelberg, Germany
e-mail: armin.volkmann@asia-europe.uni-heidelberg.de

© Springer International Publishing AG 2018

C. Siart et al. (eds.), *Digital Geoarchaeology*, Natural Science in Archaeology,
DOI 10.1007/978-3-319-25316-9_3

identified. Compared to isotope- and dendrochronological palaeoclimate investigations, the general functionality was verified. Digital maps with the relevant geoecological information (soil type, groundwater, topography, vegetation, etc.) have been the basis for this study. They were evaluated in a GIS database together with the archaeological site data. The focus therefore lies on hints of microregional humidity and thermal conditions at the settlements. It becomes apparent that prehistoric (From the Oder river region are almost no written sources before the high Middle Ages up to the thirteenth century when history starts. The high Middle Ages settlements are not under investigation.) settlements from the Iron Age to the early Middle Ages have different preferred positions with characteristic geoecological potentials in relation to time- and culture-specific needs. The areas have been chosen for suitability for agrarian-oriented settlement under a certain palaeoclimate with increasing and decreasing humid conditions. Agrarian-oriented settlements could be confirmed because of the analysis of the archaeological discoveries, like longhouses with parts for residents and stable ('Wohnstallhäuser') indicators for ridge and furrow ploughing and Celtic field systems ('Hakenpflugspuren') [about settlement structures of the investigation area, see Leube (2009, 7–17, 119–127)].

Keywords
Site catchment analyses • GIS • Settlement environment • Palaeoclimate • Ecological footprint • Palaeolandscape • Iron Age • Middle Ages • Oder river region

3.1 Introduction

Site catchment analysis is not a new method for the systematic recording of settlement surroundings. Indeed, initial attempts date back as far as von Thünen's model of concentric zones for agricultural goods in the nineteenth century (1826). The site catchment analysis method was fundamentally developed in the 1960s and 1970s by geologist Claudio Vita-Finzi and archaeologist Eric Sidney Higgs (Vita-Finzi and Higgs 1970) and expanded upon over the following decade particularly in numerous studies of prehistoric human-environment interactions (Jarmann et al. 1972; Roper 1979). The sometimes quite arbitrary concentric zones were however utilized less and less frequently in order to better take into account considerations of accessibility to geoecological resources with regard to the landscape (cf. Birkett 1985).

The increased use of computers and associated development of powerful geographic information systems (GIS) in geography as well as (after an adoption lag) archaeology inspired the methodological and theoretical development of 'new archaeology' or 'processual archaeology'.[1] This theory and paradigm shift changed the previous traditional approach of qualitative investigations to the study of cultural history by scientific, evidence-based approaches for large data analysis (e.g. Gordon and Phillips 1958; Eggert 1978). Proponents of 'processual archaeology', as it came to be known, attempted to identify and record all possible aspects of prehistoric human-environment interactions through excavation[2] so

[1] See Sect. 2.2 (Verhagen 2018).

[2] In the meaning of processual archaeology, an excavation can be defined as a translation of archaeological data to objective and comparable data.

that they might subsequently undertake complex analytical study ideally based on objective data. The limitations of this methodology were very quickly made clear as the ritual context of a feature, state of preservation and subjectivity of each archaeologist were little considered. This resulted in the 'post-processual' countermovement within which the descriptive approach to research was again advocated.[3] Processual archaeology was, with regard to its assumption of the ability to gather objective archaeological data, criticized. The interpretation of information is almost concluded during excavation and documentation. 'You only see what you know' is a common saying for field archaeologists. It is furthermore difficult to grasp and document the mindsets and mentalities of past populations from prehistory and so involve these factors in analysis and subsequent interpretations. Likewise the environmental determinism of geoecological site factors, such as those around a settlement, was brought into question by the expression of a prehistoric individual's free will. Finally, various archaeological cultures were not generally comparable with one another as they are only recognized in pieces. Of central import to post-processual archaeology was again the individual feature within the context of the prehistoric individual's influence on the specific transformation of local entities and actions.

These theories are however opposed to partly diametrically quantitative studies like GIS analyses in landscape archaeology.[4] As a result, only a few studies have emerged in the previous two decades which systematically performed site catchment analyses on a sufficiently large dataset. Fortunately, however, in recent years, the use of this analysis has been returning and its methodology further refined. In spite of all theoretical discussions within these paradigm shifts, the behaviour of prehistoric humans towards their environment can be succinctly comprehended with this approach.[5] It therefore had to be considered how one could capture, as representatively as possible, past economic zones around prehistoric settlements, whereby simple concentric rings represent only one simple model (see Fig. 3.1). In determining the accessibility of potential agricultural zones near a settlement, more complex models will take the surrounding topography into account to assess cost benefit (as in Herzog and Posluschny 2011), operating on the basis of least-cost path analyses. This is a plausible premise particularly for prehistoric, agrarian-oriented societies. Areas very distant from settlements could however be directly sought out if they possessed special natural resource potential, as in numerous third-fifth centuries AD iron smelting sites located far from contemporary settlements in Niederlausitz (Volkmann 2012). In this article, only the main features of site catchment analysis methods will be discussed and illustrated by a case study in the Oder region.

3.2 Methods of Site Catchment Analyses

Geoarchaeology as a discipline has increased in significance in recent years. In general it is concerned with the transformation of the natural landscape into the cultural landscape (cf. Zimmermann et al. 2005). In particular this transformation is researched on two levels: micro- and macroscale. On the former, the micromorphology of individual archaeological features is analysed in minute detail. At the other end of the spectrum, there is a focus on quantitative GIS studies like environmental analyses where the focus is not on an individual feature, but rather a large dataset consisting of numerous features, appropriate for statistical analysis to confirm apparent patterns.

[3] Trigger (1989, 2006); in German see Bernbeck (1997) and Eggert and Veit (1998).

[4] Cf. Kluiving (2010); for general remarks on GIS, see Bill (2010) and Burrough et al. (2015).

[5] Cf. Ducke and Kroefges (2007), Posluschny et al. (2012), Ullah (2011) and Volkmann (2013)

Fig. 3.1 Soil map of Brandenburg (BÜK300) and in the northeast, on the lower Oder in the Polish Voivodeship Zachodniopomorskie, and the geological-geomorphological map (GK25) in GIS showing settlements from the Early Roman Period phases A (*green*) and B (*yellow*), Late Roman Period phase C (*red*), to the Early Migration Period phase D (*blue*), orientated North. Also the settlements of the Late Migration Period and Early Middle Ages were studied within the systematic site catchment analysis (GIS author; BÜK300 Brandenburg State Office for Mining, Geology and Raw Materials—Landesamt für Bergbau, Geologie und Rohstoffe Brandenburg and archaeological site information from the Brandenburg State Office for Monuments and the Archaeological State Museum—Brandenburgisches Landesamt für Denkmalpflege und Archäologisches Landesmuseum)

The interplay between natural and human-influenced landscape development over time is therefore a central research tenet of geoarchaeology. As such, the changeable influence of prehistoric features like settlements, graves, etc. is treated as influencing factors on the landscape as is the influence of the landscape on the position of these features. The effects of human activity are not only limited to specific position of settlements but also the types of settlement with internally individual houses. Externally this is further applicable in the placement of fields and pastures in the immediate surroundings, as well as that of gathering and forestry zones in the broader environment. This influence has clear effects on the landscape, and the methodology of landscape analysis has actively evolved to address these interactions as an adaptation to increased amount of internal and external variables.

The underlying concept of analysis of surrounding zones is based on the idea that the location of a settlement in a given landscape is based on a combination of various ecological units of the site which collectively seem to make it attractive or unattractive. These geoecological-related site features are particularly essential for agrarian-oriented prehistoric societies although artificial, social or psychological reasons can also be crucial in settlement siting. Aspects of free will, political and socio-economic structure of prehistoric societies or cultural activities can only be taken into account partially or not at all here, although this is largely rendered irrelevant by the methods employed. When performing site catchment analyses, these aspects must be considered and discussed within the frame of the results. This is not about a simple reduction in complexity of a causal analysis of the primary reasons for site selection, which is due only partly to human reasons and partly to geoecological ones, but rather understanding of human activities as complex interactions with their geoecological surroundings. Thus, site catchment analyses can provide an explanation to better understand exactly these relationships between prehistoric human activity and natural conditions. An adaptation of the site catchment analysis is the site location analysis, which respects also socio-political factors of human behaviour.[6]

Generally within a site catchment analysis, the geoecological data from at least one archaeological site are evaluated. Where a low number of geoecological units are involved, these surrounding geodata can also be collected and evaluated in the frame of a survey. In the confines of a quantitative study, mostly up to 50 or several hundreds of archaeological sites are processed for statistical validity as most geodata[7] are not collected directly in the field but frequently extracted and analysed from respective maps with already extracted field information. A combination of geodata collection with simultaneous confirmation through corresponding field data in thematic maps is especially necessary. If appropriate thematic maps are not available or of a quality too low for analysis, perhaps because units are too generalized, a direct survey of the relevant region is probably necessary. Site catchment analyses require field data or highly detailed maps which must be available for the observed study area. In a theoretical, ideal case, these are drawn to at least a 1:25,000 scale without gross generalizations and exist without gaps for a larger study area, because the respective typical location details from the map must be standardized for all settlements (see Hunt 1992, 288). Therefore, in addition to the validation of the available geodata during collection, a method must be adapted for each site catchment analysis specific to the questions asked and appropriate for the study area. In the frame of site catchment analyses, there is no universal standard procedure, because both the context of archaeological features and the characteristics of the specific environment must be addressed.

3.2.1 Concentric Zones

The simplest conception of site catchment analysis consists of the evaluation of geodata from a feature's surroundings in concentric rings (Fig. 3.2, above left). The operating radius of prehistoric settlers can be estimated to be, depending on archaeological culture and associated intensity and significance of agriculture and animal husbandry, between 1 and 10 km.[8] Archaeological cultures whose primary subsistence strategy is based on field agriculture often have a higher local concentration in which the said fields are largely located in the immediate vicinity of the settlement. In such cases, a radius is defined between 1.5 and 6 km.[9] Where economic income is more dependent on animal husbandry, a higher daily mobility is presupposed, and an operating radius of 5–8 km is considered more realistic, because, e.g. cattle needs large areas with fresh grass. In contrast highly mobile societies like hunter-gatherer cultures roam widely, and for these an operating radius of 8–15 km is appropriate (see Renfrew and Bahn 2005, 172). The larger the radius, however, the more difficult it is to obtain significant data: A great deal of geodata are collected for potential economic zones even though not all may be utilized by the prehistoric farmers. It is therefore essentially necessary to identify prehistoric economic zones which were actually utilized, as only data from these sites are relevant. These must be collected as precisely as possible in order to avoid inaccuracy in their evaluation.

[6] See Sect. 2.3.1.

[7] Except for spatial planning, potentially present topology, as well as past topology through layers and drilling, etc.

[8] Based on cost-benefit ratios cf. Bintliff (1999, 505ff)

[9] For the Iron Age in SW Germany, '. . .the land used for agricultural needs, mainly for ploughing, is usually not further away than 1 km which is approximately a 12 or 15 min walk from the settlements' cf. Posluschny et al. (2012, 461).

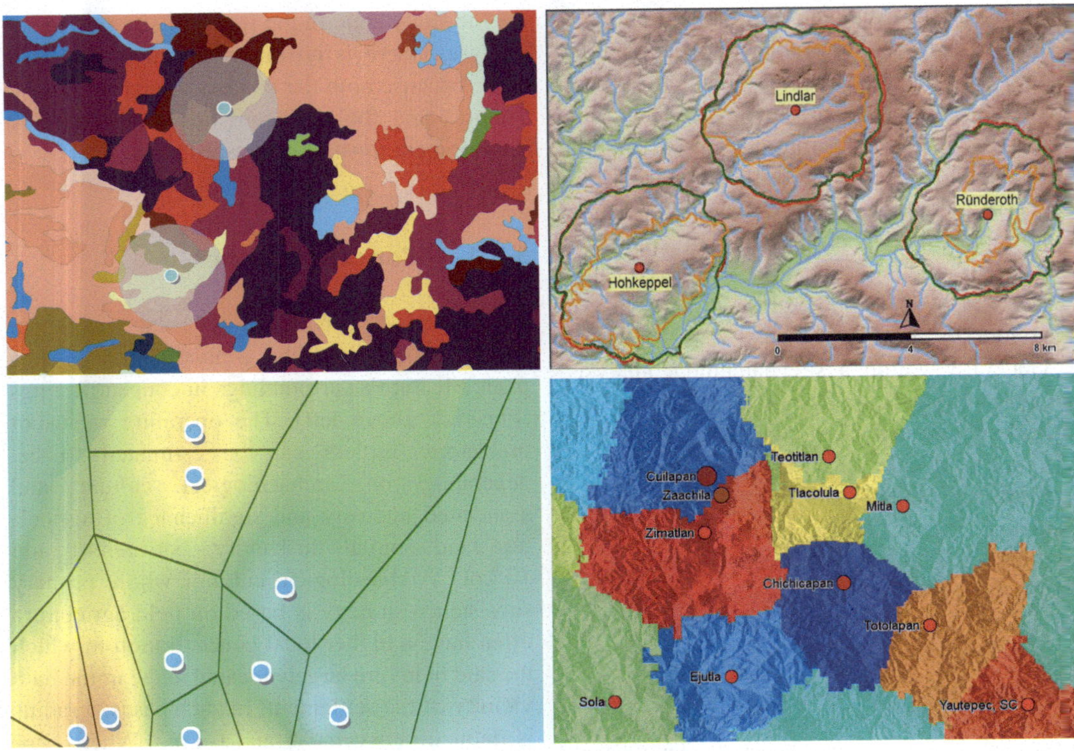

Fig. 3.2 Comparison of zones surrounding prehistoric features. *Clockwise from the top left*: concentric rings (author); accessible areas (from Herzog 2014); XTENT model of expanded accessible areas (from Ducke and Kroefges 2007); Thiessen polygons in a Voronoi diagram (author)

3.2.2 Accessible Areas

Instead of simple concentric rings, surrounding zones based on estimated accessibility can also be determined in GIS. In these areas, a zone which would have been topographically accessible to prehistoric settlers for a defined period of time is calculated around a given feature like a settlement (Fig. 3.2, top right). These calculations require the existence of surface information with terrain contours or a digital terrain model (in GeoTIFF or other grid formats) in order to determine the distance around the feature with regard to the topographic altitude differential in GIS. These computational processes are also called least-cost or path distance analyses (cf. Herzog 2014; Herzog and Posluschny 2011). Depending on terrain, foot speed lies between 6 and only 0.2 km per hour, respectively, for flat and very steep ground (Posluschny et al. 2012, 416, Fig. 3.3). Based on the consideration of least-cost analyses, it can be assumed that economic zones accessed daily were sites within approximately 1 h travelling distance. If digital terrain models are available for a study area, the use of accessibility calculations can reliably identify areas which were, with a very high probability, actively used in the past.

3.2.3 Thiessen Polygons

As unfortunately digital terrain model data are not available for all regions which could serve as the basis for the calculation of accessible areas, Thiessen polygons in Voronoi diagrams can be determined between individual feature sites in GIS. Consequently, zones can be created from GIS based on Thiessen polygons (Fig. 3.2,

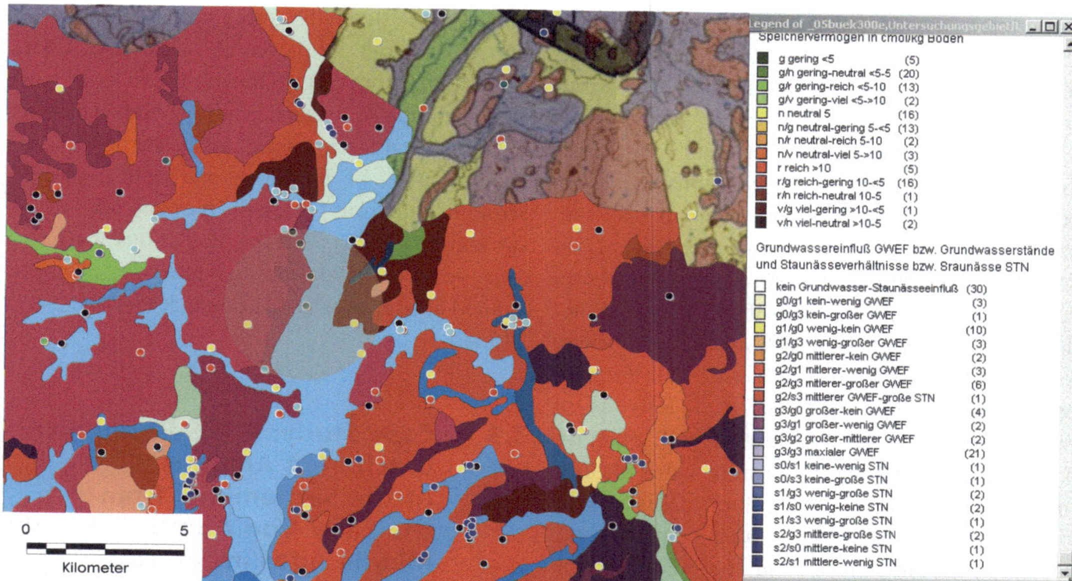

H	Gemarkung	fZN	FPL	Höhe_Nl	Exposition	Relief	GeWabstand	Nutzung	Gefährdung	Geomorp	Kleineinheit	Bodenart	Bodengüte	Bodentyp
)	Friedrichsthal	E	1	45,6	3	1-3	150	3-7	1	1	11a-11b?	das	S7Al	27-16-80-28
)	Berkholz bei Boitzenburg	B2+D	1	75,5	5	2-3	190	1	4	3	28-9-18	sl-sm	-	65-80-40-54
)	Grünow bei Prenzlau	D	11	50	2	1-3	80	1	3	3	28-9-11	dm-hsl-lm	-	64-78-61-70
)	Lützlow	D	12	63	4	2	210	1	3	3	6-8	dm	IS2D	64-71-65-72
)	Prenzlau Exerzierplatz?	C-D	32	42	7	2	120	1	3	4	8	dm sl-sm	-	55-65-67-80
)	Prenzlau Sabinenkloster?	D?	119	54	6	2-3	55	1	3	3	9-28	dm l-m	-	65-71-64-67

Fig. 3.3 Geoinformation was transferred from the various thematic maps (coupled with settlement sites in GIS) into a standardized cartographic key geosignature in the geoarchaeological database. This database is utilized for further qualitative and statistical/quantitative evaluation of the data collected. On the map, the surroundings of the Late Migration Period settlement at Friedrichsthal are marked on the digital soil survey map BÜK300 (see Fig. 3.1). In the database (*bottom*), the corresponding entry of the four soil type signatures nearest to the settlement at Friedrichsthal is outlined in *red*. Legends for digital soil fertility (*top right*) and soil moisture (*bottom right*) are also displayed

bottom left). Thiessen polygons are composed of georeferenced spatial points (seeds) around each site and accompanying planar valuation. A Thiessen polygon is a demarcation surrounding a set of all seeds in a zone which lie nearer to the feature than to any other seed belonging to a neighbouring set (Thurmaier 1999, 23–24). The collection of all Thiessen polygons is a Voronoi diagram. This structure embodies a geometric plane which precisely represents the neighbourhood relationships of the original feature's location.

Due to the triangulated irregular network, connections between seeds are generated by GIS in the course of polygon creation with the aim of keeping triangles as uniform as possible, out of which polygons are composed. This is done with the calculation of Delaunay triangulations which determine that the circumscription of a given triangle contains no seeds belonging to another triangle. In this way the stipulation to maximize the minimum angles of the triangle sides is satisfied, and the requirements for approximately equilateral triangles meet. The lines between seeds, the triangle sides, now represent the starting point for the creation of Thiessen polygons. The perpendicular bisector is then determined for each triangle edge, and the nearest intersections of these bisectors comprise the polygon vertices, the bisecting lines themselves and the edges. In this way, the triangulated irregular network becomes a smoothed rectangular network (Müller 2000), and a table listing polygon attributes is simultaneously and automatically created in GIS. The polygons have the values of their internal seeds as evaluable attributes for the following analysis.

A distance calculation between seeds and their respective Thiessen polygons can be performed whereby a specific average distance value is calculated and which generated yet another evaluable polygon attribute. From these various attributes, it is possible to interpolate thematic maps with coloured zones appropriate to the specific value attributes, e.g. with Kriging performed in this Fig. 3.2 (top left) as a two-dimensional gridded illustration. Care must be taken for the logical division of value classes (based on minimum and maximum values as well as proportionally weighted value distributions) by colour scheme. Instead of the creation of a surface modelled grid (as with Kriging), thematic contour line maps could also be generated as they would similarly outline individual territories and their areas of influence.

3.2.4 XTENT Model of Expanded Accessible Areas

The GRASS toolbox, now implemented in numerous open-source GIS software packages, includes an enhanced and improved version of the almost classically used Renfrew and Level algorithm for the calculation of accessible areas (see Ducke and Kroefges 2007). This formula, called XTENT, offers various advantages because it recognizes the characteristics of the landscape as the basis for the mobility of prehistoric settlers as in the accessibility calculations discussed above. Constraints and hierarchical relationships between sites (like settlements, local and regional centres) and their relationships with one another (see Nakoinz 2009) are however also involved, about which an expanded accessibility calculation can be taken into consideration. Furthermore, hierarchical weighting of individual features, as with the settlements and their Thiessen polygons in the Voronoi diagram, can be calculated to distinguish between centres with greater areas of influence and less important settlements with smaller ones, which serve as regional units in the collection of data (Fig. 3.2, bottom left). The combination of various methods like accessibility and Thiessen polygons enables a general reconstruction of prehistoric areas of influence and their potential economic regions through the calculation of surrounding zones. These calculations are however very complex and so have been rarely applied in the past (Ducke and Kroefges 2007, 245 ff.).

3.3 Case Study: A GIS-Based Site Catchment Analysis of the Settlements from the Early Iron Age to the Early Middle Ages in the Oder Region

A comprehensive and, above all, a systematic environmental analysis of sites from the Roman Iron Age to the Early Middle Ages was applied (Volkmann 2013). A methodological concept was developed for this purpose, whereby data for the subsequent cartographical and statistical analysis were collected within a geoinformation standard. Highly detailed digital datasets (e.g. georeferenced aerial photographs, soil maps, digital ground models, etc.) and collections (e.g. historical and topographic maps) were evaluated as a database in a standardized manner in collaboration with partners from the appropriate technical and regional authorities (e.g. 'Brandenburgisches Landesamt für Denkmalpflege' and 'Geobasisinformationen Brandenburg') (see Fig. 3.3). In addition, the archaeological site information in the archives and in the relevant technical literature was consolidated across boundaries as a geoarchaeological synthesis: a database and a catalogue. In the course of this endeavour, an intensive examination of the source material was carried out, as many sites from the Roman Iron Age or Migration Period were often not recognised as such. Thus, an entirely new and comprehensive picture of the archaeological sites emerged. It was also possible to gain access to several thesis catalogues which could be included for purposes of comparison. However, for all sites, the type of site and its dating were checked thoroughly, as the most finely detailed dating possible was an imperative foundation for the subsequent environmental analyses.

Thus, a revised chronology system was developed in Periods I–III for the Early Iron Age (EIA), Periods A–C for the Roman Iron Age, Periods D–E for the Migration Period and Period of the Early Middle Ages (EMA) for the Early Middle Ages which also permits analogies to be made with reference to the technical literature on the area being studied (e.g. Leube 2009). Besides the very important dating of the sites, the place and circumstances of their discovery were critically examined and corrected for accuracy, as these selective factors have a powerful effect on the profile of the site in the section of the study on settlement archaeology. Within the framework of this critical assessment of sources, the research history, which fundamentally shaped the status of research, was examined (Volkmann 2013, 30–39).

3.3.1 Methodological Approach

As a GIS-based study, the site catchment analysis was performed through the evaluation of numerous thematic maps on topography, soil, vegetation and other geoecological parameters to at least a 1:25,000 scale so that detailed geoinformation from the surroundings of settlement features could be evaluated. The study exclusively investigated settlements in order to ensure comparability, as other sites, such as cemeteries, require other site characteristics not identical to those required by settlements. In one early step, the geodata were collected from within a probable operating radius around respective settlements dating to the individual periods and then statistically analysed. These radii were determined on the basis of least-cost considerations with a high probability of containing the economic zones of the respective settlements (van Leusen 1998, 2002; Verhagen 2010). Not only randomly sampled geodata from a hypothetical radius but rather accessibility analyses with reference to the terrain and watercourses as potential access routes were tested in GIS (Herzog 2014). Furthermore, the deciding site characteristics of the soil and geoecological settlement surroundings as well as artificial and natural superimpositions with distorting effects are discussed (Volkmann 2013, 111–218). The ecological indicator values were fundamentally recombined into concise classes by transformative processes with regard to their applicability and reliability for issues of prehistoric, agrarian-oriented cultures and checked for climate indicators. Based on the large number of settlements studied, it was possible to highlight statistically significant climate proxies for relative humidity and palaeoclimate temperature variation.

A total of 18 different thematic maps were analysed in GIS for this study which are indicative of the geoecological potential of a settlement site. The soil survey map of Brandenburg (BÜK300) from the State Office for Geosciences and Natural Resources is based on GK25 which provides further detailed information on specific soil type and differentiates between 99 soil types with a colour key and official terminology (Fig. 3.3). For the subsequent analysis, these very different soil types were grouped by archaeological aspects into specific classes defined by potential uses. This process had the advantage of transforming many units into quite divergent, smoothed microregional soil-type signatures which were more easily comparable with one another. Thus, not only is the soil type or geomorphological unit within/upon which a settlement site is located a potential deciding factor, but so are the three nearest-lying soil types and geomorphological terrain forms from around the said settlement, copied over from GIS into a table for evaluation in a bar graph (cf. Fig. 3.4).

In Fig. 3.3, a 2.5 km radius plausible for agrarian-oriented cultures is marked around the Late Migration Period (LMP) E settlement at Friedrichsthal in the Uckermark (see Volkmann 2013, 328, Catalogue number 67). This radius encompasses the nearest-lying soil forms for analysis of the surroundings. The geoinformation

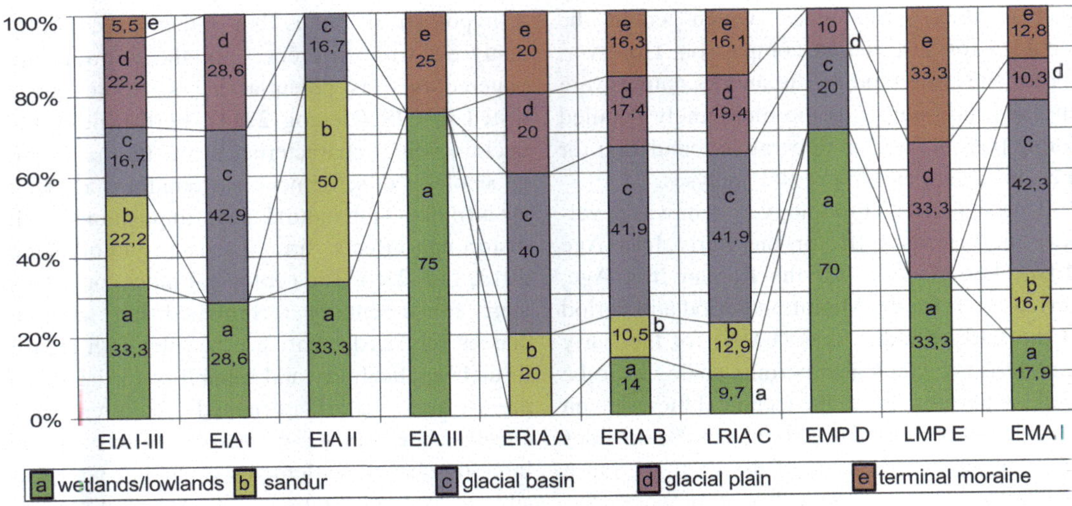

Fig. 3.4 In addition to the 17 other mapping variables used in this study, this example shows percentages of geomorphological units in the settlement surroundings. The drastically varying quantities of settlements in the humid floodplains are clearly visible; these were appropriate settlement sites only in dry conditions. These indirect indicators of palaeoclimate, e.g. of the relative increase or decrease of the humidity index, were utilized to calculate curve and weight by significance (see Volkmann 2013, 213–218); (**a**) wetlands/lowlands, (**b**) sandur, (**c**) glacial basin, (**d**) glacial plain, (**e**) terminal moraine; *EIA* Early Iron Age phases I–III, *ERIA* Early Roman Iron Age phases A–B, *LRIA* Late Roman Iron Age phase C, *EMP* Early Migration Period phase D, *LMP* Late Migration Period phase E, *EMA* Early Middle Ages (author credit)

of the nearest-lying, potential economic areas for the settlement were entered into the database for archaeo- and geoinformation. This database contains data of the surroundings of a total of 682 settlements and was the basis for the following statistical evaluation. It indicates, for example, that Late Roman Iron Age (LRIA) C settlements were preferably sited on brown earth-chernozem (red and dark red surface signatures) which was no longer of interest or played a random role in LMP E site selections. In addition to changes in economic strategies, this is plausibly explained by climate changes.

In the first step of statistical evaluation, geodata on settlement surroundings from the pre-Roman Iron to Early Middle Ages were analysed in simple bar graphs, one for each mapped region as well as map theme, as for the data on geomorphology (see Fig. 3.4). The bar percentage ratios of each time period were ordered by a degree of relative moisture, increasing from top to bottom, derived from altitude and the influence of groundwater on the respective geomorphological units.

3.3.2 Evaluation of Climate Indicators from Geofactors of the Site Catchment Analysis

The values from the bar graphs were ordered by weight in Fig. 3.4 (see formulas below) into wet and dry indices and displayed in interpolated temporal curves. The summary shows the curves within the spectrum of the respective geofactors of geomorphology, topography, precipitation, vegetation, soil type and classification, humidity and distance from a body of water. These encompass the Early Iron Age (EIA) phases I–III, Early Roman Period (ERP) phases A and B, Late Roman Period (LRP) C, Early Migration Period (EMP) phase D, LMP and Early Middle Ages (EMA). The left is the timeline in absolute period values; the right is the average curve of the relative wetness index (blue line) without insignificant distance-from-water values. Left of the diagram are the absolute values for the time periods of varying duration.

The somewhat collectively divergent curves of individual geofactors which are indicators of relative moisture increase or decrease show surprisingly large similarities simultaneously in different areas, particularly from the Roman Period to the EMA. These periods are based on the standard record and clearly indicate the necessity of reviewing dates and feature types (settlement, grave, etc.) garnered from archives as an important basis for further analyses. Because a comparable view should however be rendered from the EIA to EMA, this could only be performed on the basis of existing catalogues from the study area; otherwise, the framework of the study would be overwhelmed.[10] Naturally, the proposed dates were also checked and corrected as necessary. The precision of the evaluation results is fundamentally based on the data collection of accurately dated settlements in a sufficiently large number for the subsequent statistical analyses.

An expected lack of clarity is apparent in the EIA phases II–III, ERP phase A and LMP E due to a relatively low number of dateable settlements in these periods. This problem could however be overcome by the large number of investigated geofactors so that significant representative statements could also be made for these periods. In EMA I, the earliest phase of Slavic immigration is not apparent as the primary source of pottery finds is quite vaguely dated. Within the spectrum of finds, those from the early Slavic settlements and middle Slavic pottery types comprise small parts. In the Oder region, the early Slavic settlements typically either were in use into the middle Slavic period, or a small proportion of so-called middle Slavic pottery types already exist in the early Slavic period. Due to the parallelism of pottery, like the early Slavic undecorated Sukower Type present in low numbers in the middle Slavic pottery record, a considerable insecurity of dating is created for these settlements from the seventh to tenth century AD. Furthermore, the richly decorated Feldberger Type is still attributed to the early Slavic period, but even this is mostly associated with the middle Slavic types (sometimes wheel-thrown and combed ware like Tornower Type, which are often carinated vessels with concentric furrowed band shoulder decoration) in the domestic pottery record. During the investigation of the early Slavic period and in addition to its securely dated settlements, all generally late Slavic settlements were grouped with respective pottery types by Brather's chronology system (Brather 2000, 117–119, Fig. 70).

In the relative wetness index correlation (Fig. 3.5) for settlements dated as finely as possible, human influence on the somewhat varied economic pathways for each time period was apparent in the slightly divergent curves. In the development of the averaged standard moisture curve which touches on the analysis of multiple specific geofactors (Fig. 3.5, far right), the curve was further calibrated with the varying proportions of agricultural and pastoral economies (Fig. 3.6). For this calibration (cf. Fig. 3.7, far right), cultural historical factors of trade or fortified features with varying relevance were also considered (Fig. 3.6). Within the standardized research method procedure, the congruent or divergent curve areas could be checked for significance. The curves are to be

[10] The most important catalogues for the region are, among others, 'The Göritzer Group' (Die Göritzer Gruppe) by Griesa (1982) for the EIA in the Oder-Neisse region and 'Slavs and Germans in the Uckermark' (Slawen und Deutsche in der Uckermark) by Kirsch (2004) on regional transformative processes of the Middle Ages. The ERP and LRP were investigated by Leube in his habilitation dissertation 'A Study of Economy and Settlement by the Germanic Tribes in Northern Central Europe during the first to fifth–sixth Centuries AD' (Studien zu Wirtschaft und Siedlung bei den germanischen Stämmen im nördlichen Mitteleuropa während des 1. bis 5.–6. Jahrhunderts n. Chr.) and published in 2009. For the EMP and LMP, only two dissertations with complete catalogues in the study area were available: Schach-Dörges published 'Features from the third to sixth Centuries AD between the Lower Elbe and Oder' (Die Bodenfunde des 3. bis 6. Jahrhunderts nach Chr. zwischen unterer Elbe und Oder) in 1970, and Voß presented his doctoral dissertation 'Investigations on History of the Germanic Settlement between the Elbe/Saale and Oder/Neisse in the third to seventh centuries' (Untersuchungen zur Geschichte der germanischen Besiedlung zwischen Elbe/Saale und Oder/Neiße im 3. – 7. Jahrhundert) in 1986 although it remains unpublished. The former study considers the state of research for numerous new finds from the archives at the regional archaeology state offices in Brandenburg and Lubusz Voivodeship in a modern-day context.

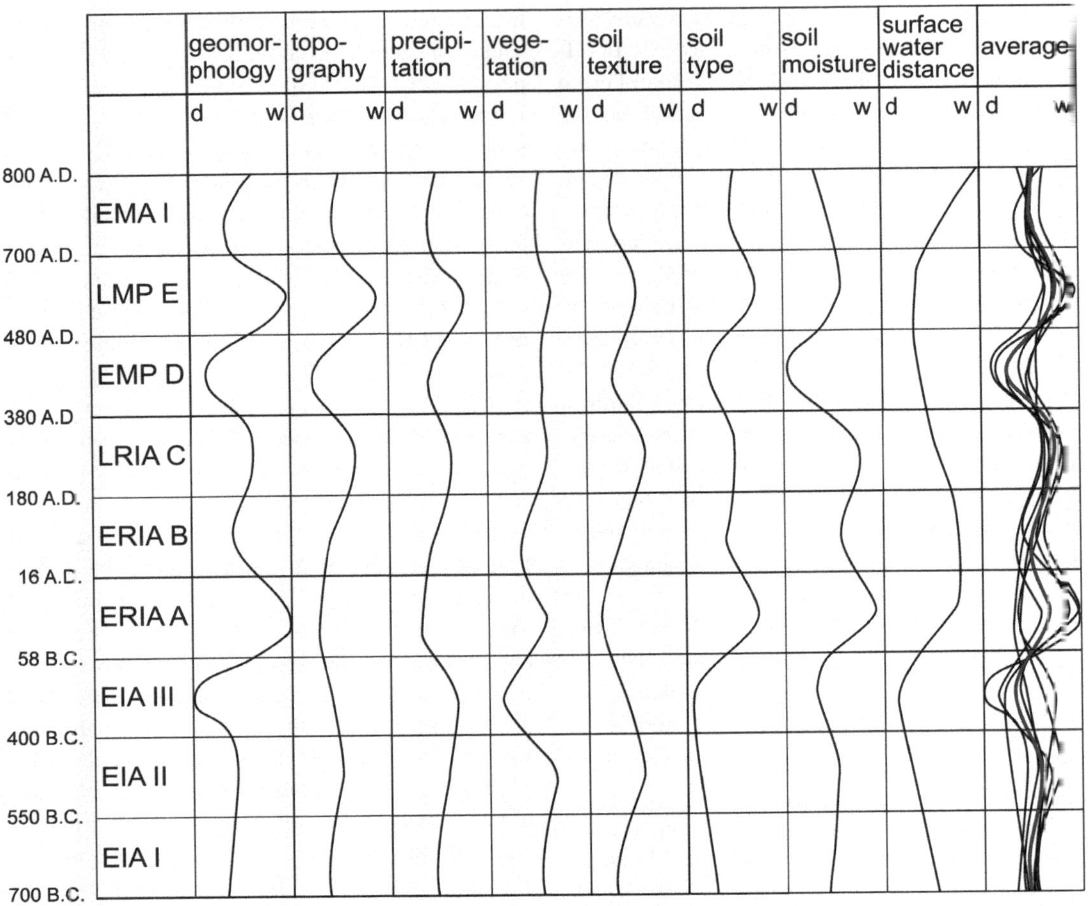

Fig. 3.5 Curve interpolation of the spectrum of various geofactors (see above) from their respective *bar graphs* (cf. Fig. 3.4) with notes on palaeoclimatic moisture activity from 700 BC to 800 AD (*from the bottom to the top left*) and their relative decrease (d = dry) or increase (w = wet)

understood as relatively comparable increases or decreases in moisture (wetness index). They do not represent absolute values such as total precipitation or temperatures. For each period only one relative value was interpolated into the curve.

In the frame of the complex correlation of relative values for moisture increase or decrease from the geofactor analyses of the settlements (and their surroundings of potential economic zones), surprisingly consistent results emerged in the curves. These were even in spite of very different underlying evaluation media from the available maps (vector-based or bitmap) with sometimes different scales and therefore general map units. The functioning of the applied methods is essentially made possible by heterogeneous data sources as here the risk of circular reasoning (that vicious circle) is lower due to independent data collection. It is however entirely apparent that the developed methods for site catchment analysis lose precision as the researched settlements increase in age as is here the case for the EIA (cf. Figs. 3.5 and 3.6, right of the average curves). In addition to the difficult dating of these settlements in the Oder region, similarly to those of the EMA, interference from subsequent human activity increases considerably as time passes, complicating the evaluation. It is understandable that the recent

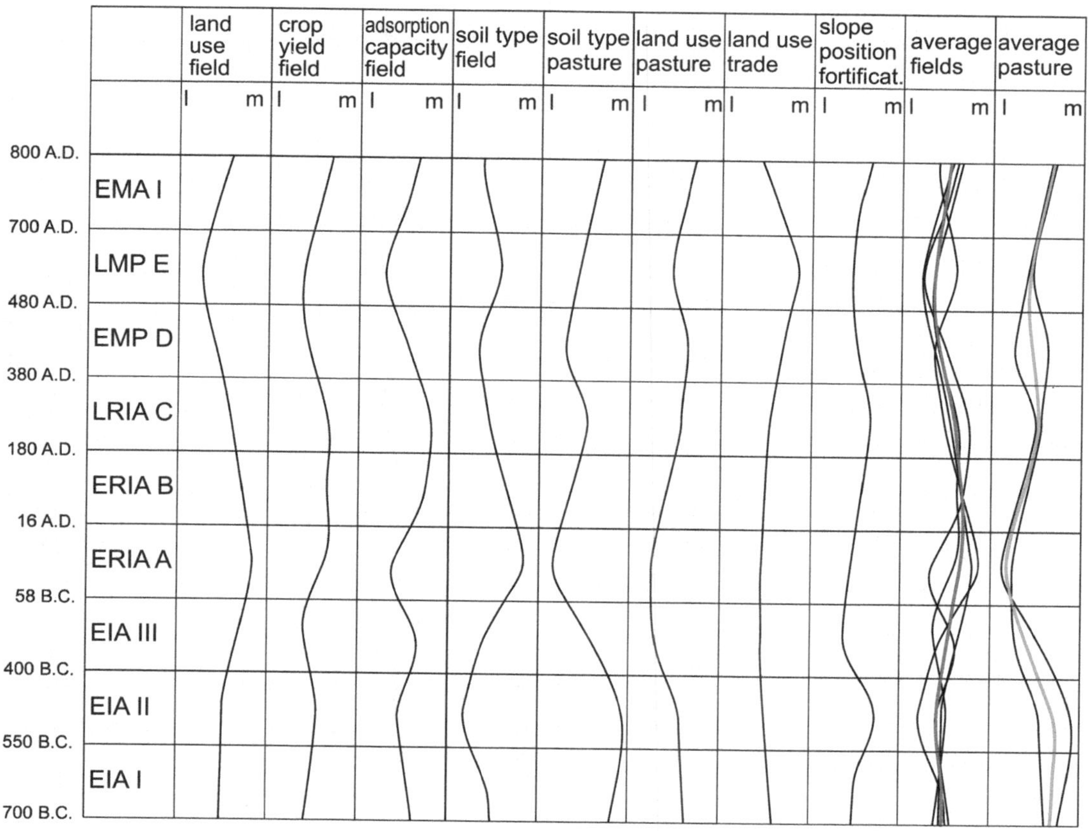

Fig. 3.6 The summary shows *curves* from the spectrum of the respective geofactors of land-use, soil yields, soil water storage, soil type and slope. *Left*: time periods. *Right*: averaged curves (*brown and green lines*) for relative decrease (l = low) or increase (m = more) of agricultural and pastoral economies. *Top*: indicator of processed geofactors from analysis of a thematic map. *Bottom*: relationship to indicators for agricultural, pastoral, trade or fortification aspects

source map units nullify the reflection of 'prehistoric reality' but, in spite of major changes to the cultural landscape over the course of time, geoecological indicators in settlement regions which were decisive in site selection and in the generation of climate indicators remain evaluable.

The wetness/dryness index is calculated as a single value per period and corresponds to a time period in the transferred curve for each peak. It is not methodologically possible to interpret values from the interpolated curve in between peaks. The formula utilized for the calculation of the wetness/dryness index is

$$I_{\mathrm{wd}} = \frac{\sum \text{totalof wetness indicators} - \sum \text{totalof dryness indicators}}{n(\text{total number of indicators})}$$

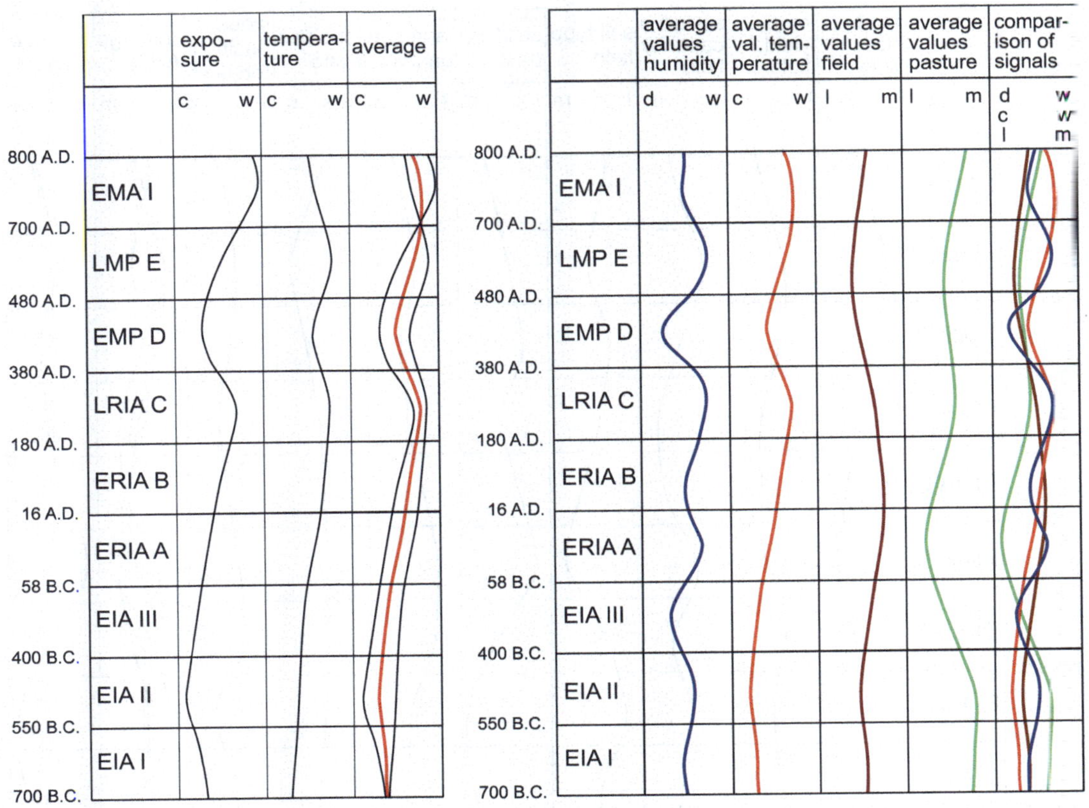

Fig. 3.7 *Left*: average temperature curve (*red*) derived from indicators of exposure and microregional temperature (c = cold, w = warm). *Right*: summary of individual curves of all palaeoclimate proxies in a calibrated moisture curve (*far right*; see Figs. 3.5 and 3.6)

In this way, the sums of individual wetness (w) and dryness (d) indicators are weighted in order to consider their greater or lesser significance:

$$\sum nw = \frac{\sum_i xi \times gi}{\sum_i gi}$$

and

$$\sum nd = \frac{\sum_i xi \times gi}{\sum_i gi}$$

For example:

$$\sum nw = \frac{x1 \times g1 + x2 \times g2 + x3 \times g3 + x4 \times g4 + \ldots + xn \times gn}{g1 + g2 + g3 + g4 + \ldots + gn}$$

This calculates the wetness/dryness index as follows:

$$I_{\mathrm{wd}} = \frac{\sum nw - \sum nd}{n}$$

The values of the wetness/dryness index lie within the range between −1 as a minimum and 1 as a maximum which corresponds to the respective bar length of the individual geofactors in Figs. 3.3, 3.4 and 3.5.[11] To ensure compatibility, the above formula was the standard applied to all geofactor analyses in order to show the changes to the calculated values over the course of the curve (Fig. 3.5). Similarly the formula was also utilized in an appropriately modified form for the development of land-use (Fig. 3.6) and temperature (Fig. 3.7).

In Fig. 3.5, the summary of curves representing the spectrum of geofactor exposure and temperature from the EIA to EMA is shown on the left with indicators of relative coolness (*c*) or warmth (*w*). On the right is the averaged curve (red line) of the averaged relative warmth index. In Fig. 3.5, right, the comparison of the averaged values for temperature indicators and agricultural and pastoral economies which influence the wetness index can be seen. In the calibration of the moisture curve (far right), the relationship with economic pathway is quite clear, but it also is the temperature (cf. Figs. 3.5, 3.6 and 3.7).

3.3.3 Palaeoclimate Development Around the Oder

The results of the GIS analysis can now confirm climate fluctuations based on the relative wetness index, reliably reconstructed and reproducible in the catalogue (Volkmann 2013). Starting with EIA phase I (Fig. 3.7, right), the comparison with phase II indicates a small increase in climatic moisture (cf. Fig. 3.5). In phase III these indicators shrink considerably. In this distinct peak, a large decrease in average annual precipitation as well as water levels above and below ground is reflected (Fig. 3.5). At the same time, the temperature curve rises (Fig. 3.7, left). After a relative warm phase in EIA phase I in spite of a generally lower temperature, phase II (Fig. 3.7, right) is recognizable by a striking drop in temperature (with concurrent increase in precipitation rates). Starting from the low temperature in the EIA, one can see an almost linear increase in temperature until the LRIA phase C where an optimal temperature (with accompanying high precipitation) is achieved.

Similarly apparent is a sharp and drastic reversal of hydrologic behaviours at the beginning of the Early Roman Iron Age (ERIA) phase A with a very pronounced wet phase (Fig. 3.5) together with a distinct change in economic pathway at the beginning of the same period. The importance of a pastoral economy sinks to a minimum as, conversely, agriculture increases considerably and becomes the most important industry (Fig. 3.6). A continuous increase of the wetness index, based on the hydromorphologic potential of water levels above and below ground as well as precipitation rates, seems credible for the mapped settlements from ERIA phase B to LRIA phase C. Concurrently, the temperature increases linearly beginning in the EIA and reaches its highest point in the LRIA phase C (Fig. 3.7, left). The indicators of land-use, soil yields and soil water storage (Fig. 3.6) show an increase in pastoral economies until phase C while simultaneously expressing a maintained high proportion of agriculture as the primary economic pathway and subsistence strategy.

The beginning of the EMP (phase D) is marked by drastic climate change which abruptly gives way to a dry phase and accompanying temperature lows. Absolute value in degrees Celsius during this cool/dry phase remains unclear from the averaged relative temperature curve (Fig. 3.7, left) although would apparently have been on a similar low level as in the EIA phase II. Simultaneously, both agricultural and pastoral economies decreased in use in the EMP although their minimums were not reached until the LMP (Fig. 3.6). Instead trade of gathered goods (such as honey) and resources (like salt) as well as,

[11] For more on weighting mathematical averages cf. Bahrenberg et al. (2010, 85ff)

perhaps, slaves and/or soldiers became a very important economic pathway supplement to agriculture and animal husbandry subsistence economies.

Of further interest is that what was the quite significant fortification aspect of settlements (judging by their frequent siting on land spurs) during LRIA phase C plays a secondary role for the entirety of the Migration Period, so that military conflicts in phases D and E only rarely occurred in the study area. The unstable phase of a strongly oscillating climate with very changeable weather also continued into the LMP phase E, although the values now indicate a distinctly wet phase. With the early Slavic period came the climatic stability in the form of again balanced, dry, but not arid weather. At the beginning of the EMA, an increase in both agricultural and pastoral economic pathways is apparent, whereas interregional trade starkly decreased. In the EMA (phase I), the possibilities of settlement defence again became important aspects of site selection as a significant increase in features such as hilltop and peninsular sites become clearly apparent (Fig. 3.6). As early as the LMP, there was a detectable relative increase in temperature, but this did not continue into the EMA phase I (Fig. 3.7, left).

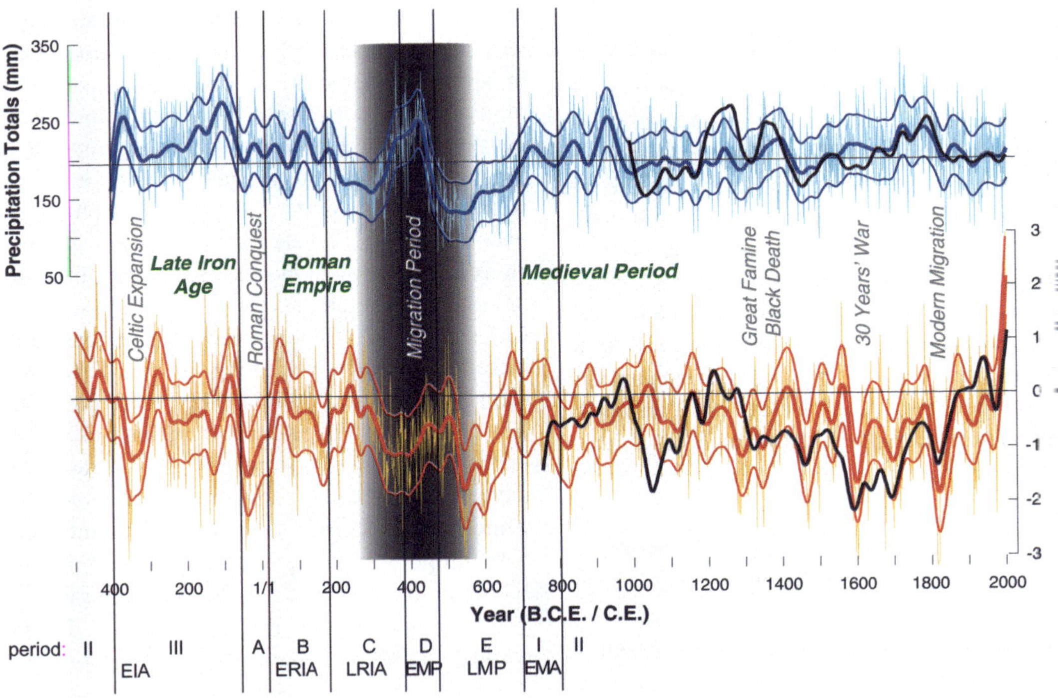

Fig. 3.8 Palaeoclimate humidity and temperature from the early Iron Age to present. The reconstructed precipitation totals (*top*) and temperature anomalies (*bottom*) are based on actual recorded values only from 1901 to 2000 AD (from Büntgen et al. 2011; author in preparation). The *thin bars* show considerably oscillating annual fluctuations in precipitation and temperature derived from values of the dendrochronological calibration curve of oak in Central Europe. The *thick black lines* show the precipitation and temperature reconstructions from independent palaeoclimate studies in Germany and Switzerland beginning in the year 1000 (*top*) and 750 AD (*bottom*). The *thick blue* (*top*) *and red* (*bottom*) *lines* represent smoothed curves of precipitation and temperature in 60-year averages; these lie within the range expressed by the *thin blue and red lines* (not including the peaks of statistical outliers in the bars). Apparently periods of demographic expansion, economic prosperity and social stability as well as of political turmoil, cultural change and population were in close relationship with phases of, respectively, stable or unstable climates (see Volkmann 2013, 213ff.)

3.3.4 Comparative Palaeoclimate Studies in a Central European Context

Compared to the scientific climate research, which was first published after the completion of the work upon which this article is based, agreement with the humidity and temperature changes of the Oder region as determined by the site catchment analyses is apparent (Fig. 3.7), clearly supporting the basic functioning of the methods applied here. More specifically however there are some contradictory assertions about the humid or dry phases in the Migration Period for regional and multiregional contexts. So in a direct comparison of both studies, there remain partly contradictory statements as in the Early Migration Period (EMP) D for which the Oder region was in a dry phase but is depicted as humid according to the averaged data for Central Europe (Fig. 3.8). Similarly in opposition are the findings on palaeoclimate of the Late Migration Period (LMP) E. This disagreement is however not necessarily irreconcilable for two reasons: firstly, the respective databases are completely different. One is composed of absolute dates derived from calibrated C^{14} dendrochronological data from Central Europe, whereas the other is based on relative dating of archaeological finds from the Oder region. Therefore, problems remain in the absolute dating of climate fluctuations in the Oder region as there are only very few scientifically measured absolute dates available. Secondly, the regional factor must also be considered, as the averaged values for all regions have been statistically smoothed in the above diagrams for Central Europe.

Despite these problems of scale, however, both studies show clear and distinct climate fluctuations during the Migration Period. Whether the dry phase occurred during the Late Roman Age (LRIA C) (Fig. 3.8) or the EMP D (Fig. 3.7) is largely inconsequential as it is the drastic climate change that is important and which, regardless of whether humidity increased or decreased dramatically, would have deprived agrarian communities of their livelihoods and thereby been a contributing factor for migration.

References

Bahrenberg G, Giese E, Mevenkamp N, Nipper J (2010) Statistische Methoden in der Geographie, Univariate und bivariate Statistik, Bd. 1. Schweizerbart, Stuttgart

Bernbeck R (1997) Theorien in der Archäologie. Francke, Tübingen

Bill R (2010) Grundlagen der Geo-Informationssysteme. Wichmann, Berlin

Bintliff J (1999) Settlement and territory. In: Barker G (ed) Companion encyclopedia of archaeology, vol 1. Routledge, London, pp 505–545

Birkett C (1985) A site catchment analysis: Walkunder Arch Cave, near Chillagoe, North Queensland. Qld Archaeol Res 2:132–142

Brather S (2000) Früh- und Hochmittelalterliche Keramik bei den Westslawen. In: Wieczorek A, Hinz HM (eds) Europas Mitte um 1000, vol Bd. 1. Theiss, Stuttgart, pp 114–120

Büntgen U, Tegel W, Nicolussi K, McCormik M, Frank D, Trouet V, Kaplan JO, Herzig F, Heussner KU, Wanner H, Luterbacher J, Esper J (2011) 2500 years of European climate variability and human susceptibility. Science 331:578–582

Burrough PA, McDonnell R, Lloyd CD (2015) Principles of geographical information systems. Oxford University Press, Oxford

Ducke B, Kroefges PC (2007) From points to areas: constructing territories from archaeological site patterns using an enhanced Xtent model. In: Posluschny A, Lambers K, Herzog I (eds) Layers of perception. Identifying settlement patterns and territories. Proceedings of the 35th international conference on computer applications and quantitative methods in archaeology (CAA), Berlin. Dr. Rudolf Habelt GmbH, Bonn

Eggert MKH (1978) Prähistorische Archäologie und Ethnologie. Studien zur amerikanischen New Archaeology. Praehist Z 53:6–164

Eggert MKH, Veit U (eds) (1998) Theorie in der Archäologie: Zur englischsprachigen Diskussion, Tübinger Archäologische Taschenbücher, vol 1. Waxmann, Münster

Gordon WR, Phillips P (1958) Method and theory in American archaeology. University of Chicago Press, Chicago

Griesa S (1982) Die Göritzer Gruppe. Veröff. Mus. Ur- und Frühgesch. Potsdam 16. Deutscher Verlag, Berlin

Herzog I (2014) A review of case studies in archaeological least-cost analysis. Archeologia e Calcolatori 25:223–239

Herzog I, Posluschny A (2011) Tilt – slope-dependent Least Cost Path Calculations Revisited. In: Jerem E, Redö F, Szeverényi V (eds) On the road to reconstructing the past. Computer Applications and Quantitative Methods in Archaeology (CAA). Proceedings of the 36th International Conference, Budapest, April 2–6, 2008. Archeaeolingua, Budapest, pp 212–218

Hunt ED (1992) Upgrading site-catchment analyses with the use of GIS: investigating the settlement patterns of horticulturalists. World Archaeol 24(2):283–309

Jarmann MR, Vita-Finzi C, Higgs E (1972) Site catchment analysis in archaeology. In: Ucko P, Tringham R, Dimbleby DW (eds) Man, settlement and urbanism. Duckworth, London, pp 61–66

Kirsch K (2004) Slawen und Deutsche in der Uckermark: Vergleichende Untersuchungen zur Siedlungsentwicklung vom 11. bis zum 14. Jahrhundert. Forsch. zur Gesch. und Kultur des östlichen Mitteleuropa 21. Franz Steiner, Stuttgart

Kluiving S (ed) (2010) Landscape archaeology between art and science – from a multi- to an interdisciplinary approach. International Landscape Archaeology Conference, Amsterdam

Leube A (2009) Studien zu Wirtschaft und Siedlung bei den germanischen Stämmen im nördlichen Mitteleuropa während des 1. bis 5.–6. Jahrhunderts n. Chr. Jh. u. Z. Ethnogr Archäol Z 33:130–146

Müller J (2000) Zur räumlichen Darstellung von Radiokarbondaten: Zwei Beispiele aus dem Endneolithikum. Online-Publ. der Freien Univ. Berlin, Sem. Ur- und Frühgeschichte. http://www.jungsteinsite.uni-kiel.de/2000_mueller/14c_raum.htm

Nakoinz O (2009) Zentralortforschung und zentralörtliche Theorie. Arch Korrbl 39:361–380

Posluschny AG, Fischer E, Rösch M, Schatz K, Stephan E, Stobbe A (2012) Modelling the agricultural potential of early iron age settlement hinterland areas in southern Germany. In: Kluiving SJ, Guttmann-Bond EB (eds) Landscape archaeology between art and science – from a multi- to an interdisciplinary approach. Landscape & Heritage Series. Proceedings. Amsterdam University Press, Amsterdam, pp 413–428

Renfrew C, Bahn P (2005) Archaeology: the key concepts. Site catchment analysis. A review. Routledge, London, pp 172–176

Roper DC (1979) The method and theory of site catchment analysis: a review. Adv Archeol Method Theory 2:119–140

Schach-Dörges H (1970) Die Bodenfunde des 3. bis 6. Jahrhunderts nach Chr. zwischen unterer Elbe und Oder. Offa-Bücher 23. K. Wachholtz, Neumünster

Thurmaier C (1999) Entscheidungsunterstützung für die Wertermittlung in der Ländlichen Entwicklung – Ein neues Einsatzgebiet für GIS? Online-Publikation der Technischen Universität München Geodätisches Institut, München

Trigger B (1989) A history of archaeological thought. Cambridge University Press, Cambridge

Trigger B (2006) A history of archaeological thought. Cambridge University Press, Cambridge

Ullah IT (2011) A GIS method for assessing the zone of human-environmental impact around archaeological sites: a test case from the late neolithic of Wadi Ziqlâb, Jordan. J Archaeol Sci 38:623–632

van Leusen M (1998) Viewshed and cost surface analysis using GIS (cartographic modelling in a cell-based GIS II). In: Barceló JA, Briz I, Vila A (ed) New techniques for old times. Computer applications and quantitative methods in archaeology. Proceedings of the 26th conference. BAR International Series 757, Barcelona, pp 215–224

van Leusen M (2002) Pattern to process: methodological investigations into the formation and interpretation of spatial patterns in archaeological landscapes, line-of-sight and cost surface analysis using GIS, chap 6. Rijksuniversiteit Groningen, Groningen, pp 1–21. http://dissertations.ub.rug.nl/faculties/arts/2002/p.m.van.leusen/

Verhagen P (2010) On the road to nowhere? Least cost paths, accessibility and the predictive modelling perspective. In: Contreras F, Farjas M, Melero FJ (eds) Proceedings of the 38th annual conference on computer applications and quantitative methods in archaeology (CAA). Archaeopress, London, pp 383–389

Verhagen P (2018) Spatial analysis in archaeology: moving into new territories. In: Siart C, Forbriger M, Bubenzer O (eds) Digital geoarchaeology: new techniques for interdisciplinary human-environmental research. Springer, Heidelberg, pp 11-26

Vita-Finzi C, Higgs E (1970) Prehistoric economy in the Mount Carmel area of Palestine: site catchment analysis. Proc Prehist Soc London 36:1–37

Volkmann A (2012) Eisenproduktionswerkplätze der späten römischen Kaiserzeit (3.–5. Jh. AD) im inneren Barbaricum. Forschungen zur Völkerwanderungszeit und zum Frühmittelalter Europas. Würzburg. http://opus.bibliothek.uni-wuerzburg.de/frontdoor.php?source_opus=7442

Volkmann A (2013) Siedlung – Klima – Migrationen: Geoarchäologische Forschungen zur Oderregion zwischen 700 vor und 1000 nach Chr. mit Schwerpunkt auf der Völkerwanderungszeit. Habelt, Frankfurt

von Thünen J (1826) Der isolirte Staat in Beziehung auf Landwirthschaft und Nationalökonomie, oder Untersuchungen über den Einfluss, den die Getreidepreise, der Reichtum des Bodens und die Abgaben auf den Ackerbau ausüben. Wiegand, Hempel & Parey, Berlin

Voß HU (1986) Untersuchungen zur Geschichte der germanischen Besiedlung zwischen Elbe/Saale und Oder/Neisse im 3.–7. Jahrhundert. Unpublished Dissertation, Humboldt Universität, Berlin

Zimmermann A, Richter J, Frank TT, Wendt KP (2005) Landschaftsarchäologie II – Überlegungen zu Prinzipien einer Landschaftsarchäologie. Ber RGK 85:37–95

Point Pattern Analysis as Tool for Digital Geoarchaeology: A Case Study of Megalithic Graves in Schleswig-Holstein, Germany

4

Daniel Knitter and Oliver Nakoinz

Abstract

In this contribution, we apply different methods of spatial and geomorphometric analysis in order to present a general approach of data exploration in areas where detailed local information is absent. Our data are based on locations of megalithic graves from Funnel Beaker societies (3700–2800 BCE) in the area of Schleswig-Holstein, Germany. Using these locations, we apply methods of point pattern analysis in order to reconstruct the spatial processes that created the sample: We use density-based measures to show the influence of first-order effects on the dataset. While first-order effects are related to the underlying areal characteristics of the point locations and hence are determinant of their intensity, second-order effects are the result of interactions between points. We conduct distance-related approaches, e.g. focusing on nearest-neighbour characteristics, in order to investigate the interaction between the points. The point pattern analyses are complemented by integrating geomorphometric measures that are indirectly indicative for some general environmental conditions, even in prehistoric times. This helps (a) to relate first-order effects to societal or environmental features and (b) to understand the specific pattern of interactions between the points. The necessary raw data in the form of digital elevation models are freely available for large parts of the globe. All analyses are conducted using free and open-source software in order to provide their limitless application.

Keywords

Spatial data analysis • Point pattern analysis • Funnel beaker • Megalithic tombs • Geomorphometry • Landscape

D. Knitter (✉)
Department of Geography, Physical Geography—Landscape Ecology and Geoinformation, Christian-Albrechts-Universität zu Kiel, Ludewig-Meyn-Str. 14, 24118 Kiel, Germany
e-mail: knitter@geographie.uni-kiel.de

O. Nakoinz
Institute of Pre- and Protohistoric Archaeology, Christian-Albrechts-Universität zu Kiel, Johanna-Mestorf-Straße 2–6, 24118 Kiel, Germany
e-mail: oliver.nakoinz@ufg.uni-kiel.de

© Springer International Publishing AG 2018
C. Siart et al. (eds.), *Digital Geoarchaeology*, Natural Science in Archaeology,
DOI 10.1007/978-3-319-25316-9_4

4.1 Introduction

In this study, we aim to present different tools for spatial data analyses that are especially useful in areas where detailed local information is absent, for instance, areas of short research history or areas that are not accessible due to political reasons. Our objective is to present a simple workflow that integrates tools from point pattern as well as geomorphometric analyses in order to get a better understanding of the processes that influenced the spatial pattern we observe. Furthermore, we want to discuss some general characteristics of spatial data that make them special.

In order to check the effectiveness of our spatial analytical tools, we use a dataset of megalithic graves from Funnel Beaker societies located in the area of Schleswig-Holstein, Germany. These are well investigated and allow to prove whether the conclusions we draw from our spatial analyses are a valuable complement of digital geoarchaeological research.

4.1.1 Why Are Spatial Data Special?

Spatial data are special because they do not fulfil one of the most common prerequisites of conventional statistical analyses: they are not random, i.e. stochastically independent. This leads to different *problems* (collection after O'Sullivan and Unwin 2010, 34):

- **Spatial autocorrelation** is a measure of the importance of a location. It measures to which degree the characteristics at a certain location—or in the study area as a whole—are indicative for other locations in the area. The concept is closely related to Tobler's first law of geography (at least for positive autocorrelation): "(...) everything is related to everything else, but near things are more related than distant things" (Tobler 1970, 236). This means that it is more likely that points next to each other have similar or comparable characteristics of, e.g. elevation than points that are distant. Local similarities are used to describe and differentiate space. For instance, an area of high concentration of people may be called settlement; a wetland area of low pH values, dense vegetation and high organic carbon content may be called swamp; etc. The *law* also indicates that this holds true for all spatial data. If spatial phenomena would vary randomly through space, spatial data would be meaningless (O'Sullivan and Unwin 2010, 35). There are different techniques that allow to assess the importance of a location—hence spatial autocorrelation—in an analysis, i.e. *Moran's I* as well as *Geary's C* (e.g. Lloyd 2011, 80–82).
- The **modifiable areal unit problem** arises when spatial data are compiled or acquired on a certain level of detail but are analysed in aggregated, areal-modified form (O'Sullivan and Unwin 2010, 36–38). For instance, ceramics are counted as single finds, but their distribution is reported in terms of ceramics' sum per survey grid cell. This grid cell is a modifiable areal unit that is *arbitrary* in terms of the investigated object. This can lead to problems in subsequent analyses because the unit of aggregation, i.e. our arbitrary survey grid cell, influences the outcome of the analysis. This kind of problem arises also by comparing archaeological sites, like megalithic graves, with environmental features, like natural regions (see Fig. 4.1). The comprehensive discussion of this issue by Openshaw (1984, 4–5, 10–11) shows that different aggregation schemes—e.g. different grid cell sizes or shapes—have a very strong effect on correlation measures.
- The common statistical problem of **ecological fallacy** is often related to modifiable areal units. It occurs when a statistical relationship at one level of aggregation is assumed to be present because it is present at another (O'Sullivan and Unwin 2010, 39). Thinking of some archaeological sites that might occur more frequently at higher elevated locations, this observation does not allow us to conclude that those sites are located there because these locations seek higher visibility. It might also be due to the need of a lower groundwater table, to flood risk reduction or creation (Schütt et al. 2002). Hence, data on occurrence of archaeological sites in different

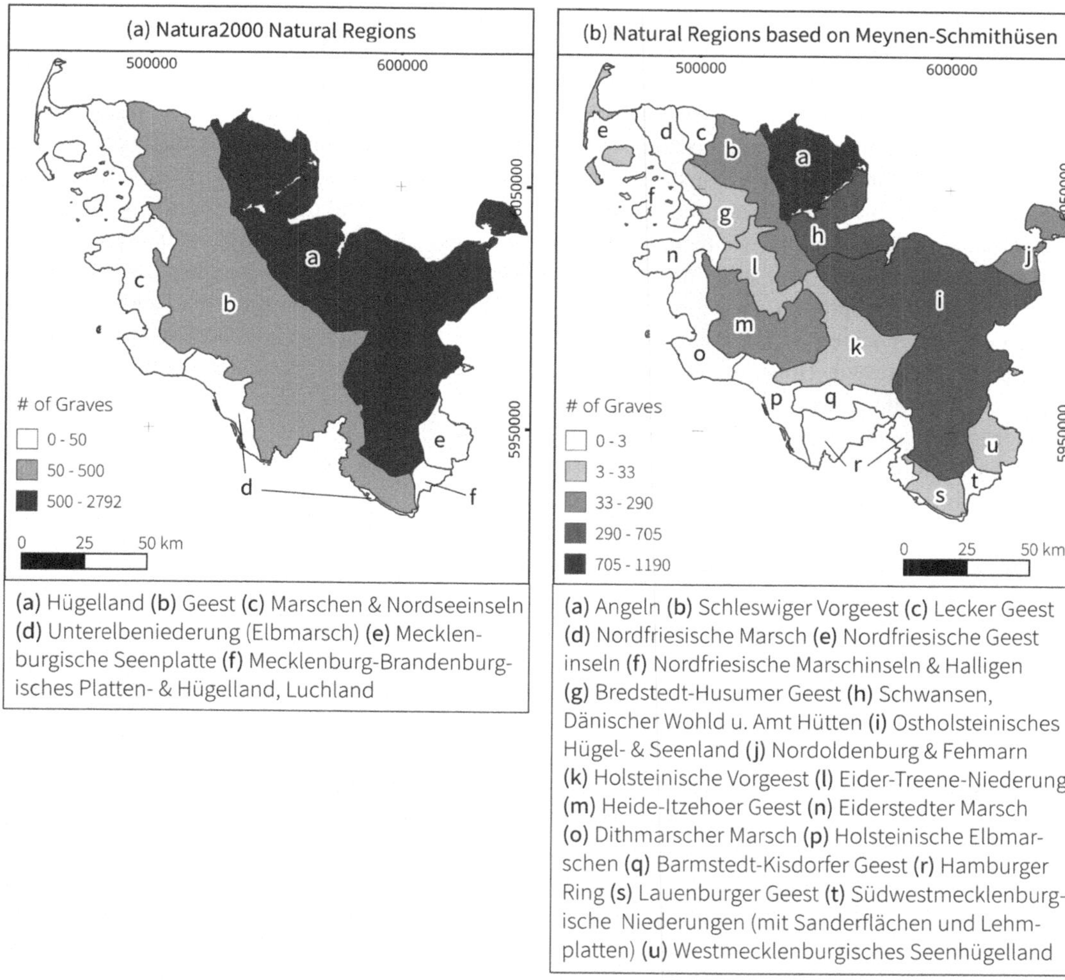

Fig. 4.1 Natural regions in Schleswig-Holstein and the number of megalithic graves. Most of megalithic graves are located in the northeast and east. The *modifiable areal unit problem* gets obvious in comparing (**a**) and (**b**). The absence of megalithic graves in the marsh-related natural regions corresponds to the fact that these areas were not inhabitable during occupation period of Funnel Beaker societies [data kindly provided by Federal Agency for Nature Conservation; (**a**) based on Ssymank (1994); (**b**) based on Meynen and Schmithüsen (1962)]

altitudes can only support the conclusion that these are often more elevated in relation to their surroundings.

- Before the start of a spatial analysis, it is necessary to decide on which geographic **scale** the analysis will be conducted because this affects what we are able to observe. The sample of this study is megalithic graves, represented as points. This already implies that this scale is too small to, for instance, investigate the orientation of the megalithic monuments. Furthermore, investigating the characteristics of megaliths as points only gives one measure—e.g. their altitude—although they cover a certain area, i.e. a certain range of altitudes.
- **Space is not uniform**; accordingly, processes measured in space can be heterogeneous although their characteristics did not change. This is an induced spatial dependence (Borcard et al. 2011, 229). In terms of megalithic graves, this may be seen when one

considers a higher population density in a certain area, therefore more settlements and as a result more megalithic graves. Their occurrence is now linked to the higher population density and not, as one could suspect, better circumstances for their construction.

- **Edge effects** are related to the issue of non-uniformity and arise when an artificial boundary is imposed on a study area (Diggle 2013, 9). In the present case, the artificial boundary is the area of the modern federal state of Schleswig-Holstein in Germany. This can produce asymmetries in the data since points in the centre of the study area can have more nearby observation than those at the edge of the study area.

Many of these points may sound trivial. Nevertheless, it is important to be aware of them since they directly influence the results. Spatial data are the result of processes. In analysing them, it is possible to detect functional relationships. But these do not infer causality (see Ahnert 2003, 19–20). Hence, it needs to be discussed continuously, whether these processes are the actual reason of the configuration of spatial data or just an artefact of the analytical approach.

In the frame of this article, we use two types of spatial data: One type are point pattern data, as represented by megalithic graves. In this case—like many others in reality—we have nothing more to describe and analyse than the graves' locations, since chronological, functional, cultural, etc. information are absent or homogeneous.

Point patterns are the result of processes that are influenced by (a) first-order effects, i.e. the point's location is influenced by the underlying area's structure but not by the location of other points (Wiegand and Moloney 2004, 210), and (b) second-order effects that occur when the location of a point is influenced by the presence or absence of other points (Wiegand and Moloney 2004, 210). Point pattern analyses are common in ecological studies (e.g. Legendre and Legendre 2012; Wiegand and Moloney 2013); up to now there are only few applications in archaeological contexts (e.g. Knitter et al. 2014).

The other type of spatial data used here are geomorphometric data, i.e. different derivatives of a digital elevation model (DEM). These geomorphometric parameter raster describe the topographic characteristics in spatially continuous form and allow us to draw indirect conclusions on the boundary conditions of location selection in the study area. The integration of DEM-related parameters in (digital geo)archaeological investigations is very common since the 1990s and often used in predictive modelling approaches (e.g. Verhagen 2007; Kvamme 2006).

4.1.2 Case Study

4.1.2.1 Natural Characteristics of the Study Area

The natural characteristics of Schleswig-Holstein are mainly the result of the last two glaciations, i.e. the Weichselian glaciation (115,000–11,600 a BP) and the Saale glaciation (300,000–130,000 a BP) (Litt et al. 2007, 34, 46). In general, three natural regions can be separated: (1) the marsh regions, (2) the geest regions, and (3) the young, heavily undulating regions of east Schleswig-Holstein (Stewig 1982, 18; Schmidtke 1995).

The marsh regions are the youngest areas, formed during the Holocene by continuous up-silting of the intertidal zone—a, in general, natural process that was accelerated by human intervention due to the creation of polders. These areas, as we see them today, did not exist during the time of Funnel Beaker societies (Stewig 1982, 35–42; Schmidtke 1995, 86ff). Since these areas were not inhabitable, no megalithic graves can be found there (Fig. 4.1a, category c; Fig. 4.1b, categories d, f, n, o, p).

East of the marsh area the geest regions are located (Fig. 4.1a, category b; Fig. 4.1b, categories b–c, e, g, k, m, q, s). These are characterised by glacial deposits and can be subdivided in *Hohe Geest* areas comprised of terminal and ground moraine deposits from the Saale glaciation as well as *Niedere Geest* areas, incised into the geest and formed by glacio-fluvial activities as a result of the melting of the

Weichselian ice sheet in the east, leading to widespread sandur deposits in the glacial meltwater valleys draining to the west (Liedtke and Marcinek 2002, 438–440). The *Niedere Geest* sinks gently under the marsh areas. In contrast, marked cliffs can occur between the marsh and the *Hohe Geest* regions due to former abrasion at the seashore or during storm tide periods (Stewig 1982, 27–34). Due to periglacial processes, *Hohe Geest* regions were levelled, and marked relief differences are present only at the border to the marsh as well as *Niedere Geest* regions (Schmidtke 1995, 78).

The eastern part of Schleswig-Holstein was covered by the ice of the Weichselian glaciation that created a very heterogeneous and complex topography with diverse series of terminal moraines, interrelated by lakes or local sandurs, eskers and kames (Fig. 4.1a, category a; Fig. 4.1b, categories a, h, i–j, t–u). The soils on moraine material are the most fertile of the entire region (Stewig 1982, 18–26; Schmidtke 1995, 40ff; Liedtke and Marcinek 2002, 433–435).

4.1.2.2 Megalithic Graves of Funnel Beaker Societies

Megalithic graves are above-ground visible monuments consisting of upright and capstones (Müller 2014, 182). Starting in the fifth millennium BCE, the tradition to build megaliths started in South-West Europe and was distributed along the coasts to reach the Baltic Sea in the fourth millennium BCE. At this time, the sea reached its present level (Schmidtke 1995, 72). In the study area megalithic graves are present since 3650 BCE, i.e. about 500 years after Neolithisation commenced and connected with the occurrence of early Funnel Beakers (Müller 2011, 8, 10). Based on ceramic style and decoration, different societal groups can be distinguished. Altogether these create the complex of Funnel Beaker societies spreading from Southern Scandinavia to Central Europe (Fig. 4.2).

The research area belongs to the North Group of Funnel Beaker societies where the *megalithic phase* corresponds to the Middle Neolithic period (3300–2800 cal BCE). The main phase of megaliths' construction was around 3100 BCE (Müller 2011, 15, 17). Megalithic graves in the Funnel Beaker societies of the North Group are attended to the construction of causewayed enclosures (Müller 2011, 29). They can be typologically sequenced into *Urdolmen*, extended *dolmen* and passage graves (Müller 2011, 31, 34). Since megaliths solely rely on slabstones and dry masonry walls, their construction necessitates a supportive mound that might have also served as a ramp during the building phase (Müller 2011, 34).

The dataset consists of 2792 individual megalithic graves in the study area of Schleswig-Holstein. In general it is assumed that more than 3000 of them occur in the area, although most of them are already destroyed. From the known locations, only 867 offer more detailed archaeological information, and only 188 of them can be seen as well preserved (Lüth 2011, 117). An exact dating of the monuments is difficult since they served as collective cemeteries where burials from different periods occur. Furthermore, many of the megalithic graves were destroyed and reused as burial places during the Late Neolithic, especially by the Single Grave culture. As a result, the role of megalithic graves within the Funnel Beaker societies is not yet clear (Lüth 2011, 117, 119). Nevertheless, they seem to be more than just cemeteries. First of all, they are clear indicators for former settlements since they were sited in close vicinity to them. Local site-specific studies also revealed some insights about their locational characteristics (after Lüth 2011, 119–122):

- They are concentrated on fine-grained, fertile soils.
- They are located along streams and fords, near lakes or at the coast; all these locations are preferable for traffic.
- They are located at exposed positions on hills, high banks of rivers and cliff coasts; all these locations have the character of prominent topographic lines.
- Building material, in the form of glacial erratics, needs to be available.

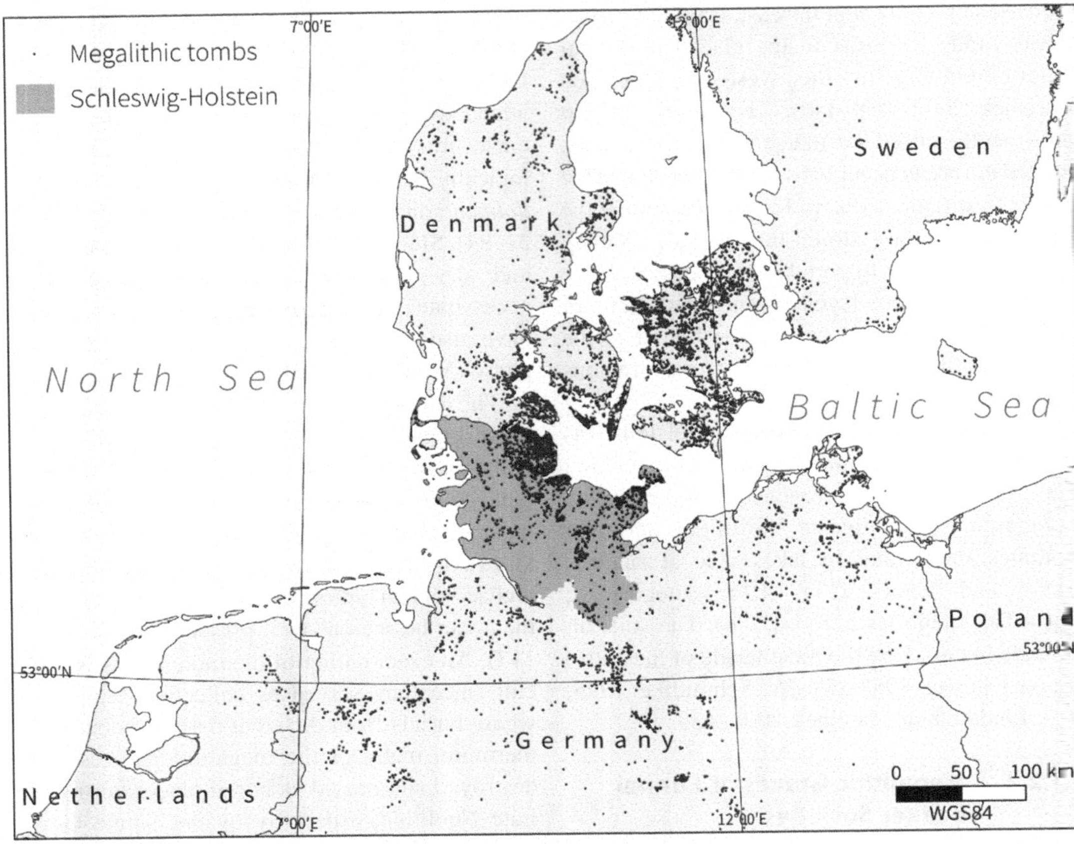

Fig. 4.2 Megalithic graves of Funnel Beaker societies in northern Europe and southern Scandinavia (based on Fritsch et al. 2010; Müller 2014)

These conditions are present in the young, strongly undulating regions in the east of Schleswig-Holstein. Nevertheless, there are also high concentrations of megalithic graves in the *Hohe Geest* regions that only partially offer the same conditions as compiled for the young moraine areas (compare Figs. 4.1 and 4.2).

4.2 Methods

4.2.1 Point Pattern Analyses

Focusing on spatial data without special attributes, like size or amount of ceramic shards, there is only the location left to analyse. The locations of the megalithic graves form our point pattern.

A point pattern consists of a set of *events*,[1] $S = (s_1, s_2, \ldots, s_n)$, in a defined research region A of size α; an event s_i describes the occurrence of objects of interest at a specific location by geographic coordinates (x_i, y_i). The definition of the research region—that corresponds to a modifiable areal unit—influences subsequent analyses. Furthermore, in defining a point pattern, it is assumed that all relevant entities are included in the pattern. Both aspects are

[1] Note that the term *event* is used here in the technical sense of spatial point pattern analysis: In a spatial point pattern, locations are referred to as *events* in order to distinguish them from other points of the region in question (Diggle 2013, 1).

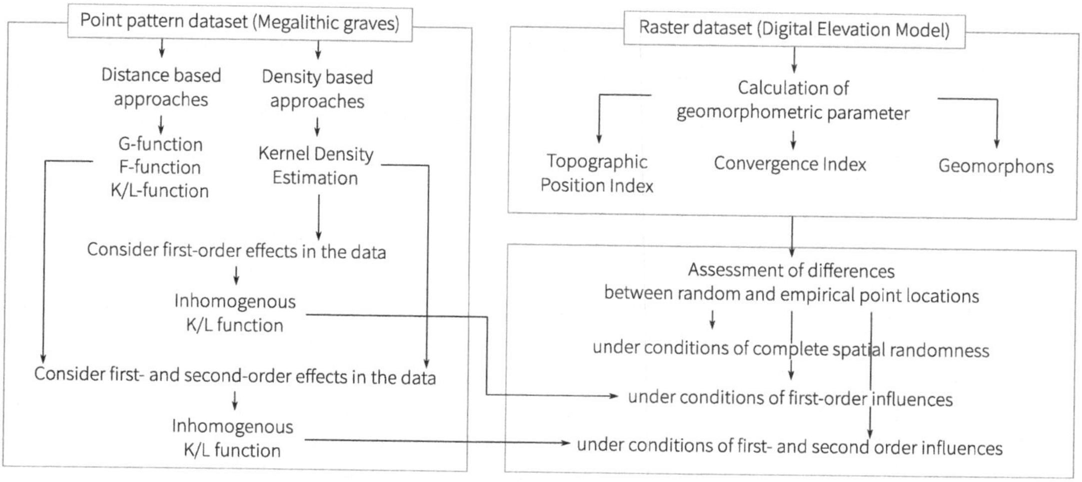

Fig. 4.3 Workflow of the applied methods; it is based entirely on following open-source software packages: *left part and lower right part*: R Core Team (2013) with packages, `sp` (Pebesma and Bivand 2005), `rgdal` (Bivand et al. 2015), `spatstat` (Baddeley and Turner 2005), `raster` (Hijmans 2015), `ggplot2` (Wickham 2009); *upper right part*: SAGA GIS (Böhner et al. 2006) and GRASS GIS (Neteler et al. 2012)

problematic in archaeological studies due to the fragmentary nature of the source material.

The events constituting a point pattern might be influenced by *first-order* as well as *second-order effects*. A point pattern analysis (Fig. 4.3) allows to a certain degree to separate these influences, though an exact and definite differentiation is not possible (O'Sullivan and Unwin 2010, 124).

4.2.1.1 Density-Based Approach: Kernel Density Estimation

First-order effects become evident in the event density, i.e. the process' intensity (Diggle 2013, 10). *Absolute location* is determinant and a clear variation across space shall be visible. The methods applicable to detect *first-order* effects are density-based measures of point patterns. In the present case, we conduct a kernel density estimation (Bivand et al. 2008, 165–168). The basic idea behind this approach is that there is a density at any location, not just at the location of events. This creates a continuous representation of intensity from a set of events allowing (a) to detect high-density spots and (b) to check whether the process is *first-order nonstationary*, i.e. it shows only local variations from the mean intensity but not an overall trend across the region, as well as (c) to link point objects to other spatially continuous geographical data (O'Sullivan and Unwin 2010, 68–71).

The density is estimated by applying a kernel function on events in a region that is centred at the location where the estimate is to be made. We use a Gaussian function which weights nearby events more strongly than distant ones. The resulting kernel density estimation is an estimate of the intensity function of the process that generated the phenomena observed (Baddeley and Turner 2015, 278):

$$\widehat{\lambda}_p = e(p)\sum_i k(x_i - p); k = f(x, \mu, \sigma)$$
$$= \frac{1}{\sigma\sqrt{2\pi}} e^{-\frac{(x-\mu)^2}{2\sigma^2}} \tag{4.1}$$

where k is the Gaussian smoothing kernel—with standard mean μ and variable kernel bandwidth, i.e. standard deviation σ—$e(p)$ is an edge correction factor, and p is the kernel's location.

It gets obvious that the kernel bandwidth σ is crucial for the resulting intensity: a too large value will blur the result and omit local details, while a too small bandwidth will omit a potential general trend. Hence, for every research question, the most suitable bandwidth values need to be discovered—a process that in general requires experimentation (Bivand et al. 2008, 166). In this study, we use three different kernel sizes:

- A large kernel with bandwidth of 10,000 m in order to show a potential trend within the data
- An *empirical* kernel whose size equals three times the mean nearest-neighbour distance of the megaliths, i.e. approximately 1500 m
- A small kernel with bandwidth of 500 m that will highlight areas where megalithic graves are clustered

The calculations are conducted using the `density` function in the `spatstat` package (Baddeley and Turner 2005).

4.2.1.2 Distance-Based Approach: G, F, K, and L Function

Second-order effects refer to distance-dependent interactions between events; hence, *relative location* is important (Diggle 2013, 57f). In order to investigate potential interactions between events, distance-based approaches are applied. The simplest is the G-function that investigates the cumulative frequency distribution of nearest-neighbour distances (O'Sullivan and Unwin 2010, 132). More specifically it calculates what fraction of all nearest-neighbour distances $d_{\min}(s_i)$ in the pattern is less than d (O'Sullivan and Unwin 2010, 132):

$$G(d) = \frac{\#\{d_{\min}(s_i)\} \leq d}{n}. \quad (4.2)$$

The function's shape provides information about the way the events are spread. If events are clustered, $G(d)$ increases rapidly at short distances; if events are evenly spaced, $G(d)$ increases slowly up to the distance where most events are spread and increases rapidly afterwards. While the G-function is useful to investigate interactions between events, it does not provide information about the event characteristics in relation to the research region. To get this, the F-function is used (O'Sullivan and Unwin 2010, 133). In applying the function, point locations anywhere in the study region are randomly selected, and the minimum distance of these to any event in the pattern is determined:

$$F(d) = \frac{\#\{d_{\min}(p_i, s)\} \leq d}{m}, \quad (4.3)$$

where $d_{\min}$ is the minimum distance from location p_i in the randomly selected set to any event in the point pattern s. The F-function behaves different than the G-function for clustered and even patterns: In a clustered pattern—where a large area of the research region is empty—$F(d)$ rises slowly for short distances but stronger for larger distances. In case of an evenly spaced pattern, there is a sharp increase of $F(d)$ at the beginning, because the proportion of empty space is small.

There is a shortcoming in applying the G- and F-functions: they only take nearest-neighbour distances into account (Bivand et al. 2008, 161–162). This is a problem especially in clustered point patterns where the nearest-neighbour functions are very short and might mask a more general structure within the data.

To overcome these restrictions, the K-function, known as Ripley's K, is employed that uses all distances between events (Ripley 2004, 159). It draws circles C of different radius d around events s_i. Within each circle, the events are counted, and the mean count for all events is calculated and divided by the overall mean of the research regions' event density (O'Sullivan and Unwin 2010, 135):

$$K(d) = \frac{\sum_{i=1}^{n} \#\{S \in C(s_i, d)\}}{n\lambda}. \quad (4.4)$$

Since the K-function implements more information than the G- or F-function, the resulting plots can offer more insights in the event interactions—like event separation in a regular pattern.

In order to stabilise the variance and to make visual comparisons easier, it is common to conduct a square root transformation of the K-function, known as L-function (Illian et al. 2008, 95):

$$L(d) = \sqrt{\frac{K(d)}{\pi}}. \tag{4.5}$$

Furthermore, we modify the L-function by subtracting the radius d from all values of $L(d)$ (O'Sullivan and Unwin 2010, 147). Due to this, the theoretical distribution corresponds to a straight line at level zero. Positive values are indicative for aggregation, since more events than expected by the theoretical model occur at corresponding distances. Negative values indicate the opposite, since less events as expected are present.

The calculations of the G-, F-, K- and L-functions are conducted using the `envelope` function in the `spatstat` package (Baddeley and Turner 2005) with default configuration and 99 simulations. The null model of the simulations is a complete spatial random process, i.e. a process without first- and second-order effects.

4.2.1.3 Data Influenced by First- and/or Second-Order Effects

Up to this stage, all distance-based approaches assumed the intensity function to be constant and the events in the point pattern independent from the spatial characteristics and location of other events. Nevertheless, due to the non-uniformity of space, such conditions practically do not occur in geographic reality. If solely first-order effects influence the location of events, an inhomogeneous or nonstationary point pattern is present (O'Sullivan and Perry 2013, 42; Bivand et al. 2008, 165).

In order to estimate the characteristics of the point pattern under the influence of first-order effects, we use the inhomogeneous K-function that is a direct generalisation of a nonstationary point process with non-constant intensity (Baddeley et al. 2000, 332). As in the stationary case, we calculate the L-function, i.e. the square root transformation of the inhomogeneous K-function. The parameter of non-constant intensity equals the result of the kernel density estimation with large bandwidth since it is representative for an overall data-inherent trend.

In general it is difficult to decide whether first- or second-order effects caused the observed point pattern. In order to differentiate between first- and second-order effects, artificial point patterns are used that correspond to different null models (Diggle 2013, 99ff):

- Including no first- or second-order effects—also called complete spatial randomness—i.e. event locations are independent of area conditions and the location of other events; we use the function `rpoispp` from the `spatstat` package (Baddeley and Turner 2005) with an intensity that is similar to the overall intensity of the megalithic graves.
- Including first- but no second-order effects; in this case the random points are created using the same intensity function that was used as determinant in the inhomogeneous case, i.e. the first-order effects mirror the overall trend in the data distribution; as in the previous case, the `rpoispp` function from the `spatstat` package (Baddeley and Turner 2005) is used, but now the intensity is a raster representing the results of the kernel density estimation with large kernel.
- Including first- and second-order effects; first-order effects are the same than in the previous case, but now there is also interaction between events due to a tendency to cluster. This is realised using the Thomas cluster process, a generalisation of the Neyman-Scott cluster process, where *parent* points are replaced by clusters of *offspring* points whose locations being isotropic Gaussian displacements from the cluster parent location (Thomas 1949). The artificial point pattern is created applying the `rThomas` function from the package `spatstat` (Baddeley and Turner 2005) with the same intensity as in the previous

case and a homogeneous offspring with intensity 2 that is comparable to the intensity of megalithic grave clusters (Fig. 4.4).

4.2.2 Raster Analyses

In order to get an understanding of the topographic characteristics of the megaliths locations, we use a digital elevation model (DEM) with a resolution of 90 × 90 m, derived from SRTM data that is processed to provide a seamless continuous topography surface (Jarvis et al. 2008). These data are the least error prone for the research area, as compared to other free available digital elevation models that offer a higher spatial resolution like Aster or the recently available SRTM1 data (both types of data are available at `http://earthexplorer.usgs.gov/`).

In analysing DEMs, we employ methods from geomorphometry, the science of quantitative geometric land-surface analysis (Pike et al. 2009, 3). We use these techniques in order to get spatially continuous information on the surface characteristics that help us to describe the location of the megalithic graves in the context of the natural environment. The larger a landform the longer its existence (Ahnert 1981). Based on this rule, we do not focus on the very local scale but on the regional scale where the *general* character of topography is comparable between the time of Funnel Beaker societies and today. Nevertheless, this generalisation is not valid for large parts of the marsh areas since these were made by humans later. Therefore, they are not considered in the subsequent analyses.

4.2.2.1 Topographic Position Index

The Topographic Position Index (TPI), developed by Guisan et al. (1999) and Weiss (2001), measures the relative topographic position of a cell as the difference between the elevation in this cell and the mean elevation in a predetermined neighbourhood, i.e. moving window (De Reu et al. 2013, 39). TPI values larger than zero represent locations whose elevation is higher than the average of their surroundings, what holds true, e.g. for hills and ridges. Negative TPI values occur at locations that are lower than their surroundings, like, e.g. valleys or pits. Flat areas as well as constantly inclined slopes have TPI values near zero.

The Topographic Position Index is inherently scale dependent. Therefore, the selection of the size of the moving window has a direct influence on the results. By integrating the TPI results for differently sized moving windows, it is possible to create slope position and landform classifications (e.g. Tagil and Jenness 2008). We focus our interest on the relative location in terms of ridges or valleys; therefore, we only use one moving window with an outer circle radius of 2500 m. TPI is calculated using the SAGA GIS module `Topographic Position Index` (Böhner et al. 2006).

4.2.2.2 Convergence Index

The Convergence Index employs the exposition of neighbouring cells in order to parameterise flow convergence and divergence. It is advantageous to plan curvature since it does not depend on absolute elevation differences (Olaya and Conrad 2009, 301). Similar to the TPI, we use it as a measure for relative topographic location since high values correspond to strong flow divergence, e.g. at ridges, while small values correspond to flow convergence, e.g. in depressions. We calculate the parameter using the SAGA GIS module `Convergence Index` (Koethe and Lehmeier 1996; Böhner et al. 2006).

4.2.2.3 Geomorphons

Besides the continuous geomorphometric measures, a landform classification based on *geomorphons* (Jasiewicz and Stepinski 2013) is employed to measure topographic characteristics in qualitative categories. Geomorphons are based on the idea that a landform element can be regarded as the specific spatial arrangement of elevation values in a selected area. Based on this analogy, the method utilises the concept of local ternary patterns, originally developed for texture

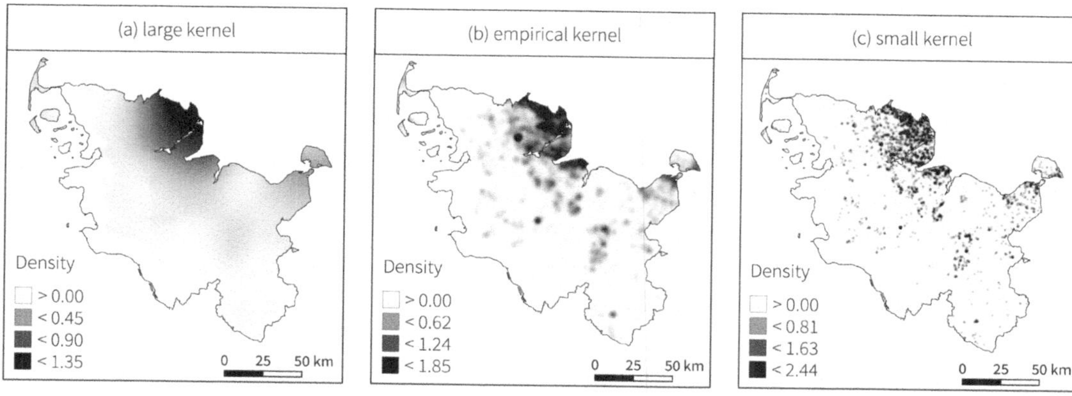

Fig. 4.4 Kernel density estimation of megalithic graves in Schleswig-Holstein; density values correspond to megaliths per square kilometre; different absolute values are due to the kernel weighting process

classification, to identify local landform elements. The resulting geomorphons are fundamental microstructures of the landscape (Jasiewicz and Stepinski 2013, 148). We use the GRASS GIS extension `geomorphons` to calculate this parameter for landforms smaller than 1 km (Stepinski and Jasiewicz 2011; Jasiewicz and Stepinski 2013).

4.3 Results

4.3.1 Density-Based Analysis of Megalithic Graves

The results of the kernel density estimation show different patterns: the large kernel (Fig. 4.4a) indicates that there is a general trend within the data that decreases from northeast to southwest. The empirical kernel (Fig. 4.4b) shows an area of high density in the northeast but also parts of moderate density that stretch throughout the region, e.g. the zone of moderate density that runs from the western edge of the high-density area in southeastern direction. This line corresponds to the transition between the geest areas (Fig. 4.1a, category b) and the young moraine area (Fig. 4.1a, category a). The small kernel (Fig. 4.4c) illustrates the location of clusters of megalithic graves. Though the majority of these is located in the northeast, they occur throughout the entire region.

4.3.2 Distance-Based Analyses of Megalithic Graves

The calculation of the distance functions provides homogeneous results: all of them show a strong deviation from the theoretical distribution of complete spatial randomness. Accordingly, the megaliths are not distributed randomly throughout space (Fig. 4.5a–d). The results of the G-function show that the distances to the nearest events are shorter than under random conditions without interactions. Nearly all events had their nearest neighbour within 1000 m. The results of the F-function illustrate that the proportion of events with a small nearest-neighbour distance between a point and an event location is smaller than in the random case. This indicates that there are larger "empty-space" areas in the megalith point pattern than under random conditions. The results of the G- and F-functions that rely on the nearest-neighbour distance between points indicate an aggregation of megalithic graves.

The results of the K- and L-functions show that the amount of megalithic graves within smaller radii around graves is higher than in the random case. Hence, events are located closer to each other as it would have been expected in the case without interactions. The K- and L-functions indicate an aggregation of events, hence interaction between the megalithic graves.

The results of the kernel density estimation indicate a general trend within the data. This

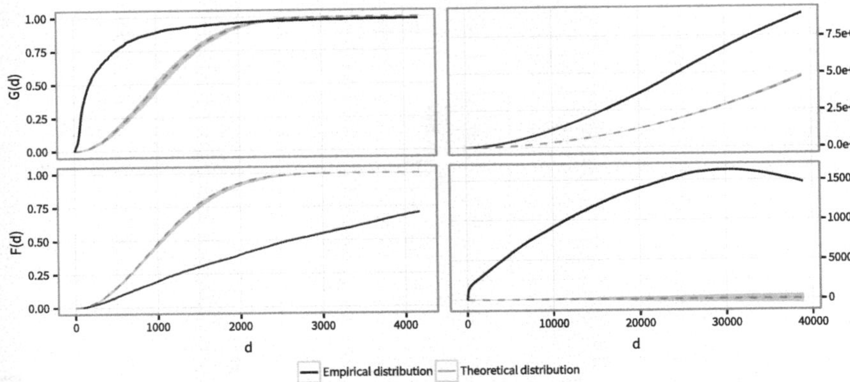

Fig. 4.5 Results of the distance function calculations for megalithic graves in Schleswig-Holstein; nearest-neighbour-based G- and F-functions (*left side*) as well as count-circle-based K- and L-functions (*right side*) show deviation from theoretical model of complete spatial randomness and illustrate that the distribution is aggregated

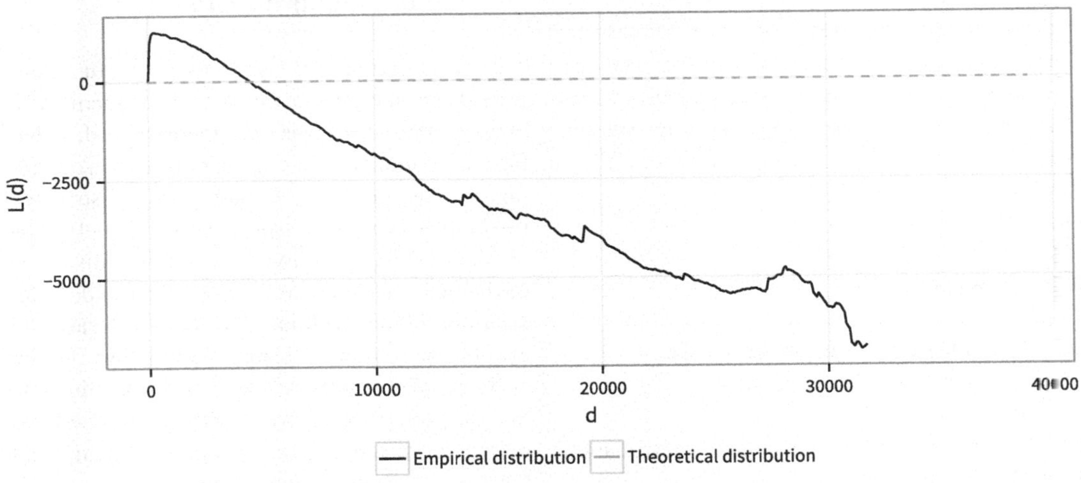

Fig. 4.6 Inhomogeneous L-function for megalithic graves in Schleswig-Holstein; at short distances, a strong clustering occurs, while starting from around 5 km, the opposite behaviour is present; first-order stationary is based on results from general trend within the data (Fig. 4.4a)

first-order effect in the point pattern influences the occurrence of megalithic graves. In order to consider this, we adapt the theoretical model to implement the general trend of megalithic graves' density throughout the area. The results of the L-function for this *inhomogeneous* pattern show still strong deviation from the theoretical model. Nevertheless, the pattern is different: at distances smaller than 5000 m, the megaliths show aggregation, while at larger distances they show the opposite trend of a more repelling distribution (Fig. 4.6).

4.3.3 Geomorphometric Parameters

We compare the locational characteristics of megaliths and artificially created points. The latter are created as resulting from (a) a complete spatial random process (Fig. 4.7a), (b) a process

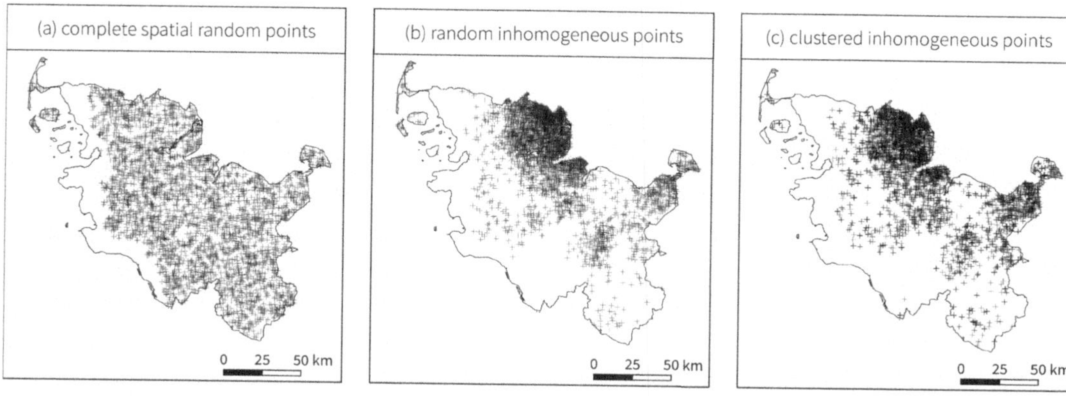

Fig. 4.7 Artificial point patterns; (**a**) complete spatial random pattern; (**b**) pattern with first-order effects; (**c**) pattern with first- and second-order effects; see text for more details

influenced by first- but not second-order effects (Fig. 4.7b) and (c) a process influenced by first- and second-order effects (Fig. 4.7c).

The values of the different geomorphometric characteristics are extracted within a buffer of 100 m of the event locations. This is done (a) to take into account that megalithic graves are no points but polygon features and (b) to enlarge the sample in order to represent locations by more than only one value. The summarising function for the extracted values was median for the continuous raster and modal for the categorical geomorphons raster.

The comparison of elevation values of artificial and empirical events shows that the megalithic graves are located to a certain degree on higher altitudes and that very low altitudes are avoided (Fig. 4.8). This behaviour is also visible in the first- and second-order influenced artificial point pattern. Besides, there are no clear differences discernible between the artificial point patterns and the megalithic graves.

In case of the Topographic Position Index, the differences in the location of artificial events are few (Fig. 4.9): although the density differs slightly, the general pattern shows a concentration at values around and below zero what is indicative for flat locations and slopes. The distribution of the megaliths is different. Although the most values are around zero, the distribution is skewed to higher values.

This pattern gets also obvious for the Convergence Index (Fig. 4.10). The distribution of the megalithic graves is skewed to higher values.

The results of landform classification geomorphons show a clear concentration of the megalithic graves on classes "ridge", "summit", "slope" and "spur". This location preference is not visible in any of the artificial point patterns (Fig. 4.11).

Altogether the results of geomorphometric parameter calculations show the tendency of megalithic graves to be located on elevated positions, i.e. on ridges and summits. Flat areas or pits are avoided as locations. The artificial point patterns do not distinctly diverge from the pattern of the megalithic graves, except for landform classification.

4.4 Discussion

4.4.1 Density-Based Analysis of Megaliths

The northeast-southwest decreasing trend in the density of megalithic graves continues to the north (Fritsch et al. 2010 as well as Fig. 4.2).

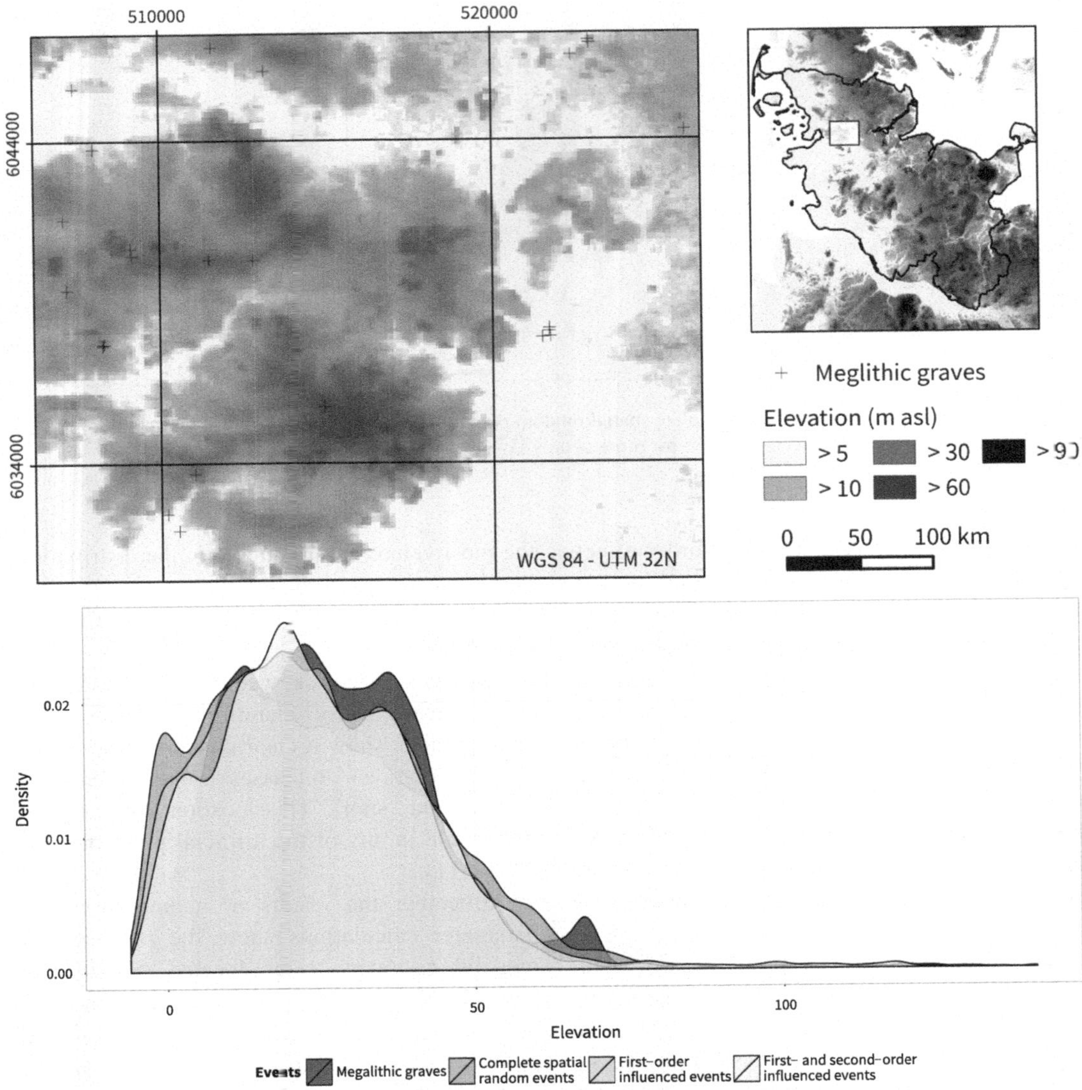

Fig. 4.8 Digital elevation model of Schleswig-Holstein (*top*) and its probability density for megalithic graves as well as artificial event locations (*bottom*)

The entire area of high density, also referred to as "megalithic heartland" (Müller 2011, 42), also influences subsequent cultural dynamics where megalithic graves played a different role (see Furholt 2012). Accordingly, it is valid to interpret the general trend reconstructed in the data as a first-order effective since it is related to a specific process of societal organisation.

Small and local clusters of high megalithic grave density occur all over the research area (Fig. 4.4c). This distribution might reflect local favourable living conditions as proposed by local archaeological studies (e.g. Lüth 2011). Nevertheless, since our methods do not integrate this kind of local-scaled but regionally specific information, no further conclusions can be drawn from this observation.

The north-south elongated line of moderate to high densities of megalithic graves (Fig. 4.4b) that corresponds to the transition between

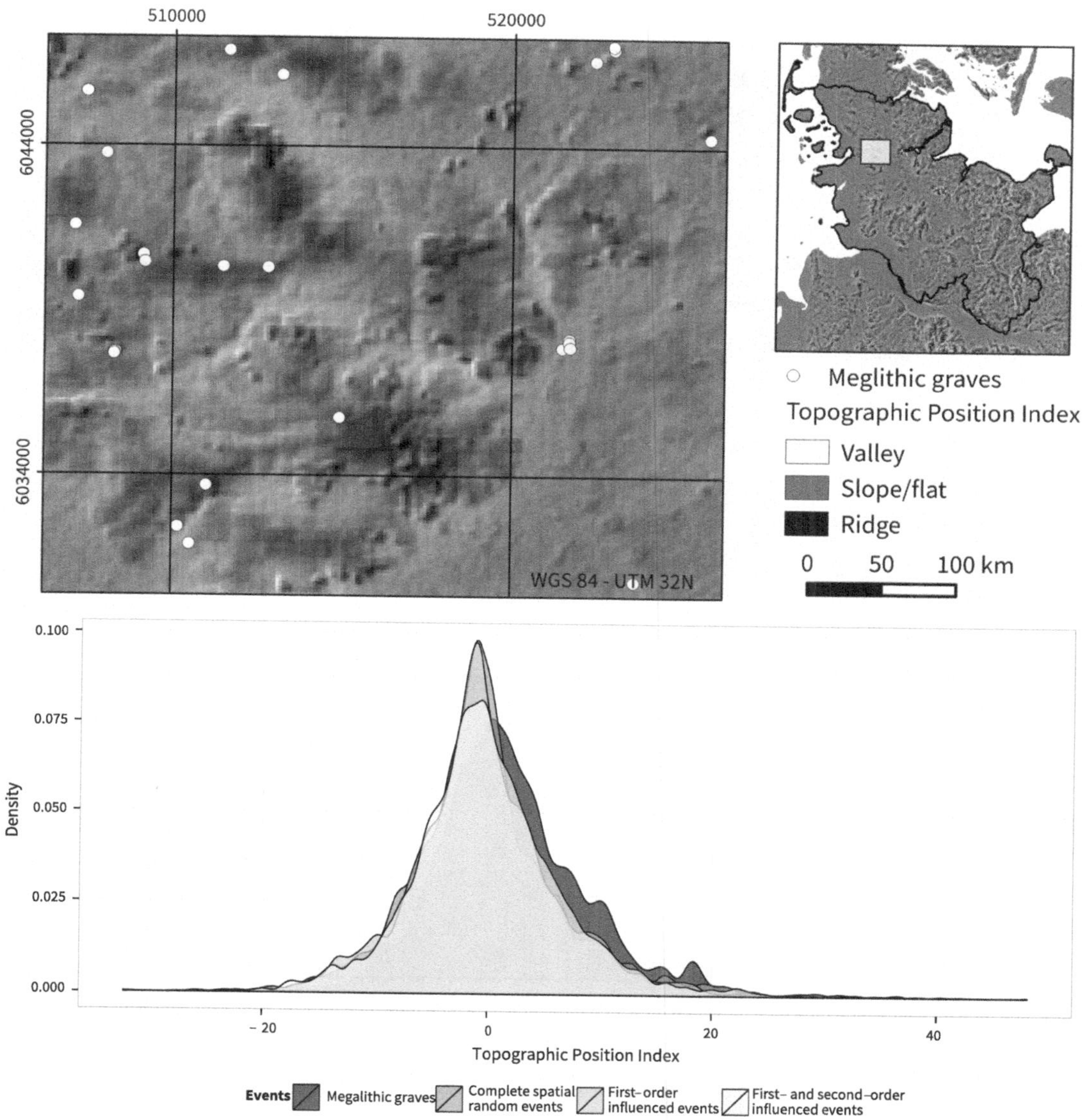

Fig. 4.9 Topographic Position Index of Schleswig-Holstein (*top*) and its probability density for megalithic graves as well as artificial event locations (*bottom*)

young moraine and geest areas can be interpreted as an exchange corridor between the north and the south. In terms of environmental characteristics, this area is favourable because of small differences in topography, while character of soils remains uniform, being light and well drained. This is the case in the sandur areas (Witt 1962, 1020). This favourable situation is also documented by the fact that an old and throughout the millennia always important road—known from mediaeval times as *Ochsenweg* (Hill 2002)—runs here. Hence, the conducted calculations also reflect the tendency of megalithic graves to be located along traffic routes.

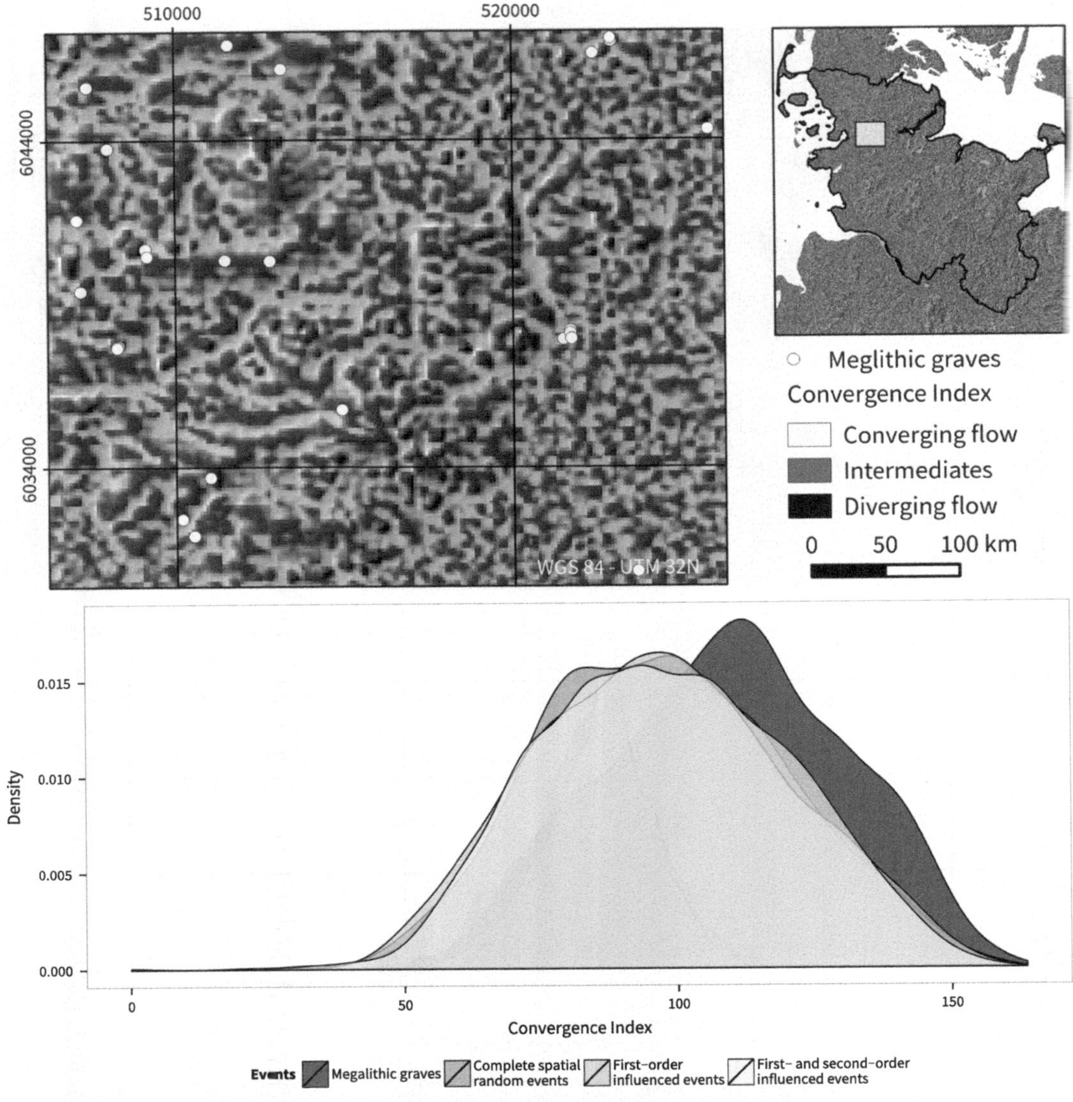

Fig. 4.10 Convergence Index of Schleswig-Holstein (*top*) and its probability density for megalithic graves as well as artificial event locations (*bottom*)

4.4.2 Distance-Based Analyses of Megaliths

The nearest-neighbour distances of the megalithic graves are mostly smaller than 1 km (see empirical curves for G- and F-functions in Fig. 4.5). This characteristic in combination with the facts (a) that the megalithic graves are found in clusters and (b) that these clusters repel one another concurs well with the observation that megalithic graves are settlement indicators (Lüth 2011, 119): Graves are constructed for more than one generation in direct vicinity to settlements. This leads to their clustered distribution. Furthermore, since each settlement needs a certain kind of space to sustain living

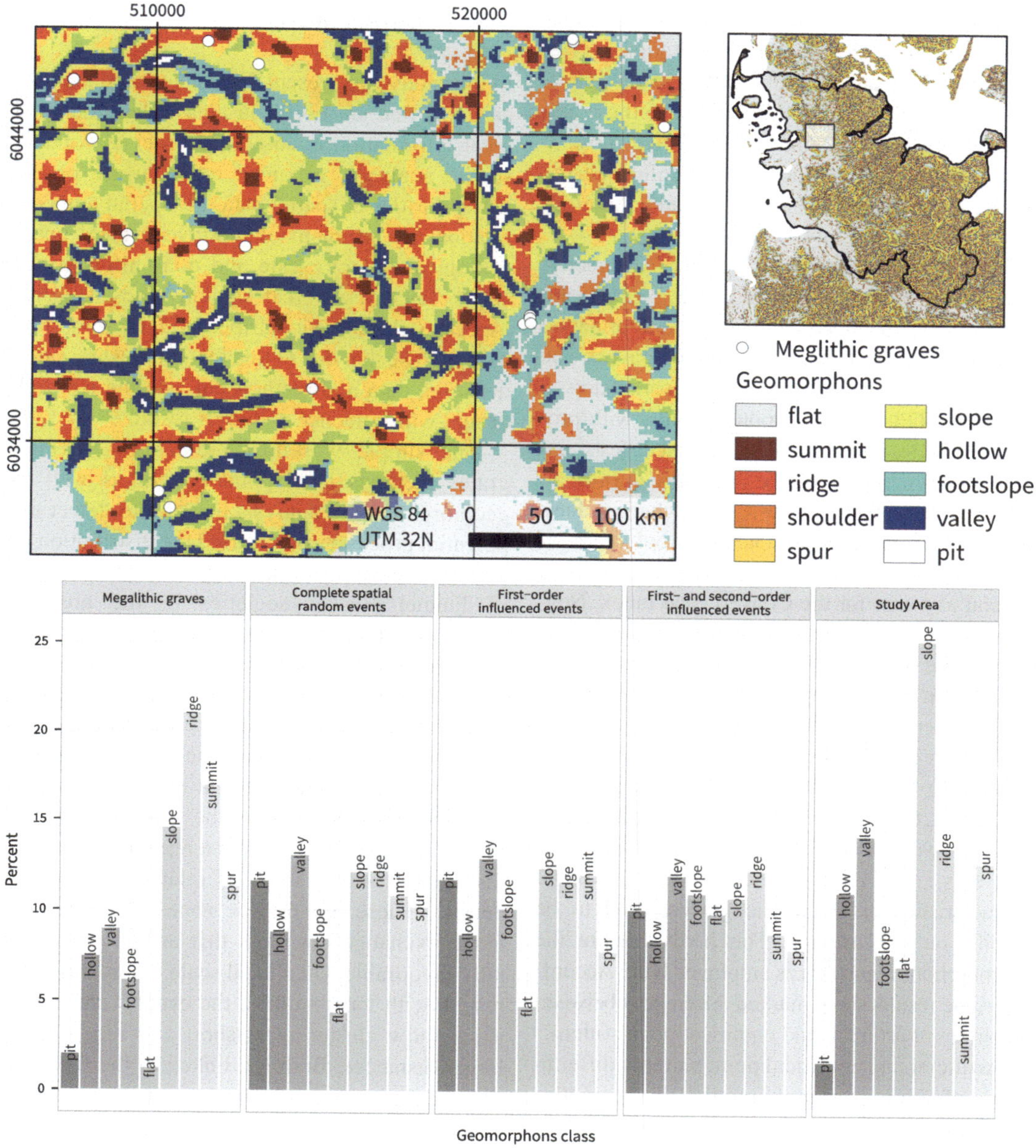

Fig. 4.11 Geomorphons in Schleswig-Holstein (*top*); the relative occurrence of megalithic graves as well as artificial events on different geomorphons classes (*bottom*)

conditions—i.e. at least 4 km as shown by empirical studies (Chisholm 2007, 146)—they tend to push potential space competitors away. Accordingly, the clusters of megaliths show a repelling behaviour at this critical distance (Fig. 4.6).

4.4.3 Geomorphometric Parameters

The results of the spatial analysis of the different geomorphometric parameters, especially those of the Topographic Position Index (TPI), are influenced by measurement errors of the radar satellite due to forest vegetation (see, e.g. Shortridge and Messina 2011, 1582, 1583). Although there are methods to remove these and other errors (e.g. Gallant et al. 2012) we chose to accept them because they do not influence the results in a way that necessitate and therefore justify additional analysis steps.

The geomorphometric parameters' characteristics of the different artificial and empirical point patterns show that not all of them yield information on the characteristics of megalithic graves. The continuous scaled geomorphometric parameters—DEM and TPI—do not produce results with pronounced differences. Detailed statistical tests and further parameters would be necessary in order to signify the slight skew to higher values, i.e. ridgelike positions, a topic that would necessitate an own contribution. The last point holds in general also true for the Convergence Index. Nevertheless, the distributions show marked differences. The Convergence Index as used here is focused on topographic features with a smaller spatial extent than the TPI. Hence, although both measure in general the same, i.e. the tendency of a location to be above or below its immediate surroundings, their explanatory value for the megalithic graves is different. This indicates that measures, aiming to describe the locational characteristics of megalithic graves, need to be specific on a local level. This holds true for the geomorphons algorithm as employed here. Accordingly, the results show marked differences between the different artificial and empirical point pattern.

Artificial and empirical point patterns differ in terms of their geomorphometric characteristics what indicates that megalithic graves are positioned on certain positions in the landscape. Although the artificial events exhibit the same patterns as the megalithic graves—clustering at short, repelling at larger distances—they are not able to reconstruct the preference of ridgelike positions with good visibility. This preference—as also proven by archaeological studies (Lüth 2011)—could be described by implementing further, more social-related, parameters: a future task.

4.5 Conclusions

In this study, we present different spatial analytical tools that are useful in areas where detailed local information is absent. In order to conduct spatial analyses, we need to be aware of the particularities of spatial data, i.e. spatial autocorrelation, the problem of modifiable areal units, ecological fallacy, scale and non-uniformity of space. Each research design involves different aspects of these particularities, and it is important to recognise that due to this fact, not just research question, data and method but also the research design will influence a study's result.

The simple workflow presented here integrates methods from point pattern as well as geomorphometric analysis in order to detect the potential processes that caused the distribution of points in space. Our points are megalithic graves of Funnel Beaker societies in the area of Schleswig-Holstein, Germany. The results lead to insights on different scales. On a supra-regional scale, the graves' distribution follows a trend of decreasing density from northeast to southwest. Thanks to archaeological investigations, we can conclude that this trend reflects the increasing distance to the *core area* of the Funnel Beaker societies. On a regional scale, the distribution shows strips of relative high event density. These prove to be zones of interaction along exchange corridors that are environmentally favourable. On a local scale, comparing our results with local studies, the events' characteristics show clustering at short and repelling at larger distances. Both are linked to the fact that they are located next to settlements. On the one hand, they show the intention to minimise distance between grave and settlement; on the other hand, they are an indirect proxy of the settlements' need to sustain a sufficient large hinterland for their subsistence.

The comparison of the geomorphometric characteristics of megalithic graves with artificial events to identify certain spatial processes shows that the location of the megalithic graves focus on specific topographic features, i.e. elevated positions with high visibility. This spatial interdependency is also known from local studies, which proves the results from the analyses of geomorphometric site characters.

Altogether methods of point pattern analysis in combination with geomorphometric measures gain multiple insights into the processes that caused an observed distribution of points. The high correspondence between the results of archaeological analysis and ours documents that the presented methodological workflow is especially useful where such information are absent. Hence, it is a promising tool for digital geoarchaeology.

Acknowledgements Daniel Knitter is grateful to the Excellence Cluster Topoi—The Formation and Transformation of Space and Knowledge in Ancient Civilizations—for supporting this study.

References

Ahnert F (1981) Über die Beziehung zwischen quantitativen, semiquantitativen und qualitativen Methoden in der Geomorphologie. Z Geomorphol Supplement Band 39:1–28

Ahnert F (2003) Einführung in die Geomorphologie. Eugen Ulmer, Stuttgart

Baddeley A, Turner R (2005) Spatstat: an R package for analyzing spatial point patterns. J Stat Softw 12 (6):1–42

Baddeley A, Turner R (2015) Spatstat manual. CRAN. http://www.spatstat.org

Baddeley AJ, Møller J, Waagepetersen R (2000) Non- and semi-parametric estimation of interaction in inhomogeneous point patterns. Stat Neerl 54(3):329–350. https://doi.org/10.1111/1467-9574.00144

Böhner J, McCloy KR, Strobl J (eds) (2006) SAGA – analysis and modelling applications. Göttinger Geographische Abhandlungen 115

Bivand RS, Pebesma EJ, Gómez-Rubio V (2008) Applied spatial data analysis with R. Springer, New York

Bivand R, Keitt T, Rowlingson B (2015) Rgdal: bindings for the geospatial data abstraction library. r package version 0.9-2, http://CRAN.R-project.org/package=rgdal

Borcard D, Gillet F, Legendre P (2011) Numerical ecology with R. Springer, New York. http://link.springer.com/10.1007/978-1-4419-7976-6

Chisholm M (2007) Rural settlement and land use. Hutchinson, London

De Reu J, Bourgeois J, Bats M, Zwertvaegher A, Gelorini V, De Smedt P, Chu W, Antrop M, De Maeyer P, Finke P, Van Meirvenne M, Verniers J, Crombé P (2013) Application of the topographic position index to heterogeneous landscapes. Geomorphology 186:39–49. https://doi.org/10.1016/j.Geomorph.2012.12.015

Diggle PJ (2013) Statistical analysis of spatial and spatio-temporal point patterns, 3rd edn. Chapman and Hall/CRC, Boca Raton

Fritsch B, Furholt M, Hinz M, Lorenz L, Nelson H, Schafferer G, Schiesberg S, Sjögren KG (2010) Dichtezentren und lokale Gruppierungen – Eine Karte zu den Großsteingräbern Mittel- und Nordeuropas. J Neolithic Archaeol. 10.12766/jna.2010.56. http://www.jna.uni-kiel.de/index.php/jna/article/view/56

Furholt M (2012) Monuments and durable landscapes in the Neolithic of southern Scandinavia and northern Central Europe. In: Furholt M, Hinz M, Miscka D (eds) "As time goes by?" Monumentality, landscapes and the temporal perspective. Proceedings of the international workshop "socio-environmental dynamics over the last 12,000 years: the creation of landscapes II (14th–18th March 2011)" in Kiel, no. 206 in Universitätsforschungen zur prähistorischen Archäologie. Habelt, Bonn, pp 115–132

Gallant JC, Read AM, Dowling TI (2012) Removal of tree offsets from SRTM and other digital surface models. ISPRS Int Arch Photogramm Remote Sens Spat Inf Sci XXXIX-B4:275–280. https://doi.org/10.5194/isprsarchives-XXXIX-B4-275-2012

Guisan A, Weiss SB, Weiss AD (1999) GLM versus CCA spatial modeling of plant species distribution. Plant Ecol 143(1):107–122. http://link.springer.com/article/10.1023/A:1009841519580

Hijmans RJ (2015) Raster: geographic data analysis and modeling. R package version 2.3-40. http://CRAN.R-project.org/package=raster

Hill T (2002) Von Wegen: auf den Spuren des Ochsenweges (Heerweg) zwischen dänischer Grenze und Eider. No. 12 in Flensburger regionale Studien, Inst. für Geographie und ihre Didaktik, Landeskunde und Regionalforschung, Flensburg

Illian J, Penttinen A, Stoyan H, Stoyan D (2008) Statistical analysis and modelling of spatial point patterns. Wiley, Chichester, WS

Jarvis A, Reuter HI, Nelson A, Guevara E (2008) Hole-filled seamless SRTM data V4. International Centre for Tropical Agriculture (CIAT). http://srtm.csi.cgiar.org

Jasiewicz J, Stepinski TF (2013) Geomorphons—a pattern recognition approach to classification and mapping of landforms. Geomorphology 182:147–156. https://doi.org/10.1016/j.geomorph.2012.11.005

Knitter D, Nakoinz O, Del Fabbro R, Kohlmeyer K, Meyer M, Schütt B (2014) The centrality of Aleppo and its environs. eTopoi J Anc Stud 3:107–127

Koethe R, Lehmeier F (1996) SARA – System zur Automatischen Relief-Analyse. User manual. Department of Geography, University of Goettingen (unpublished)

Kvamme KL (2006) There and back again: revisiting archaeological location modeling. In: Mehrer MW, Wescott KL (eds) GIS and archaeological site location modeling. Taylor & Francis, Boca Raton, pp 2–34

Legendre P, Legendre L (2012) Numerical ecology. In: Developments in environmental modelling, 3rd english edn, vol 24. Elsevier, Amsterdam

Liedtke H, Marcinek J (eds) (2002) Physische Geographie Deutschlands, 3rd edn. Perthes Geographie-Kolleg, Klett-Perthes, Gotha

Litt T, Behre KE, Meyer KD, Stephan HJ, Wansa S (2007) Stratigraphische Begriffe für das Quartär des norddeutschen Vereisungsgebietes. Eiszeitalter und Gegenwart 56(1-2):7–65. https://doi.org/10.3285/eg.56.1-2.02

Lloyd CD (2011) Local models for spatial analysis. CRC Press, Boca Raton

Lüth P (2011) Die neolithische Besiedlung des nördlichen Schleswig-Holsteins am Übergang vom Früh- zum Mittelneolithikum. Offa Berichte und Mitteilungen zur Urgeschichte, Frühgeschichte und Mittelalterarchäologie 65/66 (2008/2009):93–133

Meynen E, Schmithüsen J (eds) (1962) Handbuch der naturräumlichen Gliederung Deutschlands, vol Band II. Bundesanstalt für Landeskunde und Raumforschung, Bad Godesberg

Müller J (2011) Megaliths and funnel beakers: societies in change 4100–2700 BC. Drieendertigste Kroon-Voordracht, Amsterdam

Müller J (2014) 4100–2700 B.C. monuments and ideologies in the neolithic landscape. In: Osborne JF (ed) Approaching monumentality in archaeology, The Institute for European and Mediterranean Archaeology distinguished monograph series. State University of New York Press, Albany, pp 181–214

Neteler M, Bowman M, Landa M, Metz M (2012) GRASS GIS: a multi-purpose Open Source GIS. Environ Model Softw 31:124–130. https://doi.org/10.1016/j.envsoft. 2011.11.014

Olaya V, Conrad O (2009) Chapter 12 Geomorphometry in SAGA. In: Hengl T, Reuter HI (eds) Geomorphometry – concepts, software, applications. Developments in soil science, vol 33. Elsevier, Amsterdam, pp 293–308

Openshaw S (1984) The modifiable areal unit problem. Geo Abstracts Univ. of East Anglia, Norwich

O'Sullivan D, Perry GLW (2013) Spatial simulation: exploring pattern and process. Wiley, Chichester, WS

O'Sullivan D, Unwin D (2010) Geographic information analysis. Wiley, Hoboken

Pebesma E, Bivand RS (2005) S classes and methods for spatial data: the sp package. R News 5(2):9–13. ftp://200.236.31.2/CRAN/web/packages/sp/vignettes/intro_sp.pdf

Pike R, Evans I, Hengl T (2009) Chapter 1 Geomorphometry: a brief guide. In: Hengl T, Reuter HI (eds) Geomorphometry – concepts, software, applications. Developments in soil science, vol 33. Elsevier, Amsterdam, pp 3–30

R Core Team (2013) R: a language and environment for statistical computing. R Foundation for Statistical Computing, Vienna. http://www.R-project.org/

Ripley BD (2004) Spatial statistics. Wiley series in probability and statistics. Wiley, Hoboken

Schmidtke KD (1995) Die Entstehung Schleswig-Holsteins: von der Eiszeit zur Kulturlandschaft, 3rd edn. Wachholtz, Neumünster

Schütt B, Löhr H, Baumhauer R (2002) Mensch-Umwelt-Beziehungen in Raum und Zeit–Konzeption eines Fundstellenkatasters für die Region Trier. Petermanns Geographische Mitteilungen 146(6):74–83

Shortridge A, Messina J (2011) Spatial structure and landscape associations of SRTM error. Remote Sens Environ 115(6):1576–1587. https://doi.org/10.1016/j.rse.2011.02.017. http://www.sciencedirect.com/science/article/pii/S0034425711000678

Ssymank A (1994) Neue Anforderungen im europäischen Naturschutz. Das Schutzgebietssystem Natura 2000 und die FFH -Richtlinie der EU. Natur und Landschaft 69(9):395–406

Stepinski TF, Jasiewicz J (2011) Geomorphons – a new approach to classification of landforms. In: Geomorphometry 2011, Redlands

Stewig R (1982) Landeskunde von Schleswig-Holstein. In: Geocolleg, 2nd edn, vol 5. Borntraeger, Stuttgart (1. Aufl. im Hirt-Verl., Kiel)

Tagil S, Jenness J (2008) GIS-based automated landform classification and topographic, landcover and geologic attributes of landforms around the Yazoren Polje, Turkey. J Appl Sci 8:910–921

Thomas M (1949) A generalization of Poisson's binomial limit for use in ecology. Biometrika 36(1–2):18–25. https://doi.org/10.1093/biomet/36.1-2.18. http://biomet.oxfordjournals.org/content/36/1-2/18

Tobler WR (1970) A computer movie simulating urban growth in the detroit region. Econ Geogr 46:234–240. https://doi.org/10.2307/143141. http://www.jstor.org/stable/143141

Verhagen P (2007) Case studies in archaeological predictive modelling. Leiden University Press, Leiden

Weiss A (2001) Topographic position and landforms analysis. In: Poster presentation, ESRI User Conference, San Diego, CA, pp 200–200

Wickham H (2009) ggplot2: Elegant graphics for data analysis. Springer, New York

Wiegand T, Moloney KA (2004) Rings, circles, and null-models for point pattern analysis in ecology. Oikos 104(2):209–229. http://www.jstor.org/stable/3547954

Wiegand T, Moloney KA (2013) Handbook of spatial point-pattern analysis in ecology. CRC Press, Boca Raton

Witt W (1962) Schleswiger geest. In: Meynen E, Schmithüsen J (eds) Handbuch der naturräumlichen Gliederung Deutschlands, vol Band II. Bundesanstalt für Landeskunde und Raumforschung, Bad Godesberg, pp 1018–1021

Visual Perception in Past Built Environments: Theoretical and Procedural Issues in the Archaeological Application of Three-Dimensional Visibility Analysis

5

Eleftheria Paliou

Abstract

In recent years, the increasing number of works dealing with the application of computational two-dimensional (2D) and three-dimensional (3D) visibility analyses to archaeological architectural spaces has demonstrated the potential of these methods to form powerful tools for exploring socio-symbolic aspects in past built environments. Although 2D methods have been adopted by archaeologists since the early 2000s, applications of visual analysis of 3D urban form are a relatively new development, and until recently they had mainly been dependent on ad hoc programming. Nonetheless, new possibilities offered by commercial GIS and procedural 3D modelling software permit nowadays the analysis of 3D visual structures using built-in tools, a development that could potentially encourage and facilitate fresh approaches to the interpretation of past urban environments. A discussion on the broader theoretical concepts and methodological concerns associated with the study of visual perception in archaeological built spaces appears, therefore, particularly timely. This chapter presents briefly the state of the art in 3D visibility analysis methods and discusses some theoretical and procedural issues linked with their application to archaeological contexts.

Keywords

Three-dimensional visibility analysis • Built spaces • Probable viewsheds • Fuzzy viewsheds, Higuchi indices • Visual perception

5.1 Introduction

Visibility studies have been very popular in archaeological computing research for more than 20 years. Throughout this time, formal analytical approaches to visual perception have been

E. Paliou (✉)
Archäoinformatik/Computational Archaeology, Institute of Archaeology, University of Cologne, Cologne, Germany
e-mail: epaliou1@uni-koeln.de

© Springer International Publishing AG 2018
C. Siart et al. (eds.), *Digital Geoarchaeology*, Natural Science in Archaeology,
DOI 10.1007/978-3-319-25316-9_5

gradually advancing in sophistication and expanding in scope, encompassing a remarkable variety of 2D and 3D methods for the analysis of landscape and urban visual structure (Paliou 2013). At present, GIS-based procedures are routinely performed by archaeologists to explore the visibility, non-visibility and intervisibility of areas in the landscape which may have affected the choice of site location, wayfinding and human interactions in geographic space. On the other hand, the use of computational visual analyses in prehistoric and historic architectural contexts has been less common, despite its steady increase in recent years (Paliou et al. 2014). By far more frequently applied to date are two-dimensional space syntax analyses (Hillier 2007; Hillier and Hanson 1984) that can currently be performed with depthmapX, an open-source software package developed at the University College London. Space syntax approaches include mainly geometric and graph-based methods, such as axial[1], isovist[2] and visibility graph[3] analysis, which seek to identify patterns of copresence and movement in the built environment mainly with the aim to investigate how spatial configuration encourages or inhibits social encounters and interaction. These techniques rely on 2D plans to capture certain fundamental elements of visual space: the longest lines of sight within a street grid, the areas that can be seen from given points in space or links connecting mutually visible locations in open building and settlement areas. Analyses based on such basic descriptions of visual structure have been productively used in a considerable number of archaeological studies to draw formal comparisons of architectural configuration over time (Bintliff 2014; Letesson 2014), identify integrated and segregated urban areas (Stöger 2014) and highlight social inequalities suggested by disparities in visual and physical access in built environments.

In contrast to space syntax theories and techniques, which have systematically been enhanced via coordinated efforts over many decades, the development of methods for the visual analysis of 3D urban space has until recently been supported by pilot works in the fields of computer science, urban geography and archaeology (Paliou 2013). These attempts have been motivated by the need to approach aspects of human perception in real-life geometrically complex 3D environments that cannot be effectively studied via 2D or 2.5D spatial representations. Thus, 3D visibility analyses aim to consider also the situations where objects would have only partially blocked an observer's view, for example, while looking through arches, door and window openings, gaps between floor levels and vegetated urban areas. They can be used to both quantitatively assess the visibility of objects in interior and outdoor spaces and study the spatial configuration of vertically arranged urban layouts, which may influence the cognitive processes involved in wayfinding (Montello 2007, iv-07). In that sense, 3D visual analyses build upon and expand the traditions of both space syntax and GIS-based visibility studies. In archaeology, these methods have mainly been used to interpret Aegean Bronze Age, Roman and Late Antique sites and buildings (Earl et al. 2013; Landeschi et al. 2016; Paliou 2011; Paliou and Knight 2013; Paliou et al. 2011; Papadopoulos and Earl 2014), but their application to other archaeological contexts (e.g. Mesoamerican) seems particularly promising (Paliou 2014). Over the years a number of different methods that aim to quantify the visual properties of 3D built environments have been proposed. Some of these are briefly mentioned below[4], while approaches adopted for archaeological analysis are discussed in greater detail.

[1] Axial analysis aims to identify most 'accessible' and therefore most widely used street segments within an urban network by connecting the longest and fewest lines of sight that traverse each outdoor space in a continuous urban plan (Hillier and Hanson 1984, 82–142).

[2] An isovist is the area visible from a given point in built space (Benedikt 1979). The 2D isovist corresponds usually to the two-dimensional horizontal slice of space at the eye level of the viewer.

[3] A visibility graph is a set of edge connections that connect locations in space that are mutually visible (Turner et al. 2001).

[4] For a more detailed review, see Paliou (2013).

5.2 Approaches to Visibility Analysis in Three-Dimensional Built-Up Spaces

The methods of visibility analysis that aim to consider the third dimension can be roughly distinguished into those that make use of 2.5D data structures and those that are employed upon fully 3D spatial representations. The former could support certain 3D entities and/or 3D volumetric tools, but they are still heavily dependent upon 2.5D spatial models, normally triangulated irregular networks (TINs) or digital elevation models (DEMs). TINs and DEMs can only store a single z value for each x and y location, and as a result, they cannot take into account vertically oriented features such as walls and wall openings, arches and caves. Among the approaches that rely upon 2.5D data models belong some of the enhanced GIS viewshed analyses that seek to allow for the presence of vegetation potentially obstructing an observer's view in landscape scenes via incorporating TIN, 3D voxel or vector models of trees/plants (Dean 1997; Liu et al. 2010; Llobera 2007). Partly 3D are also volumetric visibility analyses that use urban DEMs (i.e. DEMs that include building elevation values). Such methods at a rudimentary level aim to estimate the open visible space (i.e. volume) that surrounds a perceiver from different locations, so as to make assessments on environmental quality or evaluate the impact of urban form upon observers in a cityscape (Morello and Ratti 2009; Yang et al. 2007). Volumetric visibility tools used in combination with urban DEMs have been created as extensions to commercial GIS packages (e.g. ArcGIS Viewsphere, Yang et al. 2007) or as part of voxel analyses performed with MATLAB (Morelo and Ratti 2009) and can be time-effective solutions for urban analysis in cases where it is sufficient to use geometrically simple building representations.

Fully 3D approaches, on the other hand, make use of 3D digital models and data structures and, thus, can capture the visual properties of objects and spaces of any form. A number of studies in the field of computer science and urban geography have developed and applied ad hoc computational algorithms for visibility analysis in 3D environments. The adopted solutions vary greatly in implementation and scope, ranging from methods that aim to calculate visible pixel counts in rendered camera views to approaches that seek to generate and analyse 3D isovists and visibility graphs in open urban areas (Bishop 2003; Bishop et al. 2000; Derix et al. 2008; Engel and Döllner 2009; Fisher-Gewirtzman et al. 2003, 2013; Groß 1991; Van Bilsen and Poelman 2009). Conversely, archaeological applications of 3D visibility analysis in architectural environments have so far mainly taken advantage the built-in functionalities of 3D modelling and GIS software. The first works in this area adopted a raster-based approach that involves two stages (Earl 2005; Paliou and Wheatley 2007; Paliou et al. 2011): First, a 3D model of the spaces under study is created (Fig. 5.1), and the visibility of the target objects from different viewpoints is calculated using the standard ray tracing and texture-baking tools available in most 3D modelling programmes. The recording of the visible and non-visible

Fig. 5.1 Views from a 3D reconstruction of room 3 of Xeste 3, a prehistoric building at Late Bronze Age Akrotiri, Greece. The traget object, the wall painting of the Adorants, is seen through door openings (pier-and-door partitions) (cf. Paliou et al. 2011)

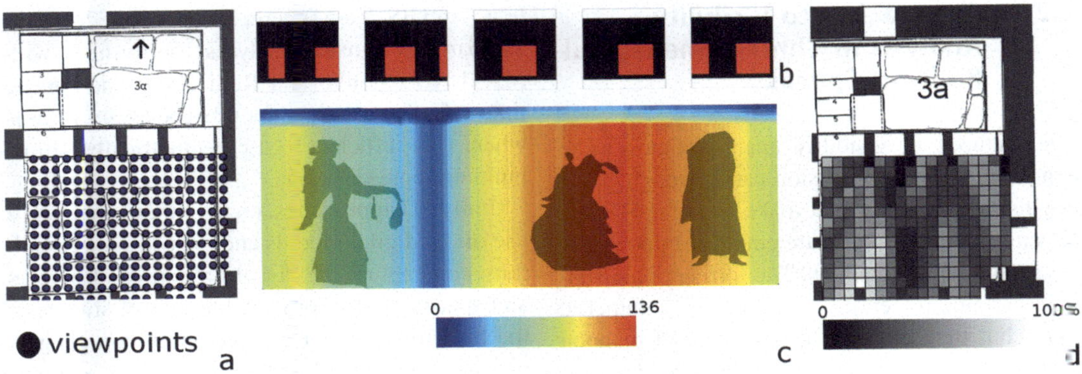

Fig. 5.2 (**a**) Observer viewpoints in equal intervals (every 20 cm) in room 3 (the arrow points to the location of the wall painting of the Adorants). (**b**) Textures of the wall surface of the Adorants holding information on illumination (= visibility) for different light locations in room 3. (**c**) A times-seen map suggesting how many times each location of the wall surface is seen from room 3. (**d**) A visible area map suggesting the percent of the total area of the three female figures that can be seen from room 3 (cf. Paliou et al. 2011)

areas of target objects in a 3D digital environment is implemented by animating a light source placed at the viewer eye level over viewpoints defined in equal intervals (Fig. 5.2a). At each light location, the texture of the target object (Fig. 5.2b), namely, a raster image that includes surface information on illuminated and not illuminated areas (corresponding to visible and non-visible areas, respectively), is extracted through the use of texture-baking tools. At a second stage, all textures are imported in a GIS to be summed up via map algebra operations using simple scripts and batch processes. The results of this analysis are usually summary maps suggesting the number of viewpoints from which the target object is seen (times-seen maps) (Fig. 5.2c) and scalar fields showing changes in the visual exposure of the target object (expressed in area units or percent) through the observers' space (Fig. 5.2d). Such maps have permitted the quantitative investigation of human experience in archaeological built environments which are redolent with socio-symbolic meaning; for instance, they have been used to explore the relationship between the visibility of prehistoric mural compositions and iconographic meaning in ritual space (Paliou et al. 2011), assess the communicative impact of elite architectural features within a Bronze Age townscape (Paliou 2011) and highlight differences in the visual experience of male and female participants in Late Antique liturgy (Paliou and Knight 2013). Furthermore, they have also been employed to quantify the visual properties of 3D prehistoric and Early Christian architectural layouts (Earl et al. 2013; Papadopoulos and Earl 2014).

More recently advances in proprietary GIS and procedural 3D modelling software have made possible the application of vector-based visibility analyses that can produce analytical outputs which are equivalent to 'times-seen' and 'visible area' maps produced with raster-based methods (Fig. 5.2c, d; ESRI ArcGIS Resource Center 2013). ESRI ArcGIS has for some time now offered users the possibility to view and to a certain degree edit and analyse fully 3D datasets that have been created with third-party 3D modelling programmes and have been imported in ArcScene as multipatch features. Multipatches are defined by a 3D vector geometry that represents the outer surface or shell of three-dimensional objects (ESRI 2012b); they enable the creation of realistic-looking 3D scenes, as they can store information on texture image, colour, transparency and lighting normal within the geometry itself. Multipatches representing buildings have been used in recent years as input for vector visibility analyses performed in a 3D GIS environment

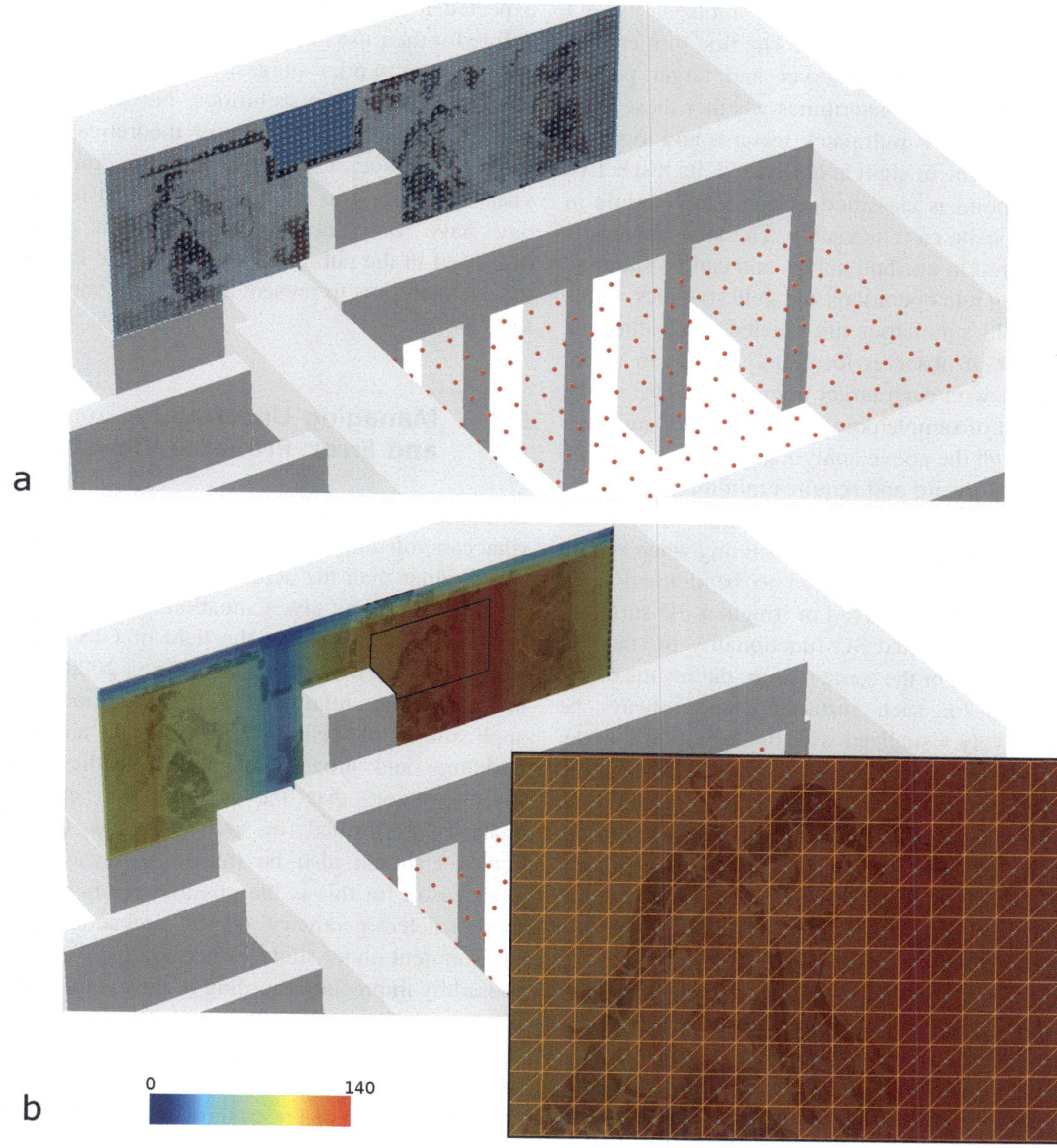

Fig. 5.3 (**a**) Observer (*red*) and target (*blue*) points in equal intervals (every 20 cm and every 5 cm, respectively). (**b**) Detail of a vector times-seen map. Multipatches representing target points and panels are used as analysis input and for the visualization of results, respectively

(ESRI ArcGIS Resource Center 2013). The analysis workflow in this case requires that both observer viewpoints and target points upon the surfaces of the objects of interest are specified (Fig. 5.3a). When the aim is to identify visible surfaces, that is, to determine how much of a wall or a building façade is visible from a set of given viewpoints, target sample points should be evenly defined at appropriate intervals upon 3D surfaces. The observer and target points, together with the multipatch features representing visual obstacles (buildings, other built structures, trees, etc.), can then be used as inputs for a vector visibility analysis performed with the *Construct*

Sight Lines and *Line of Sight* tools of the 3D Analyst ArcGIS extension. The first tool creates lines between the observer and target points, while the second determines whether these lines are blocked by multipatch features and topography. If a line of sight is obstructed, its respective target point is classified as non-visible while in the opposite case as visible. The analysis results are stored in attribute tables and can be summed up using join operations and field statistics. In the end field summaries are created suggesting the number of observer locations that are in visual contact with each target point (Fig. 5.3b) or the amount of sample points each observer can see[5]. Although the above analytical procedure is quite straightforward and requires minimum expertise from the GIS user, its implementation in a GIS can be difficult and time-consuming when evenly spaced target points need to be defined upon several vertical, curved or irregular 3D surfaces, due to the limited 3D functionality of standard GIS tools. For the same reason, the results of the analysis for such surfaces cannot always be effectively visualized with the use of GIS alone. These problems, however, can be overcome when a procedural 3D modelling programme, such as CityEngine, is integrated in the workflow (ESRI 2012a; ESRI ArcGIS Resource Center 2013). Procedural 3D modelling programmes can use shape grammar rules to create evenly sized vector panels and their centroids upon 3D surfaces of any shape; both centroids and panels can be imported in ArcScene as multipatch features (Fig. 5.3b) and used as sample target points and as surfaces for the mapping of summarized visibility values in 3D space, respectively.

Besides the above approaches, recent works in urban geography that examine the feasibility of three-dimensional visual analyses via the aid of WebGIS and Google SketchUp promise to increase the availability of tools for the study of visual urban space in the near future (Lin et al. 2015). The standardization of tools and methodologies for 3D visibility analyses is expected to offer more opportunities than ever before for their use in archaeological contexts, as some recent works suggest (Landeschi et al 2016). These new possibilities, however, also bring to the front a number of theoretical and procedural issues associated with the study of visual perception in 3D built spaces that to this day have only been sparsely—if at all—discussed in the published literature. The rest of this chapter aims to present and address some of these issues.

5.3 Managing Uncertainty and Error: Probable Viewsheds

Computational visibility analyses indicate only what can *potentially* be seen from given points in space rather than the actual views perceived by an observer in any given situation. This fact has been well established in the field of GIS landscape studies (Wheatley and Gillings 2000) but has received relatively little attention in applications of visibility analysis in ancient buildings and urban settlements; needless to say, however, that the results of visibility analyses performed in archaeological urban contexts should also be treated with caution. One reason for this is the possible errors in the reconstructed geometry of the archaeological environment under study, which are likely to be caused by imprecise recording of the extant built features on site (especially when measurements are performed by traditional methods, e.g. tape) or during the digitization of architectural plans. The main source of uncertainty, however, is the partial preservation of architectural remains and the lack of knowledge concerning features that no longer survive. On some occasions informed assumptions can be made about the presence and form of missing walls, door jambs, columns and furniture with a certain degree of confidence. Even then, however, details of such elements, for example, their exact dimensions, cannot be known with precision. Ultimately, it has to be acknowledged that a great deal of information on the geometry of past built structures, which may have affected visibility in space, will simply

[5] Equivalent to visible area maps (Fig. 5.2d)

elude us. Having said that, it is important to establish in a formal way whether lacking data can substantively affect statements and interpretations supported by visibility maps. For this purpose, probabilistic viewsheds can be created with the aim to show the likelihood that a certain target feature is exposed to a viewer as well as the propagation of possible error in the analysis output (Paliou 2009, 79). The process for the creation of probabilistic viewsheds in architectural spaces is similar to that known from landscape visibility studies (Fisher 1994): first, a number of alternative reconstructions with differences in the absence/presence, form and/or dimensions of architectural features need to be created. Binary visibility maps recorded from the same viewpoint in each proposed model should then be created, summed up and divided by the total number of alternative reconstructions (cf. Fisher 1994). The result will be a probable viewshed (Fig. 5.4):

$$p(x_{ij}) = \frac{\sum_{k=1}^{n} x_{ijk}}{n} \tag{5.1}$$

where $p(x_{ij})$ is the probability of a cell at row i and column j in the raster image being visible and x_{ij} is the value at the cell of the binary-coded viewshed in realization k such as that k takes values 1 to n.

Such probabilistic maps show the possible error distribution in visibility analysis results when solely a single viewpoint is taken into account. However, as long as interpretations are based on visual summaries that consider multiple observer locations, such as times-seen maps (Fig. 5.2c), it would be more instructive to discuss uncertainty and error propagation in relation to these more sophisticated products of analysis; in the simplest scenario between times-seen maps created from only two alternative reconstructions, the error can be described as the range of difference between the absolute values displayed in the two maps (Fig. 5.5). When more than two possible models need to be compared, as in the occasion when a number of times-seen maps are derived from alternative reconstructions among which the dimensions of certain architectural elements vary, alternative measures of dispersion can be used; for example, an indication of the possible error distribution can be given by the root-mean-square deviation (standard deviation) of the number (n) of times-seen maps considered (Paliou 2009, 80):

$$S = \sqrt{\frac{\sum (x - \bar{x})^2}{n - 1}} \tag{5.2}$$

where S is the standard deviation of the cell values of a number (n) of times-seen maps derived from an equal number of alternative reconstructions, x is the value of a pixel in a map based on one of the (n) reconstructions considered and $\bar{x}$ is the mean cell value of all times-seen maps.

This process averages the discrepancies between a number of times-seen maps that are based on a set of alternative and equally plausible

Fig. 5.4 A probable viewshed based on eight alternative reconstructions with variations in the width of the door jambs in front of the wall painting of the Adorants (marked with red circles in the ground plan)

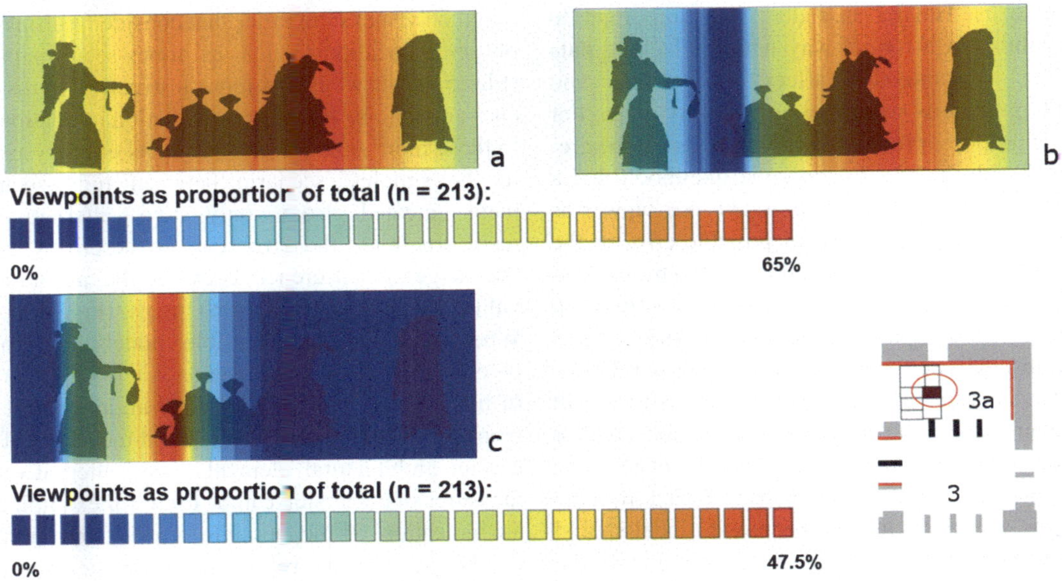

Fig. 5.5 (**a, b**) Visibility analysis results of the wall paintings of the Adorants with (**b**) and without (**a**) the wooden pier reconstructed in room 3a (marked with a red circle in the ground plan) (cf. Fig. 5.1). (**c**) The range of difference between (**a**) and (**b**) (*blue* areas are not affected by possible errors; *red* areas are those that are most affected)

reconstructions using the standard deviation of the samples. The result will be a raster (Fig. 5.6) whose cell values indicate a plus or minus error in the 'times-seen' value expressed by each cell in the map. In addition, the probability could be normally distributed around the most likely value (instead of the mean cell value), if the latter is known, for example, by sampling and statistical analysis of the dimensions of similar architectural features that have survived. This procedure provides an estimation of the range of the possible error, as well as its spatial distribution.

It is noteworthy that uncertainties in the reconstructed geometry can affect the results of the analysis in quite unpredicted ways. Sometimes, the potential error is small but widely distributed in the area of study (Fig. 5.6). In other situations the error is high but affects smaller spatial extents (Fig. 5.5). In the end, the way in which research questions have been formulated determines whether possible errors in visibility maps substantively influence suggested interpretations.

5.4 Introducing Fuzziness: Ease of Viewing and Visual Acuity in Built Environments

The above methodology aims to assess the possibility that views to a target feature are obstructed solely by the geometry of objects and built structures in a 3D scene. Nonetheless, besides geometry, visual perception is affected by a number of other factors that determine the ease with which objects are seen and are made intelligible. *Ease of viewing* is a concept that may seem obvious, and yet it can be obscure, if not explicitly defined. Publications on architectural theory have from early on attempted to formally examine the conditions that facilitate visual perception in built-up environments (Hall 1966; Märtens 1890). The influence of these works is also reflected in the list of indices proposed by Higuchi (1983, 4) to describe the visual structure of landscapes. Below, we list some of the most frequently cited factors affecting ease of viewing in built-up space and consider means to examine

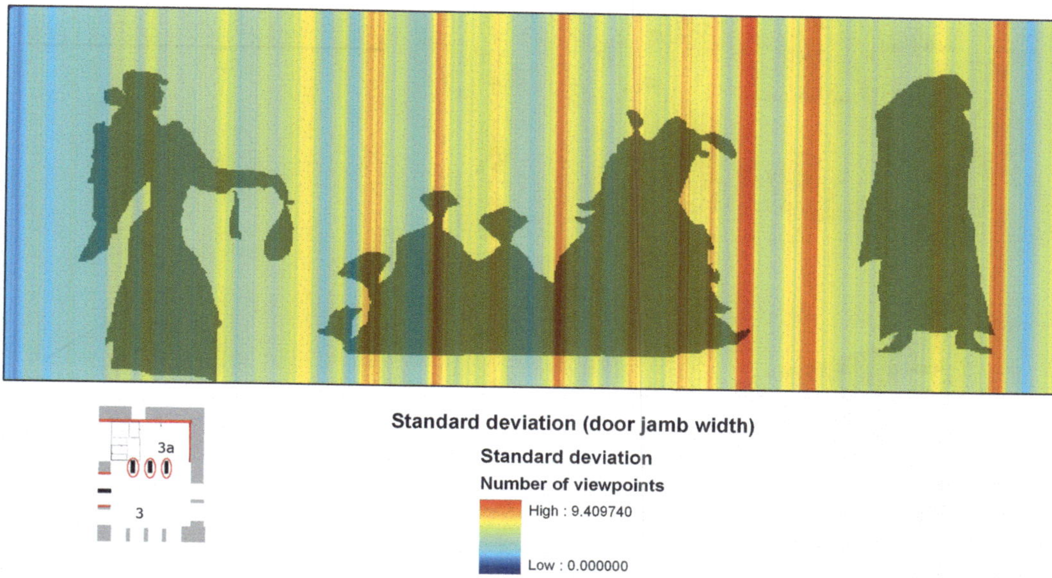

Fig. 5.6 Map suggesting the standard deviation between eight times-seen maps that were based on alternative reconstructions in which the door post width of the pier-and-door partitions in room 3 changes at the ordinal level from 18 to 25 cm

computationally their effects on visual perception:

1. *Angle of incidence*: The angle of incidence is the angle at which the line of vision strikes each surface. When the angle of incidence reaches the maximum of 90°, a surface is seen frontally and lays at about the horizon of the viewer. As the angle of incidence decreases, surfaces are seen with greater difficulty (Fig. 5.7). The angle of incidence is always larger for the frontal rather than for the longitudinal planes, and as a result, frontal surfaces are always easier to perceive (Higuchi 1983, 26). In viewshed calculations performed in a 3D digital environment, the angle of incidence can be derived by the intensity values in the texture map of a white surface that is illuminated by a single light that does not attenuate with distance. The more the light rays (i.e. sightlines) incline away from the surface's normal, the less light the surface receives and the darker it appears. High values in the lighting (texture) map of an object (approaching 255[6]) correspond to areas that can be seen from a maximum convenient angle (90° frontally), while lower values (approaching 0) to surfaces that can be appreciated with greater visual effort, if at all. It has to be noted, however, that in architectural spaces the frontal (and easier-to-see) surfaces with respect to the position of a mobile viewer entering a room or a standing observer in space are in many situations easy to identify from ground plans. Therefore, a computational analysis of the angle of incidence is not always necessary to reach an interpretation.
2. *Angle of elevation*: This is the vertical angle that determines how many degrees above or below the horizon of the onlooker a target surface can be seen. It is associated with the common experience of 'looking up' that triggers physiological and psychological responses:

[6] When the map is displayed using a stretched black and white colour ramp

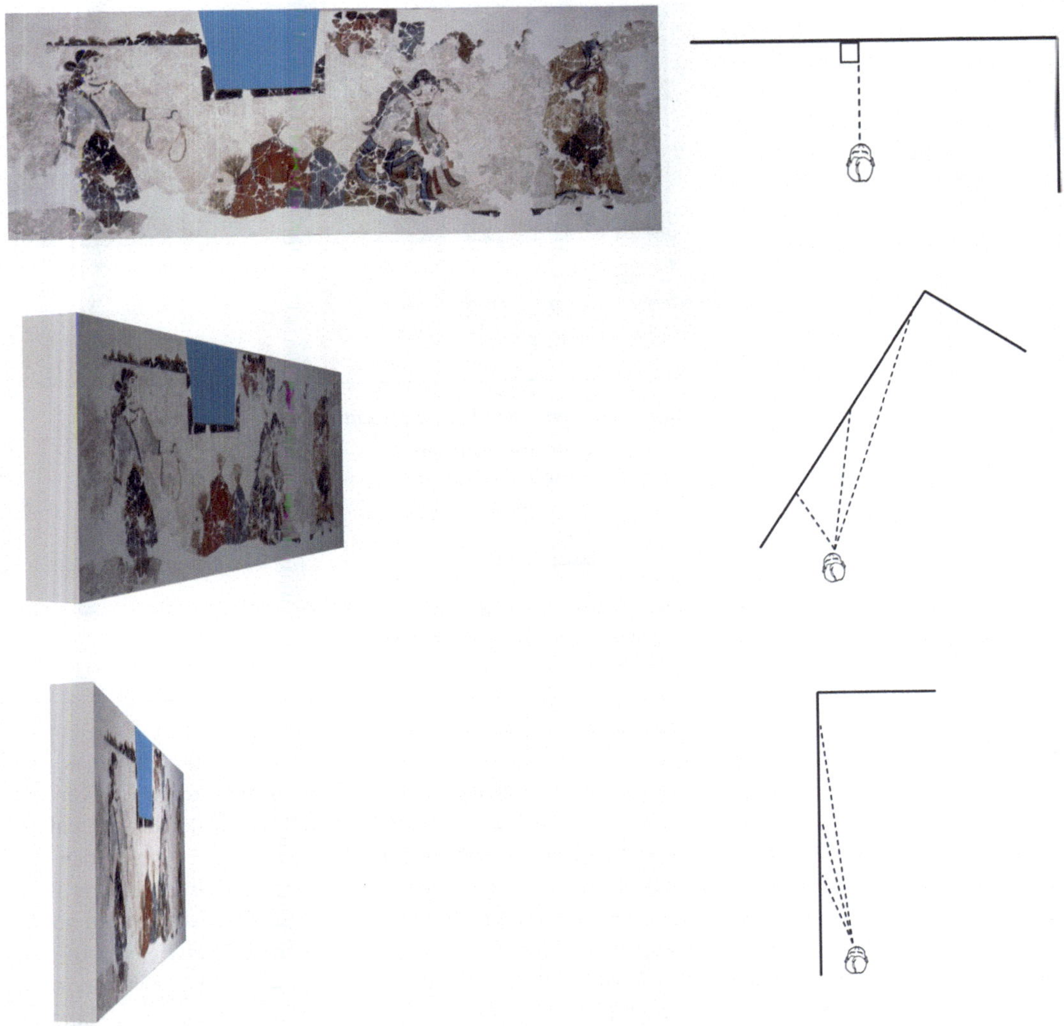

Fig. 5.7 Varying angles of incidence for a wall surface. Frontal surfaces are always easier to see

> . . .the process of looking up at an object tends to limit the mobility of the human body and to cut off the line of vision at a point above the horizontal. With the most stable line of vision for the average person being about 10–15° below the horizontal, it follows that the very process of looking up involves a certain amount of stress. Presumably this is why the term 'look up to' connotes the idea of paying respect or reverence. 'Looking up to' someone or something requires a visual effort. (Higuchi 1983, 46)

It can be argued then that the angle of elevation determines the visual effort needed to observe an object and consequently it can also be used to assess the communicative impact of symbolic features in a townscape (Paliou 2011). Considering the physiological data on vision provided by Dreyfuss (1959) (Fig. 5.8), architectural features (mural decorations, inscriptions, sculpture, etc.) that are placed up to 25° above the horizon of the onlooker can be seen with eye movements, that is, with relatively little effort and ease (Paliou 2011); thus, they are more likely to be noticed by pedestrians traversing open public spaces. On the other hand, paying attention to objects placed

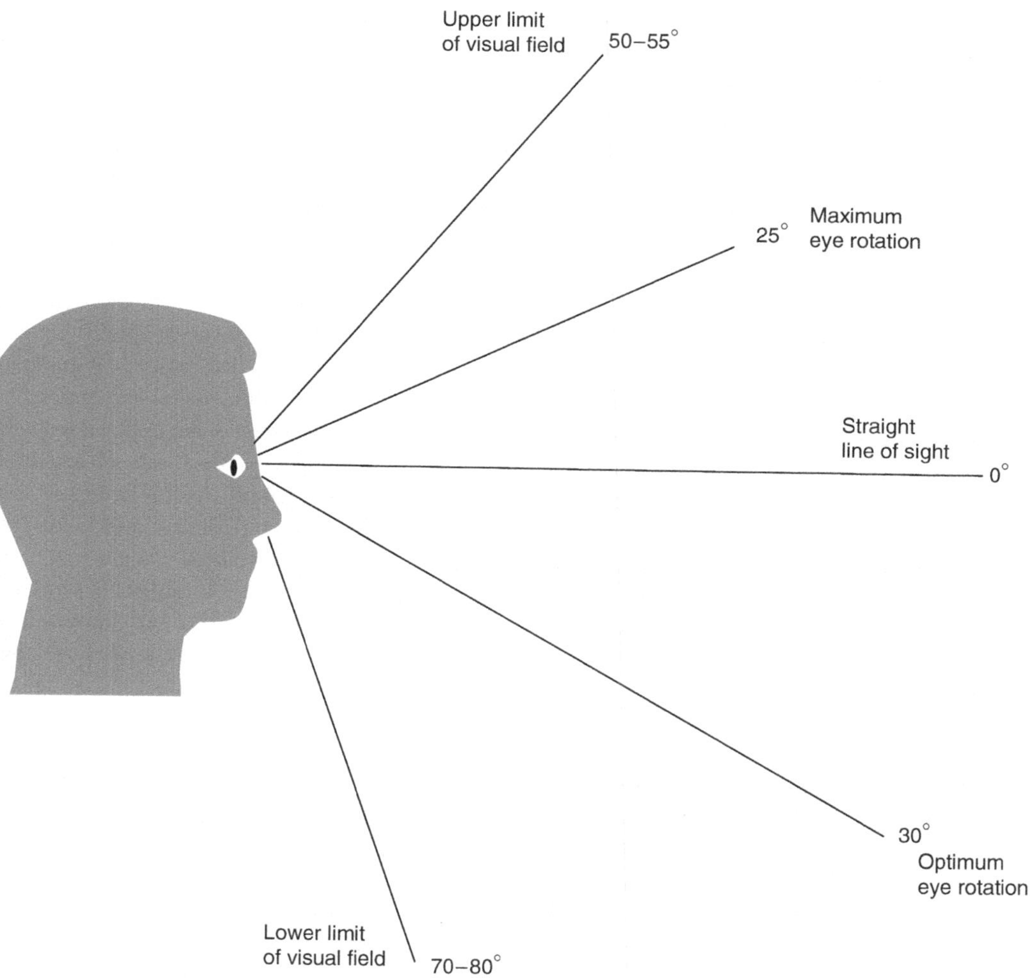

Fig. 5.8 Henry Dreyfuss (1959) basic data on vision. From Higuchi (1983, 40) (cf. Paliou 2011, Fig. 5.6)

above this angular threshold requires additional upward head movements that may force onlookers moving through densely built and populated townscapes to slow down or even stop. Pedestrians not willing to interrupt their pace of movement might not carefully observe or miss altogether objects situated way above their horizon.

The angle of elevation has also been linked with the notion of monumentality. In one of the early works on architectural analysis, Märtens (1890 cited in Higuchi 1983, 47) suggested three visual ranges at 18°, 27° and 45° at which the impact of the monument on the viewer changes significantly: at 18° a building starts to be conceived as monumental, at 27° it fills the viewers' range of vision, while at 45° it can be seen with more clarity (the viewer can distinguish small details of the building as well as see the structure as a whole).

Objects and 3D surfaces seen within critical vertical angular ranges can be easily identified computationally in raster-based and vector-based three-dimensional visibility analyses. In raster-based approaches, angular ranges can be determined by the propagation of light distributed by a photometric light source. A photometric light casts rays within a fixed set of horizontal and vertical angles defined by the user, and it can be utilized for the creation of binary viewshed maps in the same way as a light that casts rays in all directions. In this case, only areas in 3D scenes that are within the angular range specified by the user will receive light and will be classified as visible.

In vector-based analysis, the user can opt to calculate the vertical angles between the observer and target points while running the standard visibility analysis tools[7] and store the results in the attribute data table of the output shapefile. In this manner points viewed within or beyond certain significant angular ranges can be identified and visualized appropriately in the final summary maps.

3. *Distance*: Broadly speaking, the clarity with which objects in space are seen decreases with distance. Objects situated too far away from a viewer may not be intelligible despite being within his/her field of view. In the framework of visibility studies, the effects of distance decay upon visual acuity are normally taken into account by establishing different visual ranges in which the clarity of a certain feature, or the effort required by an individual to observe a particular object, changes significantly. In previous works such ranges were determined by focusing on the intelligibility of certain diagnostic features of a target. Märtens (1890 cited in Higuchi 1983, 9), for example, suggested that the maximum distance in which a person's face can be identified is 25 m, that is, the distance when the nasal bone occupies an angle of 1 min in the field of vision. Higuchi (1983) estimated three different distance thresholds (foreground, middle ground and background) in which the perception of landscape differs substantially taking into account the size of the dominant tree species[8]. Similar estimates can be made for any object or environmental feature. That said, on many occasions the effects of distance decay on the ability to recognize objects in space for what they are cannot always be easily or objectively appreciated. This is because the intelligibility of a visual target depends on many other context-dependent factors: the target's colour and contrast to the background, illumination, the familiarity of the viewer with the object, the visual acuity of the observer (20/20 vision or less), etc. The exact thresholds that mark substantive changes in the visual perception of an object are, thus, often elusive, but at least they can be discussed on the basis of arguments that are made explicit. As has been suggested on other occasions (Conolly and Lake 2006, 231), critical distance ranges in built environments can be established with real-life experiments that consider changes in the perception of similar targets as those under study.

   Visibility maps can suggest 'fuzziness' by displaying various user-defined distance thresholds between observer and target points (Earl 2005; Paliou 2009). In vector-based procedures, the distance of the observer to the target for all viewing locations can be easily taken into account by recording the length of sightlines in their attribute data table using built-in scripts[9]. Visible target points can then be classified according to their distance from the observer points and visualized appropriately (e.g. by including/excluding visibility values belonging to foreground, middle ground or background from visible area maps) to examine differences in visual acuity.

4. *Illumination*: Illumination is an important element in the definition of built space and plays a fundamental role in the perception of shape, colour and texture of surfaces that create visual impressions in the built environment. As a result, illumination is a factor that significantly affects visual acuity, human action and

[7] In the current ArcGIS version (ArcGIS 10) the ArcScene *Construct Sight Lines* tool

[8] For example, according to Higuchi the forground distance is approximately 60 times the size of the dominant tree species.

[9] The *Add Geometry Attributes* tool of ArcGIS Spatial Analyst could be of use in this case.

the comprehensibility of objects in space. In contemporary computer-aided architectural design, the effects of light upon urban form are normally examined via the application of formal lighting analysis which is implemented with the use of 3D modelling software. This approach has also been adopted in archaeological studies to foster an understanding of how the distribution of direct and indirect light could have enhanced or diminished visual perception in ancient built space. When the geometric properties and the surface materials of the environment under study are known and appropriately modelled, formal lighting analysis can give a better appreciation of distinct or subtle variations in visual experience that may have affected past architectural design, task performance and sensory engagement with space (Dawson et al. 2007; Papadopoulos and Earl 2014; Roussos 2003).

5. *Human mobility*: It should be already clear from the discussion so far that visual perception is greatly depended upon the mobility of the human body. Stasis and movement through space determine the spatial relationship of an observer's body with the target, affecting the mode in which the latter is experienced. The boundaries between these two states are rather fluid as suggested by Gibson:

> When the moving point of observation is understood as a general case, the stationary point of observation is more intelligible. It is no longer conceived as a single geometrical point in space but as a pause in locomotion, as a temporarily fixed position relative to the environment. (Gibson 1979, 75)

In visibility studies observers in built space can be assumed to be both stationary and mobile. It is often the case, however, that at times one state of the body prevails over the other depending on the nature of the practices the observers are engaged. For example, during public gatherings in controlled environments (e.g. ritual architectural spaces), individuals attend actions in space and their settings from fixed points for the greater part of the unfolding events. Stasis in space restricts visual access allowing only partial but persistent views to the environment surrounding the perceiver and the objects situated in it. On the contrary, pedestrians traversing a street network may observe their surroundings within the course and pace of their movement, which could have shaped a rapidly changing and fleeting experience of the environment. Even in this case, however, features placed in prominent locations would have probably dominated the visual perception of passers-by.

When interpreting visibility maps, the state of the human body and practices performed in space should be taken into account. While times-seen maps suggest the areas that are overall more likely to be exposed to a viewer, visible area maps could indicate changes in the visual experience of the mobile perceiver, when intersected with possible trajectories of movement in space.

5.5 Other Theoretical Issues on Visual Perception

Computational approaches to visibility have been criticized in the past because of the way they conceptualize visual perception. These criticisms have been well reviewed elsewhere (Wheatley 2014; Wheatley and Gillings 2000) and due to space limitations will not be discussed again here. There is one aspect of the critique, however, that is relevant to the scope of this chapter and is worth mentioning both because it recurs in current archaeological debates and because it could be addressed—at least to a certain degree—computationally. For some years now, concerns have been raised with regard to the dominant role of vision in visibility analyses which frequently disregard other sensory modalities that shape human perception. Such a prioritization of vision comes in contrast with current understandings of human experience (cf. Hamilakis 2014) and anthropological works that tend to emphasize the synergistic relationship between the senses, arguing for a multisensory approach to the study of past landscapes and built spaces. Interestingly, archaeologists have attempted to respond to these criticisms by developing computational approaches that aim to allow for more than one sensory modality. The solutions proposed include sensory envelopes, namely, circular catchments that define areas in which all the senses are engaged (Frieman and Gillings 2007), as well as maps that suggest the blending of the senses, for example, vision and sound, by combining visual and acoustic datasets via data fusion methods (Paliou and Knight 2013). Although such approaches do not fully address the problems involved in a multisensory approach, they at least adopt a critical perspective that acknowledges the complexity and richness of lived experience.

5.6 Summary and Conclusion

This chapter gave a short account[10] of the state of the art in three-dimensional visibility analyses in archaeology and presented briefly raster- and vector-based workflows that have been used in recent years to examine aspects of visual communication in three-dimensional prehistoric and historical built environments. This overview highlighted a current trend towards the development of analytical approaches that make use of widely available software programmes and require minimum or no programming skills by the user. The standardization of 3D visibility analysis tools and methodologies, in combination with the increasing availability of 3D datasets produced by photogrammetry, laser scanning and procedural modelling, are likely to facilitate the application of visual analyses to archaeological three-dimensional spaces in the near future. In light of these developments, this chapter also aimed to address some relevant critical theoretical and procedural issues, especially those pertaining to quantifying uncertainty and error in visibility calculations, ease of viewing and visual acuity. Nonetheless, it should be emphasized that not all aspects of human perception can be subjected to formal quantitative analysis. An embodied approach to the experience of past architectural spaces could unravel a host of additional symbolic associations that can inform archaeological studies (Hamilakis 2014). Thus, the combination of computational visibility analysis with other spatial technologies, such as augmented reality, is likely to offer a richer framework within which to interpret prehistoric and historic built environments.

[10] This review did not cover all the measures that are used to describe the visual properties of open urban areas. For example, skyline, openness and enclosure in built environments (Fisher-Gewirtzman and Wagner 2003; Lin et al. 2015) were not mentioned here as these measures have yet to be discussed within the context of archaeological and historical disciplines; nonetheless, the possibility that such concepts may prove useful heuristics for archaeological analysis in the future cannot be precluded.

Acknowledgements I would like to thank the editors, Christoph Siart, Markus Forbriger and Olaf Bubenzer, for inviting me to contribute to this volume. Most of the ideas discussed in this chapter were developed during my studies as a PhD student at the University of Southampton. For this reason, I owe special thanks to David Wheatley and Graeme Earl, who acted as my thesis supervisor and advisor, respectively, for their advice and comments on this research.

References

Benedikt ML (1979) To take hold of space: isovists and isovist fields. Environ Plann B 6:47–65

Bintliff J (2014) Spatial analysis of past built environments: houses and society in the Aegean from the early iron age till the impact of Rome. In: Paliou E, Lieberwirth U, Polla S (eds) Spatial analysis and social spaces: interdisciplinary approaches to the interpretation of historic and prehistoric built environments. De Gruyter, Berlin, pp 263–276

Bishop DI (2003) Assessment of visual qualities, impacts, and behaviours, in the landscape, by using measures of visibility. Environ Plann B 30:677–688

Bishop DI, Wherrett RJ, Miller RD (2000) Using image depth variables as predictors of visual quality. Environ Plann B 27:865–875

Conolly J, Lake M (2006) Geographical information systems in archaeology. Cambridge University Press, Cambridge

Dawson P, Levy R, Gardner D, Walls M (2007) Simulating the behaviour of light inside Arctic dwellings: implications for assessing the role of vision in task performance. World Archaeol 39:17–35

Dean DJ (1997) Improving the accuracy of forest viewsheds using triangulated networks and the visual permeability method. Can J For Res 27(7):969–977

Derix C, Gamlesæter Å, Miranda Carranza P (2008) 3D isovists and spatial sensations: two methods and a case study. In: Haq S, Hölscher C, Torgrude S (eds) SFB/TR 8 report no. 015-05/2008. Movement and orientation in built environments: evaluating design rationale and user cognition. Universität Bremen, Bremen, pp 67–72

Dreyfuss H (1959) The measure of man: human factors in design. Whitney, New York

Earl GP (2005) Texture viewsheds: spatial summaries of built archaeological spaces derived from global light mapping. Proceedings of the 11th international conference on virtual systems and multimedia, Ghent, Belgium VSMM, October 2005, pp 303–312

Earl G, Porcelli V, Papadopoulos C, Beale G, Harrison M, Pagi H, Keay S (2013) Formal and informal analysis of rendered space: the basilica Portuense. In: Bevan A, Lake M (eds) Computational approaches to archaeological spaces. Left Coast Press, Walnut Creek, pp 265–305

Engel J, Döllner J (2009) Approaches toward 3D visual analysis for digital landscapes and its applications digital landscape architecture proceedings, pp 33–41

ESRI (2012a) 3D analyst – feature and volumetric analysis. http://video.esri.com/watch/1662/3d-analyst-feature-and-volumetric-analysis. Accessed 20 Oct 2015

ESRI (2012b) The multipatch geometry type. https://www.esri.com/library/whitepapers/pdfs/multipatch-geometry-type.pdf. Accessed 26 Oct 2015

ESRI ArcGIS Resource Center (2013) 3D urban analysis. http://video.arcgis.com/watch/2208/3d-urban-analysis

Fisher P (1994) Probable and fuzzy models of the viewshed operation. In: Worboys MF (ed) Innovations in GIS: selected papers of the first international conference on GIS research UK. Taylor & Francis, London, pp 161–175

Fisher-Gewirtzman D, Wagner AI (2003) Spatial openness as a practical metric for evaluating built-up environments. Environ Plann B 30:37–49

Fisher-Gewirtzman D, Burt M, Tzamir Y (2003) A 3-D visual method for comparative evaluation of dense built-up environments. Environ Plann B 30:575–587

Fisher-Gewirtzman D, Shashkov A, Doytsher Y (2013) Voxel based volumetric visibility analysis of urban environments. Surv Rev 45:451–461

Frieman C, Gillings M (2007) Seeing is perceiving? World Archaeol 39:4–16

Gibson JJ (1979) The ecological approach to visual perception. Boston, Houghton Mifflin

Groß M (1991) The analysis of visibility-environmental interactions between computer graphics, physics and physiology. Comput Graph 15:407–415

Hall TE (1966) The hidden dimension. Doubleday, New York

Hamilakis Y (2014) Archaeology and the senses: human experience, memory, and affect. Cambridge University Press, Cambridge

Higuchi T (1983) The visual and spatial structure of landscapes. The MIT Press, Cambridge, MA

Hillier B (2007) Space is the machine, e-book. Space Syntax, London. http://www.ninsight.at/ak_stdb/SpaceIsTheMachine.pdf

Hillier B, Hanson J (1984) The social logic of space. Cambridge University Press, Cambridge

Landeschi G, Dell'Unto N, Lundqvist K, Ferdani D, Campanaro DM, Leander Touati AM (2016) 3D-GIS as a platform for visual analysis: investigating a Pompeian house. J Archaeol Sci 65:103–113

Letesson Q (2014) From building to architecture: the rise of configurational thinking in bronze age Crete. In: Paliou E, Lieberwirth U, Polla S (eds) Spatial analysis and social spaces: interdisciplinary approaches to the interpretation of historic and prehistoric built environments. De Gruyter, Berlin, pp 49–90

Lin T, Lin H, Hu M (2015) Three-dimensional visibility analysis and visual quality computation for urban open spaces aided by Google SketchUp and WebGIS. Environ Plann B 44:1–29. https://doi.org/10.1177/0265813515605097

Liu L, Zhang L, Ma J, Zhang L, Zhang X, Xiao Z, Yang L (2010) An improved line-of-sight method for visibility analysis in 3D complex landscapes. SCIENCE CHINA Inf Sci 53:2185–2194

Llobera M (2007) Modeling visibility through vegetation. Int J Geogr Inf Sci 21:799–810

Märtens H (1890) Optisches Mass für den Städtebau. Max Cohen & Sohn, Bonn

Montello DR (2007) The contribution of space syntax to a comprehensive theory of environmental psychology. In: Kubat AS, Ertekin Ö, Güney Yİ, Eyüboğlu E (eds) 6th International space syntax symposium proceedings, Istanbul, ITÜ Faculty of Architecture, pp iv-1–iv-12. http://geog.ucsb.edu/~montello/pubs/SpaceSyntax_invited.pdf. Accessed 28 May 2012

Morello E, Ratti C (2009) A digital image of the City: 3D Isovists in Lynch's urban analysis. Environ Plann B 36:837–853

Paliou E (2009) The visual consumption of mural painting in late bronze age Akrotiri (Thera, Greece): a computational approach to visibility analysis in three-dimensional built environments. Unpublished thesis, School of Humanities, University of Southampton, Southampton

Paliou E (2011) The communicative potential of Theran murals in late bronze age Akrotiri: applying viewshed analysis in 3D townscapes. Oxf J Archaeol 30:247–272

Paliou E (2013) Reconsidering the concept of visualscapes: recent advances in three-dimensional visibility analysis. In: Bevan AML (ed) Computational approaches to archaeological spaces. Left Coast Press, Walnut Creek, pp 243–264

Paliou E (2014) Visibility analysis in 3D spaces: a new dimension to the understanding of social space. In: Paliou E, Lieberwirth U, Polla S (eds) Spatial analysis and social spaces: interdisciplinary approaches to the interpretation of prehistoric and historic built environments. De Gruyter, Berlin, pp 91–113

Paliou E, Knight DJ (2013) Mapping the senses: perceptual and social aspects of late antique liturgy in San Vitale, Ravenna. In: Proceedings of CAA 2010, computer applications and quantitative methods in archaeology, international conference, Granada 6–9 April 2013, pp 229–236

Paliou E, Wheatley D (2007) Integrating spatial analysis and 3D modelling approaches to the study of visual space: late bronze age Akrotiri. In: Figueiredo A, Velho GL (eds) The world is in your eyes – proceedings of the XXXIII computer applications in archaeology conference, 21–24 March 2005. Tomar, Portugal, pp 307–312

Paliou E, Wheatley D, Earl G (2011) Three-dimensional visibility analysis of architectural spaces: iconography and visibility of the wall paintings of Xeste 3 (late bronze age Akrotiri). J Archaeol Sci 38:375–386

Paliou E, Lieberwirth U, Polla S (2014) Spatial analysis and social spaces: interdisciplinary approaches to the interpretation of prehistoric and historic built environments, Topoi. Berlin Studies of the Ancient World. De Gruyter, Berlin

Papadopoulos K, Earl GP (2014) Formal three-dimensional computational analyses of archaeological spaces. In: Paliou E, Lieberwirth U, Polla S (eds) Spatial analysis and social spaces: interdisciplinary approaches to the interpretation of historic and prehistoric built environments. De Gruyter, Berlin, pp 135–166

Roussos I (2003) High fidelity lighting of knossos. In: Arnold D, Chalmers A, Niccolucci F (eds) Conference proceedings, 4th VAST international symposium on virtual reality, Archaeology and intelligent cultural heritage 2003, pp 47–56

Stöger H (2014) The spatial signature of an Insula neighbourhood of Roman Ostia. In: Paliou E, Lieberwirth U, Polla S (eds) Spatial analysis and social spaces: interdisciplinary approaches to the interpretation of historic and prehistoric built environments. De Gruyter, Berlin, pp 297–316

Turner A, Doxa M, O'Sullivan D, Penn A (2001) From isovists to visibility graphs:a methodology for the analysis of architectural space. Environ Plann B 28:103–121

Van Bilsen A, Poelman R (2009) 3D visibility analysis in virtual worlds: the case of supervisor. In: Proceedings of construction applications of virtual reality (CONVR) 2009 conference Sydney, 5–6 Nov 2009

Wheatley D (2014) Connecting landscapes with built environments: visibility analysis, scale and the senses. In: Paliou E, Lieberwirth U, Polla S (eds) Spatial analysis and social spaces: interdisciplinary approaches to the interpretation of prehistoric and historic built environments. De Gruyter, Berlin, pp 115–134

Wheatley D, Gillings M (2000) Vision, perception and GIS: developing enriched approaches to the study of archaeological visibility. In: Lock GR (ed) Beyond the map: archaeology and spatial technologies. IOS Press, Amsterdam, pp 1–27

Yang PPJ, Putra S, Li W (2007) Viewsphere: a GIS-based 3D visibility analysis for urban design evaluation. Environ Plann B Plann Des 34:971–992

Understanding by the Lines We Map: Material Boundaries and the Social Interpretation of Archaeological Built Space

6

Benjamin N. Vis

Abstract

End users of archaeological maps are restricted in what they know about the data they are using. Mapped information is regularly used for visualisation and spatial analysis in GIS to aid interpretation. Precisely how, then, can digital spatial data best support social interpretation? Boundaries are introduced as a heuristic device to work through a series of critical observations and theoretical concepts that enable an understanding and restructuring of spatial data for social interpretation. Establishing a firm foundation for this restructuring is important to nurture a critical awareness of how archaeology can contribute to the 'new territory' of GIS approaches. While this chapter focuses on the example of built environment maps—which helps to formulate pertinent questions and to demonstrate the research process—the arguments remain valid for archaeology as social science broadly conceived.

First, I will explore some limitations associated with reading built environment survey maps as an end user and reflect on conjecturing information for spatial analyses. These observations suggest that working with spatial source data demands a deep understanding of the physical information behind archaeological evidence. Second, I will introduce the notion of *interpretive data* as a rendition of spatial data conveying *material* evidence on what matters socially about physical information. This defines a human centrist remit for social interpretation which is made explicit through the concepts of *material presence* and *agential intra-actions*. Third, I determine what social interpretation of the built environment entails by adopting an *inhabitant's perspective* and arguing the integrity of spatial analytical synchronicity in social archaeology. Finally,

B.N. Vis (✉)
Classical and Archaeological Studies (CLAS) and Kent Interdisciplinary Centre for Spatial Studies (KISS), School of European Culture and Languages, University of Kent, Canterbury, UK
e-mail: B.N.Vis@kent.ac.uk

© The Author(s) 2018
C. Siart et al. (eds.), *Digital Geoarchaeology*, Natural Science in Archaeology,
DOI 10.1007/978-3-319-25316-9_6

the chapter culminates by showing how, going forward, rigorous evidentiary understanding of spatial data grounded by an elaborate theoretical framework enables a distinct GIS approach dubbed 'interpretive GIS'.

Keywords
Archaeological evidence • Spatial theory and methods • Interpretive GIS • Built environment • Material boundaries

6.1 Introduction

In his foreword to *Setting Boundaries* (Pellow 1996), well-known anthropologist Edward T. Hall (1996) wrote: 'one can spend a lifetime on boundaries. That would be worthwhile work'. In fact, all of us spend a lifetime on boundaries. Boundaries as 'sites of difference' [a thought developed by Abbott (1995), Jones (2009, 2010)] are pervasive in the empirical reality of our material world. Philosophically speaking, a site of difference is not a thing in and of itself, but the edge along which a thing becomes distinct from its surroundings. As a concept, boundaries become the way in which differentiation, through observation, perception, and experience, allows us to recognise the matter and furnishing of the world. Mundane boundaries are the locations of encounter, shaping the objects and units the world consists of from various elements. Placing them within terms of Schutz's (1967) constitutive phenomenology, boundaries are the way in which we come to know the world (see Vis 2013, under review). Therefore, we all spend a lifetime on boundaries, and boundaries play a highly meaningful and determinant part in our empirical social lives.

In this chapter I endeavour to highlight and elucidate some significant ways in which boundaries can contribute to the analysis and social interpretation of spatial archaeological evidence on built environments. This is not to say that their relevance cannot be extended beyond the phenomenon of built environments, but by selecting a particular category of evidence, the value of boundaries can be more readily demonstrated. Furthermore, despite the focus on archaeological spatial evidence here, the ubiquity of boundaries in the empirical material world implies their validity as a research concept for human–environmental relations in general. Because the conceptual boundary is a spatial metaphor, it is only fitting that it is in the human manipulation and transformation of space that we should explore how boundaries can heuristically advance the use, analysis, and social interpretation of spatial archaeological evidence.

Ultimately, the aim is to provide the critical evidentiary reflection and theoretical backing necessary to convert digital spatial data, especially envisioned within Geographical Information System (GIS) software, into data that is structured by interpretive social meaning and primed to be analysed through its quantitative counterparts. Since one of the important advantages of quantitative empirical tools and information consists of the ability to generate comparative knowledge, the abstract and universally applicable notion of boundaries—conveying the empirical reality that gives our material world its shape—are well matched. The issues with creating qualitative or interpretive data for social analysis within GIS are steeped in theoretical depth concerning the nature of archaeological evidence and defining analytical purpose.

The research directions this chapter opens are found within the 'new territory' in GIS approaches that is informed by the archaeological–theoretical perspective, but not within the themes Verhagen (2018, this volume) identifies (i.e. cross-fertilisation with other techniques, network analysis, agent-based modelling, dynamical simulation modelling, or advanced statistical software). The universal significance of boundaries asserted in the opening statements sets out a path that here is developed from an end user perspective (i.e. analysis seeking social interpretation) on archaeological evidence on built space. Starting with the archaeological built environment survey maps which an end user may acquire, I will go through a series of observations on the way such data is presented. This brings out the complications associated with the usability of the lines on archaeological maps; moving from archaeological evidence to physical evidence, to material evidence. The theoretical implications that follow demonstrate how we can make critical use of lines on maps for spatial analysis in support of social interpretation. In conclusion, this theory is placed in the context of the requirements for developing interpretive research in a GIS environment. This has profound consequences for how we conduct archaeological GIS and how we proceed to develop new research processes with GIS. The concept of (material) boundaries is used throughout to exemplify how the staged questioning of archaeological built environment survey maps plays out. The path this sets out leads towards the theoretically and creatively enticing idea of developing 'interpretive GIS'.

6.2 Archaeological Evidence as Lines on Maps

6.2.1 Boundaries in Archaeological Survey Maps

In keeping with the geographical and landscape focus of this volume, the archaeology of built environments is taken to comprise the full scale of the relations between human constructions and the material traces of spatial transformations (i.e. developed landscapes and built-up space, such as within settlements). Except for aspects of detailed excavations, artefactual studies, and sampling of substances, archaeological information on built environment sites usually get translated into maps. The majority of such maps result from archaeological topographical surveys, employing various techniques. These may include remote sensing, geophysical and aerial surveys, and terrestrial altimetric (theodolite) surveys of the geographical distribution and shape of traces. The processed output of these is almost invariably some kind of map, showing archaeological features.

Therefore, it is in the context of the archaeological survey map that we first consider the pervasiveness of boundaries as sites of difference shaping the objects and units the world consists of. This thought has been developed in the social sciences (see Abbott 1995; Jones 2009, 2010) but can be brought to bear on the world of material objects too (Smith and Varzi 1997, 2000; Smith 2001; see also Vis 2014a). Smith and Varzi's (1997, 2000) *bona fide* boundary concept holds that entities emerge from the spatial discontinuities or physical heterogeneity along their edges. It follows that boundaries themselves do not exist as things. Rather they convey the location where physical distinctions take place. The opposite of *bona fide* boundaries is *fiat* boundaries, which Smith and Varzi define as all distinctions that are not associated with spatial discontinuity or physical heterogeneity for differentiation. In other words, *fiat* boundaries are based on ideas and conventions, and this explicitly includes maps. In archaeology, our maps are based on physical traces which we then seek to interpret.

So, if we follow Abbott (1995: 857), in that '[...] we should start with boundaries and investigate how people create entities by linking those boundaries into units. We should not look for boundaries of things but things of boundaries', we are caught in a process of twisting conversions (see Vis 2014a). By working from a mapped representation of a physical situation (a built environment), we must first work through the symbology that conveys the morphology of traces to determine the physical entities they are traces of. After we have determined the physical entities, their edges, or outlines, become our starting point for *bona fide* boundaries. In turn, we look through a theoretical and interpretive lens to define how we understand these sites of difference (*fiat*) while we continue to look empirically to acknowledge the material properties that articulate the qualities of the distinction (*bona fide*). There is no limit to the scale of *bona fide* boundaries, and therefore we could go down to the level of particle physics. For the sake of pragmatism as commensurate to our field of research, we should declare our scale of operations as that of human beings building space. The resolution of detail in information can then usefully be set at the level of discrete humanly occupiable spaces (see Vis, under review). Hence, our boundary perspective on archaeological built environment survey maps heuristically focuses on identifying the way spaces are separated to form distinctly circumscribed spatial subdivisions.

6.2.2 Reading Lines on Maps

For argument's sake, we could say that the primary source of information on archaeological topographical survey maps consists of lines. More elaborate use of symbology may exist to distinguish kinds of lines, but predominantly the occurrence and shape of archaeological traces of spatial transformations are visually represented by lines. Thus, the end user or interpretive analyst would likely encounter archaeological evidence of built space as lines on maps when acquiring spatial data. This confronts us with a twofold heuristic challenge: First, how much do we actually know about the empirical (physical) reality these lines convey as archaeological evidence? Second, how do we get to the entities shaped by linked-up boundaries on the human scale of occupiable spatial subdivisions? This will highlight limitations to the usability of information contained in archaeological spatial data which are not necessarily new, but are seldomly made explicit.

The first consideration regards the physical characteristics and condition behind the classification that is implied by any line mapped as archaeological evidence. We can commonsensically acknowledge the heterogeneity of any construction in both the technique and materials used (e.g. bricks and mortar). Even cyclopean masonry (e.g. Mycenaean and Inca architecture) does not render a constant surface. Yet, a line suggests that the physical characteristics along its course remain the same. Especially when architecture is concerned, it invokes the impression of the regular and constant vertical faces of modern construction we are used to, obscuring any specification of the physical characteristics that may afford human beings a different relationship with that spatial distinction. Lines also suggest a parity of physical conditions that applies along all full lengths. The same visual style of line can be used from archaeological feature to archaeological feature, whereas we know that preservation is rarely equal throughout a site. Beyond environmental forces and historical events acting on spatial constructions, the original characteristics of building and engineering may have influenced how spatial constructions appear as an archaeological trace. On top of this ambiguity, which is by and large intrinsic to archaeological evidence, very often the same style line can be used for multiple conditions and situations within a single plan.

Two of the most common additions to enrich a single-line style are dotted (or rhythmic) lines and (irregularly) intermittent lines. Dotted lines are intuitively used and understood as archaeological features mapped with a degree of uncertainty (cf. Hutson 2012). More often than not, it is still unknown or unqualified what this uncertainty consists of or even whether the dotted line is a projection of an educated guess or an expectation. It is useful to alert us to uncertainty, but such lines are still not straightforward to work with since it can be doubted what kind of physical condition they convey. Intermittent lines are different. Often we see bits of lines of irregular length, which in most cases suggest actually separate traces. The immediate difficulty with this is that it is, physically speaking, entirely unclear to what extent the physical condition suggests that some of such traces belong to the same spatial subdivision. This applies when we first already assume that all traces are of the same physical construction. Intermittent lines create the physical suggestion of poorer preservation than elsewhere on a site where longer continuous lines are used. However, there is also the option that visual intermittency is entirely justified by intentional physical construction (e.g. gaps for passage).

The practice of mapping is wrought with written and unwritten conventions. Beyond sustained critiques of mapping in geography in general (e.g. Monmonier 1996; Wood 1992; MacEachren 2004; Lilley 2011), Hutson (2012) offers a particularly thorough review of the conventions used in archaeological mapping of Maya architecture. It shows the potential traps of reading Maya archaeological site maps without preparation, relying on experience with other maps. The particular practice of 'prism mapping' of architectural volumes is deceptive and not even consistently applied within Maya archaeology (if the application is documented at all by the mapper). The outer lines of features often, but not always, provide the outlines of the trace, and the diagonals, often, but not always, represent simply the height of the mounds of rubble (resulting from building deterioration and collapse).

Other archaeological conventions that map the same traces (but not the same physical information) include hachures, contour lines, and outlines. Hutson (2012) demonstrates hachures provide more information about length and steepness of the slope of mounds of rubble than do prisms, while contours and outlines could arguably be seen as more objective. However, he refuses to argue in favour of any one approach. Instead, Hutson follows Galison (1998, 2000, 2010) to point out there is a difference between mechanical objectivity (removing one's interpretation from representation and automation) and judgmental objectivity (allowing one to add clarity from experience and interpretation) in maps.[1] Since all representation is interpretive, we can refer back to the *fiat* nature of maps (Smith and Varzi 1997, 2000). It remains unaltered that all these conventions of representation can use a similar visual style of line, which the end user must make sense of as physical information. However, historical examples in Hutson (2012) may remind us of alternatives to using simple lines. This may evoke impressions of 19th century urban surveys, which sometimes added symbols to lines in order to express spatial relations (see Oliver 1993). Not all walls are built the same. There is an array of ways in which walls connect inside to outside and determine degrees of spatial separation. When examining assortments of buildings, a typical example revolves around whether or not a building consists of multiple rooms. Seemingly internal arrangements could in fact be physically enforced separations, therefore composing an accretion of buildings, accessed separately.[2]

6.2.3 Documenting the Physical Information that Matters

Depending on scale, survey technique, and resolution, simplification and visual classification of lines on archaeological survey maps are completely understandable and indeed unavoidable (cf. Hutson 2012). An end user will likely not question the line(s) mapped as designated part(s) of any particular archaeological feature, assuming parity of physical information for the shape. Yet, the difference in material conditions and situations either from line to line or along a single line may provide valuable clues for the spatial structure they create. When the ultimate aim is analysing built space for social interpretation, all of this truly matters. At the same time, we can normally safely assume that information is never meaningfully obscured. In fact, the integrity

[1] Galison (2000, 2010) discusses the historical progress from 'genial depiction' before 1820, 'mechanical objectivity' between 1820 and 1920, to 'judgmental objectivity' after 1920 in scientific or knowledge-based images. The latter two intermingle, as they do in Hutson's (2012) view. In the end, all maps and mapped representations are also images (Aitken and Craine 2006) and can be evaluated in this same context, which is usefully paralleled by understanding the historical development of cartography [see Lilley (2011)].

[2] It can be admitted that even in state-of-the-art maps of the contemporary world, such as Ordnance Survey MasterMap in the UK, several polygons can be used for a single building. There is no information to know when polygonal separation implies an internal or external spatial arrangement. Naturally the purpose of each map differs. Embracing the material nature of archaeology, meticulous physical information would be justified.

of lines on a map is such that it could correctly convey the inability of the mapper to distinguish or interpret the physical situation further.

What the situation calls for is at least the equal care for making metadata available. Unfortunately, the archaeological fieldwork reports that may be able to shed some light on the physical conditions and preservation throughout a site, the environmental and visibility conditions during the work, or the conventions and pragmatic decisions made, are not always easily accessible or even available. Even if this information is available in written form and personal communication, then still it is unlikely to specify and comment on each archaeological trace that it mapped. Moreover, it is to be expected that future analytical purpose may generate questions that could not be foreseen or considered when the maps were produced. This means that even the most carefully presented and documented spatial data may ultimately result in unexpected ambiguities. Therefore, however unfortunate it may seem, after due scientific diligence, the remaining ambiguities can only be solved by consistent yet pragmatically informed rules of thumb.

This realisation must not be seen as an excuse not to produce the best possible metadata and data (re)presentation. Nor will this realisation change the fact that each specific archaeological project may allow for more information being recorded or the same information being recorded in a more useful way. Most significantly, if the archaeological survey map is the end product of a project, it would be fair to expect that it is being prepared in a best possible way to enable flexible future potential use. With the wealth of mapping conventions and symbologies available, especially in this digital age, we have far from exhausted the possibilities to improve on how physical information is conveyed on our maps.[3]

Where the onsite conditions allow, spatial–material information such as internal arrangements, separate yet associated traces, changes in preservation within the same feature, relief characteristics, etc. could be conveyed with the lines we map.

6.2.4 Complementing Lines that Stop

Until now we have been using the term archaeological traces to characterise archaeological evidence, whereas it was proposed that at the human scale information on built environments should distinguish discrete occupiable spaces. Lucas (2015) cautions against the dominant view of archaeological evidence as fragments. However, it is fair to say that there is an important discrepancy in terms of completeness between traces of a built environment and spaces of a built environment. We know that due to site formation processes, there is no perfect preservation of past situations. Yet, social interpretations in archaeology are usually concerned with understanding situations that occurred in the past. By stating 'finding remnants of an assemblage is not the same as finding an assemblage itself', Lucas (2015: 321) urges us to reflect on what survives: archaeological evidence seen as relics. In the context of going from traces to spaces, metaphorically a past lost to a past found, the resolve lies in redressing fragments as surviving traces of entities from a past situation.[4] Following the

---

[3] It may be worth mentioning here that with new recording technologies, especially high-resolution 3D photogrammetry and laser scanning or LiDAR, a cleanly presented 2D archaeological map as end product could systematically refer back to the much fuller source of information of the original digital records. Kyriakidis E (2016, personal communication) argued the advantages of a similar workflow for digitally producing the 2D map of ancient Eleusis from 3D information. If we can ensure data lineages through stable links and unique identifiers, this may prove a significant advantage in the future. Providing a data lineage still does not exonerate us from a responsibility to produce our mapped interpretations with the most complete and usable information in flexible and readable formats.

[4] It should be acknowledged that this is not an argument against the truism that interpretations of the past are a product of the present. However, it is asserted here that a different kind of knowledge is produced from a strict adherence to Lucas' (2015) alternative of seeing material evidence as relics (evidence of why things survive), which suggests a focus on the formation, meaning, and relevance of archaeological traces as entities in and of themselves.

understanding of boundaries posited earlier, entities as they occur to us should be understood by the boundaries from which they emerge. Since through site formation we tend to lose the occupiable spatial entities of the past, we must first ensure our spatial data conveys the bounded spaces of the built environment. This will require a process of, first, interpreting well-preserved traces into spatial subdivisions, then, applying critical conjectures and expert judgments on the basis of the physical information that archaeological survey maps contain.

This brings us to a further significant ambiguity about the lines we map, which is the typically impossible to answer question: why does a line stop? We have briefly touched upon the reasons for the intermittency of lines, but it is worth giving interrupted lines more explicit consideration. If distinguishing the spaces of a past built environment depends on conjecture due to the fragmentary nature of persistent traces, linking up and filling in the gaps is most reliably done by knowing why a line representing a physical trace stops. All we know when a line stops is that the feature must no longer be perceptible or measurable. Beyond that obvious reply, there are several reasons why this may be. The specific reason can impart physical information that would aid one to make reliable conjectures of spatial information. But first we should attempt to discern if it is a simple onsite, technical, or archaeological visibility problem. This could include problems with vegetation or the limited exposure of archaeological traces on the surface, limitations to the equipment used, e.g. obstruction or environmental physics (e.g. blockages and distortions of geophysical signals), or did the archaeology literally recede into the ground? Beyond compromised visibility, we must assume the physical trace did actually stop.

Both in the field and especially once recorded and mapped, it is not unusual that one can no longer distinguish why a feature or line representation appears truncated (e.g. Demarest 1997). I will list a few fairly straightforward reasons. First, what appears to be a truncation is not necessarily truncated. The features could be intentionally constructed, i.e. actually preserved and originally finished that way. This is arguably the most important distinction: do we see the representation of the finished article thanks to decent preservation or is the shape of this feature a representation of something broken? If the latter, there are still many options. Was the feature destroyed and, if so, when, by whom, and how? Did it deteriorate over time and, if so, by gradual dilapidation of the original feature after disuse or due to other site formation processes? Was it damaged by modification, reuse, or reappropriation in the ancient or recent past? Did it suffer from decay of perishable building materials or decay due to the perishing of originally incorporated natural elements (such as trees and plants)? Was (part of) the feature removed by either animals or humans after disuse or abandonment? Without a symbology for line ends, when conditions of archaeological recording allow for it, the end user will once again rely on rules of thumb to carry out the conjectures. Fortunately this can be done in critical and archaeologically knowledgeable and sensible ways (see Vis, under review, 2014b). Once the metadata of the project as well as spatial data and analogous information from historic and cultural proximity have been exhausted, one can still apply visual and morphological contrast when constructing complementary data, document the applied rules of thumb, and mark up data for easy separation of these conjectures from retentions of originally acquired spatial data.

6.2.5 Conjecturing Entities

While conjecturing is not as strong a process as fully fledged reconstruction, there is a risk of getting caught in the fallacy of perfection. In archaeological reconstruction, there is the tendency of reconstructing everything to an absolutely pristine and clean state. Buildings and surfaces are all fully functional and, in outstanding condition, their environments devoid of any form of pollution. One could argue this striving for perfection is a kind of hyperrealism. The reconstructed situation is not one that likely ever existed. A true reconstruction of the past should account for areas laying fallow and buildings in disuse or disrepair. Such occurrences would have been part and parcel of a built environment in flux. Beyond the visual impression, the critical social interpretive importance of perfection generated by conjecture or reconstruction can vary. This depends on the purpose of analysis or interpretation, i.e. what is the newly constructed data supposed to comment on or contribute to?

To illustrate this, let me err on the spatial side of the analytical spectrum. The reliability of population estimates based on buildings heavily depends on knowledge or assumptions that determine which were occupied and which had only occasional, shared, intermittent, or partial use (introducing degrees of spatial duplication or redundancy). However, a general understanding of the functional structure or spatial experience and opportunities for inhabitants does not rely on information about which house was occupied at each specific moment in time or its state of repair. From our own experience, we can accept that when a house is unoccupied, would we always know? And when it is dilapidated, it is still recognisable as a house and still poses a physical impediment to access that space. The building still has potential to be a household or to be repurposed. The space and the experience of that space is still structured in roughly the same way, even if the affective and sociocultural context may differ from particular case to particular case. The point is to alert us to the fact that when the notion of occupiable space (to understand how boundaries compose built environments) requires us to conjecture, we create an approximation of a situation in the past, not a reconstruction of any particular situation in the past. A sufficiently critical research design will be aware of both the social interpretive limitations and opportunities this offers.

Crucially, working on archaeological maps of built environment sites from the perspective of boundaries, requires us to ask pertinent questions that improve our understanding of the physical information that first generates and subsequently is represented by our spatial data. Understanding exactly how physical information is captured in spatial data helps us to (re)interpret the format and representations that spatial data is presented in. Furthermore, critically (re)assessing spatial data representation makes us reflect on which physical information exactly supports our understanding of how spatial relations and morphology are determined. Going through this process means that when we move on to spatial analysis and social interpretation of built space, we are much better prepared for data treatment. The rigorous questioning and processing of archaeological evidence up to this point also provides nonprescriptive advice for those looking to improve the way archaeological built environment maps are drafted and published with a view to future analysis. Now that our lines on maps are essentially converted into the *bona fide* boundaries of a past situation, we can only commend the creation of a precisely defined archive of physical information. The next step towards spatial analysis for social interpretation requires that we are equally critical of the theoretical assumptions such social interpretation is based on.

6.3 The Material Nature of Boundaries

6.3.1 Interpretive Data

Having arrived at this stage, we are able to examine archaeological built environment maps for the physical information they contain and identify the boundaries shaping the entities of which the built environment consists. We must now move from *bona fide* physical presence of boundaries to *fiat*, to address how we understand their presence interpretively. Analysing physical presence results in little more than dimensions: information about the geometry, topology, morphology, and topography of built space. The difficulty with such information is that it is entirely contingent whether it has any bearing on social understanding.[5] The evidence base of archaeology is distinctly material, which is arguably why the discipline has struggled to contribute more substantially to social theory and synthesis has traditionally shown little attention to material integration (Sherratt 1993; Vis 2009). What this foregrounds is that our precise understanding of empirically derived (mechanical objectivity) archaeological evidence as physical evidence (judgmental objectivity) (cf. Galison 1998, 2000, 2010) has yet to transition into *material* evidence.

This deliberate distinction of material evidence is meant to alert to us that data which captures material evidence must directly be of an interpretive nature. The next step towards social interpretation of built space is thus to construct *interpretive data.* Interpretive data redirects traditional data interpretation with precision and rigour, because it structurally links the empirical origins of spatial data to the ideas we have about them. These structural links provide the bridges that resolve the leaps between ontological registers (cf. Lucas 2015) that trouble and restrict the value of many archaeological interpretations. Rigorous use of interpretive data also carefully delimits the ontological register of interpretation, i.e. it ensures that data is commensurate with the understanding sought through analysis.

Crucially, then, recognising the interpretive nature of material evidence and the subsequent construction and use of interpretive data would contribute to more rigorous theory building and causal explanation in archaeology. Structuring evidence accordingly provides concepts to work to in the middle range of 'empirical theory' and consequently to construct better arguments on the basis of analysing this evidence (*sensu* Smith 2011, 2015; Ellen 2010). Nicholas and Markey (2015: 287) formulate the questions well: 'how do we know what we know about the archaeological record; and what types of evidence suffice for providing adequate "proof" for our interpretations?' Understanding archaeological evidence as physical evidence on the one hand and material evidence on the other provides the basis for establishing categories that enable arguments on a particular plane of interpretation.

6.3.2 Material Boundaries

How do we make our boundaries of physical information on the built environment into material boundaries? This requires a careful definition of their material nature. Positing a cogent and useful critique of archaeological theory and argumentation, Wallace (2011) proposes and defines the notion of *the material*. Critical realism offers a philosophical ontology to facilitate the development of epistemology in substantive disciplines (formed by a substantive domain) (Yeung 1997; Sayer 2013; Cox 2013). Wallace's intervention is aimed at the postmodern (or possibly post-postmodern) tribulations in archaeology that at best seem to find compromise in relativist acceptance leading to (often uncritical) eclecticism in theory and methods (see Fahlander 2012; Bintliff and Pearce 2011; Vis 2012).

Even more compelling than its critiques are critical realism's concepts and processes which link empirical and conceptual modes of research (or quantitative and qualitative social science, see Pratt 1995). Therefore, critical realism is hypothetically a particularly suitable match for the discipline of archaeology (see Wallace 2011; Vis, under review for elaborations). The critical realist focus on forming ontological entities and the categories it contains based on causality—in terms of specific causal powers emergent from internal relations—is what allows *the material* to gain a particular meaning. The physical and environmental processes operating in the matter of the built environment exist without human engagement with the built environment. The material, then, is what emerges when sociocultural and physical aspects become necessarily internally related through human interaction in

[5] Even the most abstract, and deemed objective, way of measuring is typically based on units recognised and conceptualised by humans (e.g. metres). Purely dimensional analysis expressed in such units may have value for the hypothesis that such units were used in the construction of the built environment.

time-space (Vis, under review; adapted from Wallace's (2011) original reasoning). Consequently, material boundaries refer to the understanding of the causal powers emerging from human interactions constructing or encountering the physical edges of occupiable spaces in the built environment.

6.3.3 Material Presence

A thorough understanding of material evidence is needed to characterise the kind of understanding a category of *the material* (here: material boundaries) might permit us, in this case limited to a spatial lens on built environments. This section presents a theoretical context to knowing what the material nature of our evidence is evidence of, or, what I mean by social interpretation.

It is important to recognise that material evidence is still a category of empirical information and therefore adheres to an empirical tradition of knowledge production (cf. supporting Smith's (2015) 'better arguments'). The usual mode of conduct that sees archaeology interpret data, even if we can identify various archaeological data as unwittingly interpretive data,[6] has relied heavily on correlating analogous information. Citing ethnographic, (ethno)historical, and experimental sources of information (used as analogous verification for interpretations), Nicholas and Markey (2015) develop an argument for the use of 'traditional' or 'indigenous' knowledge in archaeology. Questions can certainly be asked about to what extent treating evidence empirically is the only valid structure of reasoning. A critical realist would not deny indigenous knowledge causal power. In explanation (why something happened or occurred in that way), indigenous knowledge will play an equally valid role alongside history, personal agencies and memories, sociocultural systems and categories and human–environmental relationships, both internal and external to the situation or phenomenon being studied. What such knowledges demand is a structural linkage of that understanding to archaeological evidence. In this chapter, archaeological evidence has become material evidence, and all we know about this evidence is that it refers to a *presence* to human beings situated in the past. Therefore, the empirical information we are restricted to is *material presence* (see Vis, under review, 2016).

The interpretive perspective of *material presence*, however, seeks not to explain exactly and comprehensively why a social empirical reality occurred. Despite this limitation, the social significance of material presence (a dynamically and generatively affective and afforded causal power) will have developmental resonance. Explaining occurrence would be tantamount to knowing exactly all causal powers working towards production in some detail.[7] Instead, when presence is material we are restricted by the resolution of information invested in the social, physical, and temporal processes becoming internally related to form *the material* (cf. Wallace 2011). Interpretive knowledge creation will result from and is delimited by the definition of that entity and the causal powers of the specific category of it that forms our field of interest (here: material boundaries of the built environment). This is not to say that when studying *material presence* constitutive or supplementary causal powers have disappeared, but they have become indistinct registers of understanding. Therefore, they are simply not the purview of research operating from that perspective.

In contrast, in this chapter, the further restriction to spatially determinant characteristics is the pragmatic effect of the pervasiveness of

[6] For example, the notion of a house or residential building appearing as a unit on a map constitutes a type of interpretive data. It conflates conceptual understanding with physical characteristics. Unfortunately, such conflation is often used as a layman's term, therefore precluding knowledge on how the unit of analysis causally relates to the interpretation.

[7] For example, regular flooding in combination with a dependence on a periodically flooding water source may cause a community to build on poles. Flooding is a strongly determinant part of explaining the occurrence of building on poles. Yet, regardless of why raising buildings onto poles may have been necessary, this way of building will have social (affective and afforded behavioural and developmental) effects.

generating spatial information in past and current archaeological practice. Allowing the spatial characteristics of material evidence to speak without analogies or verification derived from other kinds of evidence may feel unnecessarily restrictive[8] to some. To me it is a fair and necessary challenge when we consider that for much of human development we do not have other types of (social) evidence available.[9]

6.3.4 Material Boundaries as Agential Intra-actions

I will concede that archaeological and geographical theorising—especially as it is progressing from discursive systems theory and agency to ANT (Actor-Network Theory) and its influence on materiality—is producing ideas that resemble critical realist emergence. Critical realism is meaningfully introduced as a frame of reference for structuring the conceptual development of our social interpretive endeavour. Wallace (2011) is right to recognise that archaeology can inject material cogency into critical realism, which in turn helps to overcome the contradictions and fallacies stemming from the disciplinary split between scientific empiricism and social interpretation (Vis, under review). The flexibility of the philosophical structure and processes this provides, as well as the clarity of language in social scientific adaptations, I find are far preferable to the imprecisions and metaphysical truisms ANT-inspired theory produces. Like Smith (2015), I argue for interpretation based on causal explanations and mechanisms. To this end, the understanding that was needed to construct and structure empirical information as interpretive data permits the researcher to organise, query, and rearrange data in an exploratory way, leading to new understanding and interpretation (see below). For this iterative research process, critical realism is more appropriately matched.

Against this backdrop I will introduce another concept that recently has been gaining traction in the social sciences (Kleinman 2012) and is now finding its way into archaeology, called *agential realism*. Albeit complex and highly abstract, the agential realist view on empirical evidence can add detail to the notion of *the material* as an emergent entity, thanks to emphasising its *performative* dimension. In agential realism, things or phenomena emerging from discursive human–nonhuman interactions are called *agential intra-actions* (Barad 2003). Meirion Jones (2015) argues to redress the archaeological perspective on material evidence in agential realism to move away from an inert object-like world and to re-emphasise interaction and constitution. This is sensitive both to the archaeological appreciation of the mutability of material and in the social interpretive sense to regarding the world as being alive and changing. In order to nurture an understanding of seeing phenomenal emergence through discursive interaction (the phenomenon central to this chapter is the built environment composed of boundaries), I quote Barad's (2003: 815) definition in full, before I will make connections to ideas originating from more familiar discourse:

> [...] *phenomena are the ontological inseparability of agentially intra-acting "components."* That is, phenomena are ontologically primitive relations—relations without pre-existing relata. The notion of *intra-action* (in contrast to the usual "interaction," which presumes the prior existence of independent entities/relata) represents a profound conceptual shift. It is through specific agential intra-actions

[8] Interpretation from this perspective—i.e. not seeking direct comparison from other casuistic evidence to draw analogies with the material evidence encountered—is necessarily subjective and self-referential. Such interpretation is in constant reference to how we understand and experience our relationship with empirical reality. Rather than drawing on any particular case, this perspective includes *all* the knowledge and experience researchers have acquired about humanness in their own lives, structured by theoretical constructions and concepts derived from rational reasoning. This is where we must acknowledge again that how we have been brought up and taught to think matters to evaluating the validity of reasoning. To avoid the suggestion of a false categorical opposition, note that Nicholas and Markey (2015: 290) indicate that indigenous knowledge, likewise, both results from 'learned experiences and explanations' and is characterised by empirical observation, repetition, verification, and inference.

[9] Working exclusively with spatially relevant evidence should not be confused with the 'spatial fetish' of spatial science [see Werlen (2005), Zierhofer (2002)].

> that the boundaries and properties of the "components" of phenomena become determinate and that particular embodied concepts become meaningful. A specific intra-action (involving a specific material configuration of the "apparatus of observation") enacts an *agential cut* (in contrast to the Cartesian cut—an inherent distinction—between subject and object) effecting a separation between "subject" and "object." That is, the agential cut enacts a *local* resolution *within* the phenomenon of the inherent ontological indeterminacy. In other words, relata do not pre-exist relations; rather, relata-within-phenomena emerge through specific intra-actions. Crucially then, intra-actions enact *agential separability*—the local condition of *exteriority-within-phenomena*. The notion of agential separability is of fundamental importance, for in the absence of a classical ontological condition of exteriority between observer and observed it provides the condition for the possibility of objectivity. Moreover, the agential cut enacts a local causal structure among "components" of a phenomenon in the marking of the "measuring agencies" ("effect") by the "measured object" ("cause"). Hence, *the notion of intra-actions constitutes a reworking of the traditional notion of causality*.

At the risk of exposing my inferior understanding of discipline-specific intricacies at play, to me, there is significant resemblance to the way interrelated causal powers in critical realism drive emergence. This logic provides the basis for phenomena and all associated entities (objects and categories) but also defines the substantive domain of research (cf. Sayer 1993; Yeung 1997; Pratt 1995). The idea of separability and exteriority resembles the logical destination of going through the implications of *autopoiesis* in systems theory. When systems are enacted, their self-generative inherent coherence contains their own distinctions towards their outside (Luhmann 1986; Arnoldi 2001). In the context of built environments, especially the incorporation of the material dimension of architectural systems into social scientific constructs is relevant as such incorporation permits the bounding of space (Koch 2005; Vis 2009).

Furthermore, the agential cut seems to represent the ability to separate oneself as an agent or researcher from the phenomenon one studies and thus co-constitutes. By the same token, it enables a distinction between human agency and nonhuman agency, even if all such 'bodies' partake in the constitutive material–discursive practices of materialisation (see Barad 2003). This connects well to agency as causal power which is suggested in Wallace's (2011) proposition that material has 'agency without intent' which follows on from Fletcher's (2004) 'actors without intent'. The realm of phenomena may naturally include things in which humans do not partake. Consequentially, I agree with Meirion Jones (2015) that nonhuman agency should not be reduced to human agency. Distinguishing types of agency is an important corrective on the ANT-inspired idea that humans and nonhumans must possess equal agency (as in symmetrical archaeology). The logical extreme of the latter is that nothing can be distinguished discretely (represented by terms such as mixtures and meshworks) (Webmoor 2007; Webmoor and Witmore 2008; Ingold 2008; for critique: Wallace 2011; Vis, under review).

6.3.5 Human Centrism in Interpretive Research

Declaring for purposes here that my aim is to equip analysis seeking social interpretation, the salient differences with a mixtures and meshworks view arise from the start. Critical realist logic serves the interests of human understanding. It holds that if something has effect (i.e. the exercise of causal power), it must exist. Taking the vantage of everyday human experience, it is virtually impossible to deny the existence of discrete objects and categories, even if through alternative lenses (e.g. scientific magnification) everything may end up as mixtures of small particles. The human frame of reference matters to acknowledge which part of reality (the field of interest) we are trying to understand. Therefore, exactly why there is a need to be guarded in explaining that scientific proceedings are different when we as human beings study human beings, I have yet to grasp (insofar as we accept that the nature of human beings is captured in the category of our species).

Especially when conducting interpretive scholarship rather than ethology, everyday human experiential knowledge[10] is an appropriate frame of reference, which applies to all members of our species.

It seems an irrefutable truth that if the notion of species bears any relevance, a different dimension to understanding is available when we are concerned with our own, and we can operationalise this in research. Seen from Barad's agential realism, the specificity of the agential intra-actions of research on human beings by human beings would suggest the same. The consequence is that we can understand phenomena involving human agency (not necessarily human agency exclusively) differently from phenomena not involving human agency. A monist philosophical stance, therefore, does not require perspectives informed by totalising mixtures and symmetry (*sensu* Webmoor 2007; Webmoor and Witmore 2008; cf. Hodder 2014). Human centrism can be a permissible mode of knowledge production even if metaphysically we can agree that human beings are not situated centre stage by default. So '[...] we do not have to assume that materials can only be understood because of their mediation by human cultural or social activities' (Meirion Jones 2015: 334).

Agential realism supports the analysis of assemblages with an active role for the material. The associated agential distinctions support a mode of understanding that is particular to when assemblages including humans are concerned. If we try to describe this 'human-centric' stance in Hodder's (2014) asymmetrical relations between humans and things, we get something like this. The situation where things depend on things (TT) is not considered separate from relations that involve things depending on humans (TH) or humans depending on TT or when things depend on things that depend on humans. Or, in critical realist terms: for the causal power of social science (or social interpretation) to occur, the participatory presence of human beings and their influence are necessary. On the basis of Bohr's physical experiments, Barad's (2003) arguments suggest that since the conjunction of machine and object of study creates a particular situation, this intra-action (of machine, object, researcher, environment) 'causes' an outcome to emerge that otherwise would not have existed. This reworking of the notion of causality again roughly pertains to how critical realist causal power can flexibly nuance the relations between cause (interrelations) and effect or, put differently, emergence. In social interpretive research, it follows therefore that beyond the judgmental objectivism in interpretive data, even the most replicable kind of automated analysis will give outcomes that intra-act with the researcher to let subjective understanding emerge that is probably unique (or endlessly complex) despite meticulous conceptual delimitation of the substantive domain.

6.4 Social Archaeology, Time, and Spatial Analysis

6.4.1 The Inhabitant's Perspective

It would appear that time is the dominant force that stands between archaeology and any aspirations of contributing to or operating as a social science. The obvious remark to make is that archaeology's subjects tend to be dead and those of social science are alive and kicking. However, the growing discourse on materiality and the influential popularity of ANT suggests the social sciences are struggling to reconcile their traditionally live human purview with the nonhuman elements of the world. As I have made explicit, there is not only a substantive but a meaningful (for research purposes) difference between human agency and nonhuman agency and whether an assemblage includes human agency. Giving material an active voice, thanks

[10] Critical realism evaluates knowledge in terms of 'practical adequacy'. Knowledge that is useful in and applicable to empirical situations is better than knowledge that is not. Knowledge and its concepts need to be revised when empirical situations are encountered to which it cannot be applied. This is the principle of iterative abstraction (Sayer 1981; Yeung 1997).

to its interrelations with human beings as conceived in *the material*, sustains an inter-human (i.e. social) mode of understanding of nonhuman things. That is, without any actual human agency currently being interrelated with nonhuman agency, we can project an understanding of the causal power (or intra-action) that emerges from a situation in which human agency would be present. While this mode of research may lose the a priori explanatory detail of socio-cultural context or personal affect and biography—as available to social science in live situations—it supports an analytical intersubjective social understanding.[11] Importantly, if social interpretive science of *the material* can be conducted without the live participation of human agency, there is in essence no difference between situations in the past and the present.

Within this chapter, I called upon the human scale and the perspective of human experience several times. As both the performativity in Barad's (2003) agential realism and causation in critical realism suggest, the presence of *the material* depends on the interrelation of human agency through participation. For the *fiat* understanding of material boundaries as a category of evidence this means that the active disposition of people in the built environment must be assumed. Or, put differently, the built environment must be actively inhabited for material boundaries to exist. For the purposes of social interpretation, we should therefore adopt an *inhabitant's perspective* (also Vis 2016) on understanding the material nature of boundaries. From this perspective we can then proceed to differentiate the diversity of the physical situation and construction of boundaries based on understanding how this influences their role in determining spatial frames and relations in human and social assemblages.

The inhabitant's perspective stands in contrast to the 'god's view' (Morton et al. 2014) or the god trick (Aitken and Craine 2006; Wood 1992; Monmonier 1996) of mapping from an impossibly high or distanced viewpoint and visualising and analysing data from a totalising perspective. Maps excel at giving data overviews and I do not argue against this. What I propose is that we should 'people the past' (or any material evidence) using a variety of dynamic spatial methods (see Morton et al. 2014) and create static data renderings which allow us ways towards understanding spatial situations from an inhabitant's perspective. It is not an argument for exact replication of experience, as indeed, in contemporary lives, we can never attain a complete replication of even our direct neighbour's perspective. Understanding the other and their time-space specific and social position and situation is inevitably always based on part projected fabrication; such is the artificiality of analysis.

One of the principal artificialities is the stasis of mapped information (even in most agent-based dynamic models, this is retained). Even if stasis nullifies the constant change and mutability as indicated by archaeological progressive time (see Meirion Jones 2015), it is actually quite consistent with inhabitants of spatial situations, such as built environments. We know the built environment can change and we are likely to have experienced such change, but in everyday life we do not expect it to. If sudden large changes occur, they are almost always disruptive (e.g. natural and human disasters, such as London 1666, and major urban renewal programmes). In general, though, we expect spatial–material situations to last, both those transformed or constructed and those naturally occurring. By and large we expect materials to be stable (see Hodder 2014): our homes and things are still there after we return from work. The fact material spaces only appear stable on the surface is actually the most pragmatic attitude in daily life.

The consequence of assuming material stability means that, in the sense of mapping material boundaries, the stasis of the dataset will assume

[11] Agency in archaeology is very often concerned with the individual as a *particular* individual. In social science, insights gained can sometimes be evaluated against an individualised context of live participants. What this appeals to is Schumpeter's and Weber's classical idea of methodological individualism, where individual actions and motivations must explain social phenomena (Heath 2015), but accepting that the individual is a generalised analytical unit.

the interrelations that constitute their causal power are actively operated. Therefore, the built environment map is now static and dynamic at the same time. The active operation of interrelations generating causal power also intrinsically carries the ability for spatial situations to change. The subtle differences in how the intra-action takes place depend on each individual case, including personal and time-space-specific dependencies [sensu the time geographical adaptation of individual biography (Schütz 1967) into life-path by Pred (1981, 1984; see Vis 2009; cf. Hodder 2014)]. In the practice of operating material boundaries, which comes down to executing interrelationships and assemblages, their material status is reconfirmed. Intra-acting the built environment maintains it. The practice of 'mending wall' (see Oles 2015) consists of the repetition of closely similar intra-actions, stabilising the physical conditions on which *the material* nature relies. These conditions are the foundation of change over time ceteris paribus, excluding the effects of ongoing nonhuman processes (cf. site formation processes or Hodder's TT). This means that when material boundaries are not operated, for a period, the potential of generating a phenomenon of similar essential spatial characteristics remains. What we can map, then, is this potential, knowing that the phenomena would exist in varying rhythms.

6.4.2 Synchronicity

Throughout I have made passing remarks related to time such as in the built environment situation that is created through conjecture (cf. reconstruction). It has just become apparent that a stable rendering of the performativity of material boundary operations generates an artificial simultaneity in our data. The foregrounds a juxtaposition of synchronicity and diachronicity that is often found in archaeology and the interests of archaeological recording and analysis. Archaeology's proclivity to emphasise time as a developmental process often results in research foci that concentrate on diachronicity. However, the static condition of spatial data or the stasis created by maps produces situations of simultaneous totalities. Consequently spatial data and spatial analysis are criticised, because they assume a synchronicity that sits uneasily with time as developmental process.

Galinié et al. (2004), Lefebvre (2009), and Rodier et al. (2009) have developed an archaeologically and historically dynamic model of information on urban fabric as an alternative to the more pervasive time-sliced (a particular or aggregate moment in time) representation, called OH_FET (based on 'historical objects'). Such approach relies heavily on the equal availability of archaeological and historical dating evidence across areas of space. In practice it is not only unusual to have equivalent dating evidence for all archaeological evidence of an area of built environment, but it is also difficult to account for the states, conditions, and iterations of persisting elements within the built environment. This is especially true when modelling 'historical objects' based on combinations of social use, space, and time (Rodier et al. 2009; Lefebvre 2009). Spatial analysis assumes the input dataset is synchronous. In contrast, analysing dynamically modelled spatial data runs the risk of 'fetishising' the formation processes of archaeological evidence instead of (social) developmental processes. However, the difficulty of incorporating temporal dynamics into spatial built environment data should not be an excuse for uncritical use of a synchronous mode of analysis.

Time-slice synchronicity should result from the judgmentally objective (*sensu* Galison 1998) representation of assembled spatio-temporal data. That data should then be converted into interpretive data through the stages informed by the concepts this chapter has introduced so far. The developmental significance in the inhabitant's perspective likely operates on a scale that differs from archaeological formation. Another sequence of concepts should be devised to identify the appropriate human scale(s) of time and how it manifests interrelations with archaeological evidence. Currently, for social interpretation, there are better concepts available for the critical understanding of synchronous analysis. This may therefore be preferable to introducing a further problem of data availability and the theoretical complications of data modelling, of course, depending on purpose and case. Time-slice synchronicity has the added advantage that it puts past built environment situations on a level playing field with analysing contemporary built environment situations.

Data that is not there cannot be analysed. The synchronicity assumed in analysing spatial datasets has the side-effect that the coincidence of all data entries makes implicit that the analysis will also assume that this coincidence is all that exists. A record of fragments thus is complete as a record in itself. Inhabitants of a past situation would not have encountered a fragmented world, and therefore I have introduced the preparatory step of conjecture. Conjecturing ensures the spatial information best approximates the social empirical reality of a past situation. The known unknown of any spatially determinant characteristics for which we do not have archaeological traces remains unresolved (the absence of evidence is not evidence of absence). Conjecture merely helps to avoid an analysis of fragments instead of past situations, even if our rendering of that situation is incomplete. Although Lucas (2015) stresses that the evidentiary fragment lacks a temporal dimension, when putting together (incl. conjecture) a time-slice of spatial data, we may pick any moment in the material presence of archaeological evidence that survives from a past situation. The incompleteness of evidence for a past situation does not rob the fragment of its duration. Complementing fragments with synchronic conjectures simply implies an analysis and interpretation focused on what archaeological evidence is evidence of.

Ultimately, these arguments improve our critical understanding of spatial analysis, but spatial analysis itself remains temporally undiscriminating. What is being analysed comprises a 'would-be' moment and situation of a (past) inhabited built environment, seeing all material boundaries as if operated at once.

6.4.3 What Is Social Interpretation of Material Boundaries?

There is nothing new about having to negotiate these limitations. However, to arrive at an 'interpretive GIS' and applying associate analytical tools, we must fully understand the extent of the opportunity for social interpretation. Many of the problems discussed so far tend to disappear into the background of 'archaeological evidencing'. It can be appreciated just how much influence the treatment of archaeological evidence has on exactly which kind of interpretive and analytical work is supported by it. At this stage we can ask, so why does it matter that our data best approximates a would-be inhabited built environment? What is its value, its contribution, its relevance? From the outset I placed the arguments in this chapter in the context of social interpretation, but until now virtually all effort has gone into preparing and seeing built environment data as material boundaries. What is the social interpretation of material boundaries?

With social interpretation of material evidence, I seek to place *the material* as a fundamental part of social science. This is not an attempt to reformulate the many guises of social archaeology (see Preucel and Meskell 2007) under a new heading. It should be seen as part of recent calls for archaeology to act as a social science and to contribute to pertinent societal issues (Smith et al. 2012; Kintigh et al. 2014; Smith 2015). In archaeology as a social science, it could be said that 'we are no longer concerned with how these materials can be interpreted; instead we are interested in how these materials intra-acted with past people' (Meirion Jones 2015: 336), specifying our interpretive process and purpose. From this vantage point, the difference between the purviews of archaeology and social science is all but gone. The synchronous mode of analysis of spatially determinant characteristics further emphasises these equal terms of operation.

The primary cornerstones of our endeavour consist of spatial and human information. Therefore, here, social interpretation should contribute to understanding how human–environmental relations create built environment situations of being in the world, how these situations function, and how they develop (cf. Graham 1999 on refocusing research concerning Maya urbanness). Social interpretation provides a performative lens on socio-spatial practice as interrelated with the material environment to analyse and study constitutive society–space relations comparatively and historically (cf. Griffiths 2013). This kind of social interpretation is of undisputable value to social scientific questions on emergence and human–environmental relations, because it places the human past as an inextricable part of the continuing processes that determine the fundamental nature of human societies (cf. Kintigh et al. 2014). In doing so, archaeology can contribute to building an evidence base of scenarios containing the widest possible sociocultural form variety and environmentally situated developmental trajectories (sensu Smith 2012). Such evidence base will further enlighten the essential relations determining the nature of human inhabitation of the world. The social empirical reality of this is necessarily spatial.

This brings us to the conclusion that social interpretation of the spatially determinant characteristics of material boundaries composing built environments will revolve around the socio-spatial opportunities involving humans introducing material presence and encountering material presence. The typical limitation to generally available archaeological physical information on space implies a focus on structure. So, after all, it is still about geometry, topology, morphology, and topography of built space, but exclusively through a socially constitutive lens on a human scale. Spatially determinant physical information is what Marcus (2017) would call a background (or context). The foreground being rather what is traditionally considered (landscape) architecture. In these terms, the structural elements of the background may depend on the architectonic constructions (perceptible results of building actions) of the foreground.

Space has a tendency of slipping into the background, ending up as something we think and act 'with' rather than think and act 'on' (Marcus 2017). However, archaeology's and architecture's developmental viewpoint clarifies immediately that the built environment is something that is *made*. Only acting on the naturally occurring physical environment (human–environmental intra-action) can create built environments, which consist of occupiable spaces constituted by things (boundaries) that specifically serve the purpose of inhabitation. To understand the essential nature of the background in terms of itself (see Marcus 2017), we must foreground the physical determinacy of the background (acting on and occupation). Consequentially, the research becomes about how space does what it does as relevant to the human perception and experience of their purpose to inhabit the world. The diversity (or differentiation) of material boundaries results from identifying the smallest intrinsically coherent elements (or operations) in which we construct and encounter spatially determinant characteristics (see Vis 2014b for an applied example).

Boundaries have been used pervasively as a heuristic reference. This conceptual example should not be confused with asserting that material boundaries have now become a prerequisite for spatial analysis in support of social interpretation. Nonetheless, the basic properties of boundaries grant spatial structure a relational plurality that classification according to discrete spaces does not. What is left is to place material boundaries in an analytically able toolkit to advance complex understanding of the situations in which they occur.

6.5 Towards Interpretive GIS: A Prospect

6.5.1 The Challenge of Interpretive GIS

It is only following this elaborate philosophical and theoretical preamble that we can confidently turn to GIS as a toolkit in aid of social interpretations of spatial data (here: built environments). Verhagen (2018, this volume) glances over 30 years of GIS applications in archaeology and emphasises the apprehension with which the more theoretically and interpretively inclined archaeologists have received it. The lack of direct theoretical engagement with GIS is still lamented during many CAA conferences (Computer Applications and Quantitative Methods in Archaeology). The outlook of this chapter is not for the archaeological 'spatial turn' to generate theory, but to generate theory for the 'spatial turn'. Though archaeological evidence is necessarily spatial, we should not allow spatial empiricism to become the driver of social insights. In this light it might be a fallacy to desire the integration of native GIS concepts in archaeological theory and interpretation. Yes, GIS could (and probably should) be part of a mixed multimethod approach, but when we dedicate ourselves to direct theoretical engagement with the spatial data GIS relies on, it becomes apparent we have to reconceptualise our data in GIS. Since we now have a concept of *interpretive data*, we should devise ways to appropriate GIS formats to become commensurable with the understanding packed into these data.

This strategy towards GIS is not an instant proponent of an eclectic use of spatial analytical tools (Hacıgüzeller 2012; Hacıgüzeller and Thaler 2014; but also in qualitative GIS: Cope and Elwood 2009). Constructive and defensible eclecticism requires a solid, that is, coherent and consistent, fundamental framework of the research process and knowledge production in which it is placed. How else can we evaluate the contribution and validity of its results? However, as material boundaries are but one aspect or operation in the constitution of built environments and used here to contribute to just one mode of understanding, this strategy does support pluralism in perspectives. What this strategy requires first, before even considering mixing and matching methods and ideas, is to work through the consequences of reconciling GIS as a toolkit with interpreting archaeological evidence. That is because the implications of following a fundamental theoretical route into GIS [as propagated by Wheatley and Gillings (2000), Gillings (2012), Hacıgüzeller (2012), McEwan and Millican (2012), Kosiba and Bauer (2013), Wheatley (2014); and following the lead of critique developed in Geographical Information Science by Kwan and Schwanen (2009), Leszczynski (2009)] are much more profound.

As I have shown, it suggests that the very structure of empirically recorded and representational spatial data needs to be questioned on top of the properties of the technology and its tools in order to avoid the 'black box effect' (Griffiths 2013). Carrying out the critical questioning that archaeological evidence is subjected to in this chapter strongly suggests that the structure (visual forms of empirical representation) in which spatial data reaches us is often suboptimal and sometimes even inappropriate for specific end user purposes. The data structure is suboptimal because data production was the work of a different ontological register, and the structure may be inappropriate because its format explicitly permits the use of measures and tools that were produced in a different ontological register.

6.5.2 Towards Interpretive Data Structures

As noted before, Nicholas and Markey (2015) point out that archaeological interpretation has relied heavily on analogous information, even in contextual archaeology (see Hodder and Hutson 2003). In the empirical practice of producing representative maps resulting from archaeological surveying and excavation, a level of interpretation is applied that identifies the (sociocultural) categories of the features we think we have seen (cf. Hutson 2015). Such identifications are typically the outcome of a form of reasoning that goes something like this: 'this' must have been 'that', where 'this' is one or multiple archaeological traces and 'that' is a choice of preconceived categories or entities derived from other (comparable) information. If an archaeological map consists of representation as a cartographic reflection of reality (cf. Hacıgüzeller 2012), then we must ask: whose (or which) reality? Whatever the categories of interpretation are derived from arguably makes up the reality the map imparts from that moment. Even though the map contains physical information shaped in space, the categories identified may not have been defined on the basis of such information at all. How we regard our evidence—even saying it 'is archaeological'—will determine to a large extent what we consider to be real (see Lucas 2012; Meirion Jones 2015).

Contrary to Lucas' (2012) recommendation to nuance the linearity from data collection to interpretation, the conceptual and preparatory processes of this chapter assume such linearity. That is not because data collection and interpretation could not be constructively blended (provided one continues working within one ontological register), but because end users of data will always depend on acquiring data prepared through different research. As recognised before, the requirements of future analyses cannot be foreseen, making data collection and recording for future use an especially tricky balance to strike. Lucas' (2012) solution of employing the concept of materialisation conveying the process of becoming of archaeological entities is unlikely to solve that problem. Critical realism teaches us that entities emerge through interrelations, but the interrelations of site formation and interpretive recording that make up Lucas' archaeological entities are not the entities archaeology as social science pursues an understanding of. When analysing *interpretive data* to advance social interpretations, we are no longer concerned with the mode of interpretation that asks what archaeological evidence is evidence of (cf. Lucas 2012, 2015; Meirion Jones 2015). With *interpretive data* we construct a spatial world of material evidence: a world of material presence.[12] This is a view from beyond any disciplinary evidence. Instead, it conjures up a live world of would-be human participants in intra-actions.

The next concern is establishing the data structure (format, elements, units, and their relations) of such a world of presence. The term geographical information system indicates that as long as phenomena are geographically located (which in a strict sense applies to all archaeology) it can store information on them. It is up to us to decide how we store information with a geospatial reference. The task at hand has been defined as pursuing understandings of the spatially determinant characteristics of occupiable spatial entities constituted from the smallest intrinsically coherent elements (or operations) as constructed and encountered by people. With this, the range of possible entities to emerge can be varied, and they have not been named yet. Any concurrence with entities presented as spatial data derived from other research would be a coincidence. For example, when working towards boundaries of spaces, it is unlikely we will find 'houses' in our finished dataset. Even though the category of a house could be an example of *interpretive data*, material boundary interrelations will disaggregate a house into distinct boundary parts. If working towards the *occupational function* of spaces, this may be quite the reverse.

The prominent critical realist research process of iterative abstraction (Sayer 1981; Yeung 1997) suggests that to get to the actual identifiable structure of *interpretive data* a sequence of contrasting 'material evidence concepts' to physical evidence (empirical edges of spatial discontinuity) follows, until they reach stability as empirically applicable material boundary concepts. Once these material boundary concepts are applied, we have *interpretive data* which are structured in an appropriate way, commensurable to the performative theory that informs the socio-

[12] It could be argued that the idea of a 'world of presence' is an example of nonrepresentational use of GIS (see Hacıgüzeller 2012). A world of presence is necessarily the complete evidence of particular occurrences or phenomena, but one should be vigilant not to mistake that to mean *all* evidence or representing the *whole* of reality. In the colloquial sense, within GIS this is still visually (re) presented. We cannot overcome ourselves as researchers to become inhabitants of the spatial situation constructed in data. This resembles the paradoxical conundrum created by the impasse of situatedness resulting from deeply acculturated emic research or indeed 'archaeological phenomenology'.

spatial significance of boundaries. For the development of boundary line type (BLT) mapping (Vis, under review, 2014b) I have gone through this process using radically contrasting spatial urban built environment datasets. It resulted in mappable concepts that to date appear stable and a data structure that posed genuine challenge to the traditional abilities of GIS software. The concepts proved entirely possible to map, albeit a laborious and complex process. However, GIS proved natively virtually incapable to query the data structure representing the concepts sensibly and commensurably without further original tool development.

The data structure that was generated by simply applying the concepts was not premeditated. In summary, this resulted from the spatial differences between material boundary operations (i.e. those enclosing several distinct occupiable spaces at once, those circumscribing one occupiable space discretely, and those specifying a characteristic persisting for only part of a circumscription of a discrete occupiable space), and the necessarily particular relations of each occupiable space to its outside (consisting of other bounded occupiable spaces). It turned out that data identification according to elementary boundary operations as experienced from the human perspective of one side of the boundary and then the other, initially produces further disaggregation into units. These units then constitute each and every occupiable space. Once mapped, this material boundary data is at once fully quantifiable (and geospatial) and contains an interpretively rich yet critically delimited description.

6.5.3 The Interpretive Advantages of GIS

Verhagen (2018, this volume) displays a perceptive awareness of the advantages of using GIS, many aspects of which are indeed advantages in the context of pursuing social interpretation. I will therefore only expand on a few, starting with the statement that GIS as a heuristic toolkit is not as reductionist as it appears. In fact, GIS is software and software is code. If we see a computer as a capsule (the hardware) with a capacity of artificial intelligence, it is very apparent that nothing is fixed. In principle GIS provides one with a complete expanse of adaptability to requirements. Naturally, to effectuate changes or develop tools within GIS software, the ability to code is an indispensable skill. The real challenge, it would appear, is to identify exactly what one's requirements are and to ground them in theory from start to finish. This is only possible through a fundamental process of critically questioning the quality and information behind any spatial evidence and meticulously defining the (interpretive or analytical) purpose of the research. Only then can interpretive data structures be created in empirical applications.

The term 'interpretive GIS' is used deliberately. This marks its distinction from the social interpretation-ridden qualitative GIS (qualGIS) and critical GIS, which are more established fields with slightly diverging remits and connotations (O'Sullivan 2006; Elwood and Cope 2009; Hacıgüzeller 2012). However, the particular theoretical–archaeological challenge I subject GIS to is not without overlap. Mixed and multimethod approaches may already qualify as qualitative GIS, and certainly this field includes the geolocation of qualitative ideas or data by adding them into GIS environments (Kwan and Knigge 2006; see examples in Cope and Elwood 2009). In fact, taking most GIS software's native abilities, the problem is not that GIS data cannot be invested with qualitative meaning at all. The attributes in the database structure allow one to attach all kinds of meaning and interpretation to any bit of spatial data. Images (usually seen as qualitative data) are raster format and can simply be imported into a GIS environment. In addition, insofar as qualitative GIS embraces the uses of GIS in qualitative research, various archaeological applications of GIS could be grouped under this umbrella without doubt. And so could historical GIS (HGIS) (see Gregory and Ell 2007). The relevance of critical GIS to archaeology is elaborately reviewed by Hacıgüzeller (2012). Classifying the research process this chapter sets out as part of an existing field is arguably a moot point in comparison to its pertinent progress. Therefore, the term 'interpretive GIS', following on from *interpretive data*, captures the purpose it serves well.

At the least I will claim that the approach I promote is a theory-laden and developmental approach to GIS that forces GIS research to operate on a higher critical plane than normally is brought to the fore, even if not exactly as Hacıgüzeller (2012) or Verhagen (2018, this volume) proposes. When following this path, it may indeed prove necessary to devise software interventions. In qualitative GIS the software development of GIS to imbue or directly import qualitative data and associated analytical techniques is also recognised as a notable advance (Elwood and Cope 2009; Jung 2009). It is on the basis of the structure of *interpretive data* that it will become apparent that new GIS abilities or tools may need to be developed. Such tools must be capable of working in respect of the new data structure in order to ask the interpretive questions this structure warrants (see Vis, under review). Innovative and purposive tool development enables spatial archaeologists to grow independent from the toolsets and research environments developed by experts external to our discipline (see Verhagen 2018, this volume). Furthermore, when theoretical understanding structures the spatial data that is queried, the interpretive value and appropriateness of the quantitative tools analysing that structure can be much better evaluated. We can ask how the tools and their measures are capable of revealing or supporting us in identifying explanatory causal powers within our dataset (cf. Smith 2015).

The conceptual implications of identifying material boundaries as spatial data have demonstrated that when it is not necessary to alter GIS software, still new theoretically informed data structures can be developed. But boundaries as a concept offer just one of undoubtedly many powerful ways in which we can reimagine and rearrange our data according to interpretive spatial requirements,[13] yet to be fully explored. Listing the rudimentary carriers of information in GIS, at face value, they may appear as a list of limitations (i.e. raster, an invariable pixel size with assigned values; vector, a combination of points, polylines, and polygons; a geodatabase supporting and attributing information to each spatial element). Instead, in terms of topography, spatial morphology, topology, and spatial relations, there is a truly vast array of possibilities to structure data before software manipulation. Many data structures will have been used in previous and current research; many structures may only exist hypothetically and have yet to find a use. This is where interpretive GIS can benefit and advance the field. GIS not only enables pluralism (Verhagen 2018, this volume), but GIS also enables experimentalism and exploration. Since it is up to the researcher which data is stored and projected in a GIS and how it appears, the sky is literally the limit.

At the low-tech level, the great advantage of digital spatial data over a printed-out map with a legend is the manipulability and comparability of mapped data. Adjusting simply how the elements on the map appear (symbology and classification) or layering several maps (an HGIS favourite), in either vector or raster format, can provide huge leaps in driving interpretive questions and spatially grounding our understanding. Following the arguments for affective geovisualisation (Aitken and Craine 2006, 2009), the traditional investigative capacities of GIS can be placed in an interpretive or qualitative research framework. When the aim is not to elicit an emotional response or understanding from the geovisualisation itself (cf. Aitken and Craine 2006), one could still conceptually map geo-affective information for visual inspection

[13] Network science, while often not geographical or even spatial, should be listed among the ways in which notable advancements in knowledge production have been achieved using new data structures and associated queries and measures [see Brughmans et al. (2016)].

or further analysis instead of geo-representations. Viewing GIS this way brings research back to maps as causal power and images, and there are many examples in the history of science of why visual exploration is a valuable resource.

Finally, when working with spatial data, our options are not restricted to GIS alone. Some graphics and design, graph, and social network analysis software may offer alternative pathways into launching spatial investigations. The advantage of GIS is that this can be done while projected onto geographical space, which is not always necessary, but in terms of underpinning comparability and the complex linking of information is very powerful.

6.5.4 Concluding Reflection

What has been demonstrated here is the culmination of a hard-fought battle to forge the alignment of the scientific empiricism and social scientific conceptualism of archaeology (regarded as social scientific conduct). It strongly suggests there is no easy way to develop interpretive GIS. Yet, for the sake of archaeological argumentation and relevance, I believe it is worthwhile. Especially for those ready to criticise GIS or computational archaeology as unwarranted reductionism and devoid of theoretical meaning, the simple but rudimentary question 'what is archaeological spatial data exactly and what can we do with it to make it serve social interpretation?' opens developmental paths of high potential. These paths may include entirely new discovery, previously impossible queries or exploration, juxtaposing and layering plural concepts and methods, or simply advancing the evidence base of current hypotheses. It truly is an open invitation for rigorous theorists to apply themselves to work through the implications of geospatially anchored ideas as necessarily related to their empirical archaeological counterparts. The outcomes and consequences will form an opportunity to shape both GIS formats and analytical research creatively.

References

Abbott A (1995) Things of boundaries. Soc Res 62 (4):857–882

Aitken S, Craine J (2006) Guest editorial: Affective geovisualizations. Directions Magazine. http://www.directionsmag.com/entry/guest-editorial-affective-geovisualizations/123211. Accessed 05 July 2016

Aitken S, Craine J (2009) Into the image and beyond: affective visual geographies and GIScience. In: Cope M, Elwood S (eds) Qualitative GIS: a mixed methods approach. Sage, Los Angeles, pp 139–155

Arnoldi J (2001) Niklas Luhmann: an introduction. Theory Culture Soc 18(1):1–13

Barad K (2003) Posthumanist Performativity: towards an understanding of how matter comes to matter. Signs J Women Cult Soc 28(3):801–832

Bintliff J, Pearce M (2011) Introduction. In: Bintliff J, Pearce M (eds) The death of archaeological theory? Oxbow Books, Oxford, pp 1–6

Brughmans T, Collar A, Cowards F (eds) (2016) The connected past: challenges to network studies in archaeology and history. University of Oxford Press, Oxford

Cope M, Elwood S (eds) (2009) Qualitative GIS: a mixed methods approach. Sage, Los Angeles

Cox KR (2013) The continuing relevance of old debates. Dialogues Hum Geogr 3(1):49–55

Demarest AA (1997) The Vanderbilt Petexbatun Regional Archaeological Project 1989– 1994: overview, history, and major results of a multidisciplinary study of the Classic Maya collapse. Anc Mesoam 8:209–227

Ellen R (2010) Theories in anthropology and anthropological theory. J Roy Anthropol Inst 16:387–404

Elwood S, Cope M (2009) Introduction: qualitative GIS: forging mixed methods through representations, analytical innovations, and conceptual engagements working from a theoretical perspective. In: Cope M, Elwood S (eds) Qualitative GIS: a mixed methods approach. Sage, Los Angeles, pp 1–12

Fahlander F (2012) Are we there yet?: archaeology and the postmodern in the new millennium. Curr Swed Archaeol 20:109–129

Fletcher RJ (2004) Materiality, space, time and outcome. In: Bintliff JL (ed) A companion to archaeology. Blackwell, Oxford, pp 110–140

Galinié H, Rodier X, Saligny L (2004) Entités fonctionnelles, entités spatiales et dynamique urbaine dans la longue durée. Hist Mes 19(3/4):223–242

Galison P (1998) Judgment against objectivity. In: Jones CA, Galison P (eds) Picturing science producing art. Routledge, New York, pp 327–359

Galison P (2000) Objectivity is romantic. In: Frye BE (ed) The humanities and the sciences. American Council of Learned Societies Occasional Paper 47:15–43. http://archives.acls.org/op/op47-3.htm. Accessed 09 July 2016

Galison P (2010) The objective image. Faculteit Geesteswetenschappen, Universiteit Utrech, Utrecht
Gillings M (2012) Landscape phenomenology, GIS and the role of affordance. J Archaeol Method Theory 19 (4):601–611
Graham E (1999) Stone cities, green cities. Archeol Pap Am Anthropol Assoc 9(1):185–194
Gregory IN, Ell PS (2007) Historical GIS: technologies, methodologies and scholarship. Cambridge University Press, Cambridge
Griffiths S (2013) GIS and research into historical 'spaces of practice': overcoming the epistemological barriers. In: von Lünen A, Travis C (eds) History and GIS: epistemologies, considerations and reflections. Springer, Dordrecht, pp 153–171
Hacıgüzeller P (2012) GIS, critique, representation and beyond. J Soc Archaeol 12(2):245–263
Hacıgüzeller P, Thaler U (2014) Three tales of two cities? A comparative analysis of topological, visual and metric properties of archaeological space in Malia and Pylos. In: Paliou E, Lieberwirth U, Polla S (eds) Spatial analysis and social spaces: interdisciplinary approaches to the interpretation of prehistoric and historic built environments. TOPOI 18, Berlin Studies of the Ancient Worlds. De Gruyter, Berlin, pp 203–262
Hall ET (1996) Foreword. In: Pellow D (ed) Setting boundaries: the anthropology of spatial and social organization. Bergin & Garvey, Westport, pp vii–viii
Heath J (2015) Methodological individualism. In: Zalta EN (ed) The stanford encyclopedia of philosophy (Spring 2015 Edition). http://plato.stanford.edu/archives/spr2015/entries/methodological-individualism/. Accessed 07 July 2015
Hodder I (2014) The entanglements of humans and things: a long-term view. New Literary Hist 45(1):19–36
Hodder I, Hutson SR (2003) Reading the past: current approaches to interpretation in archaeology. Cambridge University Press, Cambridge
Hutson SR (2012) Unavoidable imperfections: historical contexts for representing ruined Maya buildings. In: Pillsbury J (ed) Past presented: archaeological illustration and the ancient Americas. Dumbarton Oaks, Washington, pp 283–316
Hutson SR (2015) Adapting LiDAR data for regional variation in the tropics: a case study from the Northern Maya Lowlands. J Archaeol Sci Rep 4:252–263
Ingold T (2008) Bindings against boundaries: entanglements of life in an open world. Environ Plann A 40(8):1796–1810
Jones R (2009) Categories, borders and boundaries. Prog Hum Geogr 33(2):174–189
Jones R (2010) The spatiality of boundaries. Prog Hum Geogr 34(2):263–267
Jung JK (2009) Computer-aided qualitative GIS: a software-level integration of qualitative research and GIS. In: Cope M, Elwood S (eds) Qualitative GIS: a mixed methods approach. Sage, Los Angeles, pp 115–136
Kintigh KW, Altschul J, Beaudry M, Drennan R, Kinzig A, Kohler T, Limp WF, Maschner H, Michener W, Pauketat T, Peregrine P, Sabloff J, Wilkinson T, Wright H, Zeder M (2014) Grand challenges for archaeology. Am Antiq 79(1):5–24
Kleinman A (2012) Intra-actions. Mousse 34:76–81
Koch A (2005) Autopoietic spatial systems: the significance of actor network theory and systems theory for the development of a system theoretical approach of space. Soc Geogr 1(1):5–14
Kosiba S, Bauer AM (2013) Mapping the political landscape: towards a GIS analysis of environmental and social difference. J Archaeol Method Theory 20 (1):61–101
Kwan MP, Knigge L (2006) Doing qualitative research with GIS: an oxymoronic endeavor? Environ Plann A 38(11):1999–2002
Kwan MP, Schwanen T (2009) Critical quantitative geographies. Environ Plann A 41:261–264
Lefebvre B (2009) How to describe and show dynamics of urban fabric: cartography and chronometry? In: Proceedings of the 37th computer applications and quantitative methods in archaeology conference, Williamsburg. http://www.caa2009.org/articles/Lefebvre_Contribution224_a.pdf. Accessed 26 Sept 2013
Leszczynski A (2009) Quantitative limits to qualitative engagements: GIS, its critics, and the philosophical divide. Prof Geogr 63(3):350–365
Lilley KD (2011) Urban mappings: visualizing late medieval Chester in cartographic and textual form. In: Clarke C (ed) Mapping the medieval city. University of Wales Press, Cardiff, pp 19–41
Lucas G (2012) Understanding the archaeological record. Cambridge University Press, New York
Lucas G (2015) Evidence of what? On the possibilities of archaeological interpretation. In: Chapman R, Wylie A (eds) Material evidence: learning from archaeological practice. Routledge, Abingdon, pp 311–323
Luhmann N (1986) The autopoiesis of social systems. In: Geyer F, van der Zouwen J (eds) Sociocybernetic paradoxes: observation, control and evolution of self-steering systems. Sage, London, pp 172–192
MacEachren AM (2004) How maps work: representation, visualization and design. The Guilford Press, New York
Marcus L (2017, in prep) Spatial capital: measures and meanings (unpublished manuscript)
McEwan DG, Millican K (2012) In search of the middle ground: quantitative spatial techniques and experiential theory in archaeology. J Archaeol Method Theory 19(4):491–494
Meirion Jones A (2015) Meeting the past halfway: a consideration of the ontology of material evidence in archaeology. In: Chapman R, Wylie A (eds) Material evidence: learning from archaeological practice. Routledge, Abingdon, pp 324–338
Monmonier M (1996) How to lie with maps. University of Chicago Press, Chicago

Morton SG, Peuramaki-Brown MM, Dawson PC, Seibert JD (2014) Peopling the past: interpreting models for pedestrian movement in ancient civic-ceremonial centres. In: Rau S, Schönherr E (eds) Mapping spatial relations, their perceptions and dynamics: the city today and in the past. Springer International, Heidelberg, pp 25–44

Nicholas G, Markey N (2015) Traditional knowledge, archaeological evidence, and other ways of knowing. In: Chapman R, Wylie A (eds) Material evidence: learning from archaeological practice. Routledge, Abingdon, pp 287–307

O'Sullivan D (2006) Geographical information science: critical GIS. Prog Hum Geogr 30(6):783–791

Oles T (2015) Walls: enclosure and ethics in the modern landscape. University of Chicago Press, Chicago

Oliver R (1993) Ordnance survey maps: a concise guide for historians. The Charles Close Society, London

Pellow D (ed) (1996) Setting boundaries: the anthropology of spatial and social organization. Bergin & Garvey, Westport

Pratt AC (1995) Putting critical realism to work: the practical implications for geographical research. Prog Hum Geogr 19(1):61–74

Pred AR (1981) Social reproduction and the time-geography of everyday life. Geogr Ann B 63(1):5–22

Pred AR (1984) Place as historically contingent process: structuration and the time-geography of becoming places. Ann Assoc Am Geogr 74(2):279–297

Preucel RW, Meskell L (2007) Knowledges. In: Meskell L, Preucel RW (eds) A companion to social archaeology. Blackwell, Oxford, pp 3–22

Rodier X, Saligny L, Lefebvre B, Pouliot J (2009) ToToPI (Topographie de Tours Pré-Industriel): a GIS for understanding urban dynamics based on the OH_FET model (Social Use, Space and Time). In: Proceedings of the 37th computer applications and quantitative methods in archaeology conference, Williamsburg. http://proceedings.caaconference.org/files/2009/41_Rodier_et_al_CAA2009.pdf. Accessed 08 July 2016

Sayer A (1981) Abstraction: a realist interpretation. Radic Philos 28:6–15

Sayer A (1993) Postmodernist thought in geography: a realist view. Antipode 25(4):320–344

Sayer A (2013) Looking forward to new realist debates. Dialogues Hum Geogr 3(1):22–25

Schütz A (1967) The phenomenology of the social world. Northwestern University Press, Evanston

Sherratt A (1993) The relativity of theory. In: Yoffee N, Sherratt A (eds) Archaeological theory: who sets the agenda. New directions in archaeology. Cambridge University Press, Cambridge, pp 119–130

Smith B (2001) Fiat objects. Topoi 20(2):131–148

Smith ME (2011) Empirical urban theory for archaeologists. J Archaeol Method Theory 18:167–192

Smith ME (2012) The role of ancient cities in research on contemporary urbanization. UGEC Viewpoints 8:15–19

Smith ME (2015) How can archaeologists make better arguments? SAA Archaeol Rec 15(4):18–23

Smith B, Varzi AC (1997) Fiat and bona fide boundaries: towards an ontology of spatially extended objects. In: Hirtle S, Frank AU (eds) Spatial information theory a theoretical basis for GIS. Springer, Berlin, pp 103–119

Smith B, Varzi AC (2000) Fiat and bona fide boundaries Philos Phenomenol Res 60(2):401–420

Smith ME, Feinman GM, Drennan RD, Earle T, Morris I (2012) Archaeology as a social science. PNAS Early Edition, pp 1–5, doi:https://doi.org/10.1073/pnas.1201714109

Verhagen P (2018) Spatial analysis in archaeology: moving into new territories. In: Siart C, Forbriger M, Bubenzer O (eds) Digital geoarchaeology: new techniques for interdisciplinary human-environmental research. Springer, Heidelberg, pp 11–26

Vis BN (2009) Built environments, constructed societies: inverting spatial analysis. Sidestone Press, Leiden

Vis BN (2012) The death of archaeological theory? (Book review). Antiquity 86:274–275

Vis BN (under review) Cities made of boundaries: mapping social life in urban form. UCL Press, London

Vis BN (2013) Establishing boundaries: a conceptualisation for the comparative social study of built environment configurations. Spaces Flows 2(4):15–29

Vis BN (2014a) Boundary concepts for studying the built environment: a framework of socio-spatial reasoning for identifying and operationalising comparative analytical units in GIS. In: Earl G (ed) Proceedings of CAA 2012 Southampton. Amsterdam University Press, pp 820–838. http://dare.uva.nl/aup/en/record/500958. Accessed 01 June 2016

Vis BN (2014b) Mapping socio-spatial relations in the urban built environment through time: describing the socio-spatial significance of inhabiting urban form. In: Rau S, Schönherr E (eds) Mapping spatial relations, their perceptions and dynamics: the city today and in the past. Lecture notes in geoinformation and cartography. Springer International, pp 45–93

Vis BN (2016) The material logic of urban space. J Space Syntax 6(2):271–274

Wallace S (2011) Contradictions of archaeological theory: engaging critical realism and archaeological theory. Routledge, Oxford

Webmoor T (2007) What about 'one more turn after the social' in archaeological reasoning?: taking things seriously. World Archaeol 39(4):563–578

Webmoor T, Witmore CL (2008) Things are us!: a commentary on human/things relations under the banner of a 'social' Archaeology. Nor Archaeol Rev 41 (1):53–70

Werlen B (2005) Regions and everyday regionalizations: from a space-centred towards an actioncentred human geography. In: van Houtum H, Kramsch O, Zierhofer W (eds) Bordering space. Ashgate, Aldershot pp 47–60

Wheatley D (2014) Connecting landscapes with built environments: visibility analysis, scale and the senses. In: Paliou E, Lieberwirth U, Polla S (eds) Spatial analysis and social spaces: interdisciplinary approaches to the interpretation of prehistoric and historic built environments. TOPOI 18, Berlin Studies of the Ancient Worlds. De Gruyter, Berlin, pp 115–134

Wheatley D, Gillings M (2000) Vision, perception and gis: some notes on the development of enriched approaches to the study of archaeological visibility. In: Lock G (ed) Beyond the map: archaeology and spatial technologies. Nato science series A: life sciences. IOS Press, Amsterdam, pp 1–27

Wood D (1992) The power of maps. The Guilford Press, New York

Yeung HW (1997) Critical realism and realist research in human geography: a method or a philosophy in search of a method? Prog Hum Geogr 21(1):51–74

Zierhofer W (2002) Speech acts and space(s): language pragmatics and the discursive constitution of the social. Environ Plann A 34(8):1355–1372

Open Access This chapter is licensed under the terms of the Creative Commons Attribution 4.0 International License (http://creativecommons.org/licenses/by/4.0/), which permits use, sharing, adaptation, distribution and reproduction in any medium or format, as long as you give appropriate credit to the original author(s) and the source, provide a link to the Creative Commons license and indicate if changes were made.

The images or other third party material in this chapter are included in the chapter's Creative Commons license, unless indicated otherwise in a credit line to the material. If material is not included in the chapter's Creative Commons license and your intended use is not permitted by statutory regulation or exceeds the permitted use, you will need to obtain permission directly from the copyright holder.

Part II

Remote Sensing and Digital Image Analysis

Airborne and Spaceborne Remote Sensing and Digital Image Analysis in Archaeology 7

Karsten Lambers

Abstract

Remote sensing has a long and successful track record of detecting and mapping archaeological traces of human activity in the landscape. Since the early twentieth century, the tools and procedures of aerial archaeology evolved gradually, while earth observation remote sensing experienced major steps of technological and methodological advancements and innovation that today enable the monitoring of the earth's surface at unprecedented accuracy, resolution and complexity. Much of the remote sensing data acquired in this process potentially holds important information about the location and context of archaeological sites and objects. Archaeology has started to make use of this tremendous potential by developing new approaches for the detection and mapping of archaeological traces based on digital remote sensing data and the associated tools and procedures. This chapter reviews the history, tools, methods, procedures and products of archaeological remote sensing and digital image analysis, emphasising recent trends towards convergence of aerial archaeology and earth observation remote sensing.

Keywords

Remote sensing • Digital image analysis • Archaeological prospection • Object detection

7.1 Introduction

"*Remote sensing* is the science and art of obtaining information about an object, area, or phenomenon through the analysis of data acquired by a device that is not in contact with the object, area, or phenomenon under investigation" (Lillesand et al. 2015: 1; emphasis in original). This generic definition of remote sensing, a technique with many uses across a wide range of disciplines, is also valid in archaeology, where we commonly understand "device" as a sensor mounted on

K. Lambers (✉)
Leiden University, Leiden, the Netherlands
e-mail: k.lambers@arch.leidenuniv.nl

© Springer International Publishing AG 2018

C. Siart et al. (eds.), *Digital Geoarchaeology*, Natural Science in Archaeology,
DOI 10.1007/978-3-319-25316-9_7

an airborne or spaceborne platform and "object, area, or phenomenon" as a portion of a landscape with its natural and cultural components. Geophysical prospection, which is a form of near-surface remote sensing and often subsumed under that term as well (Johnson 2006; Wiseman and El-Baz 2007), is not treated here following common terminology in Europe (see Chap. 14). Furthermore, this chapter focuses on image-based remote sensing, while range-based remote sensing is treated elsewhere (see Chap. 11).

The benefits of using remote sensing as a recording technique in archaeology are manifold. For example, one of the major advantages is that sensitive archaeological objects are not touched nor otherwise affected by remote sensing. This is in line with recent trends towards non-invasive methods of investigation that help to preserve the archaeological heritage (Corsi et al. 2013). In addition, the bird's eye perspective helps to observe and understand archaeological sites and objects in their landscape context that was formed by interwoven natural and anthropogenic processes (Musson et al. 2013). Furthermore, today remote sensing data is continuously being generated in the environmental sciences, in cartography and geodesy and in the military and commercial sectors, leading to an ever-increasing quantity and quality of data that potentially hold information about archaeological contexts. These data are available at a wide variety of scales and resolutions and often with a considerable time depth. They can thus contribute to a broad range of archaeological research questions.

In what follows, the history of remote sensing in archaeology and earth observation is summarised in Sect. 7.2. Section 7.3 provides a brief overview of platforms, sensors and data and their application to archaeology. Section 7.4 discusses the archaeological analysis of digital remote sensing data, focusing on recent trends and illustrating this with a case study from own research. This is then followed by an outlook in Sect. 7.5.

7.2 A Look Back

7.2.1 Aerial Archaeology

The first aerial images of archaeological sites were taken from military balloons around the turn of the nineteenth to the twentieth century (Trümpler 2005; Verhoeven et al. 2013; Campana 2017a). Shortly after, during World War I, aerial photographs taken for military reconnaissance from aeroplanes covered many archaeological sites and ruins in Europe and the Near East for the first time. In the 1920s, O.G.S. Crawford was the first archaeologist to systematically use crop marks, i.e. observable differences in plant growth caused by subsurface archaeological remains, for archaeological site detection and mapping. While crop marks and other proxies such as soil, shadow, snow and flood marks work well in the temperate climate regions of central Europe with their extended agricultural fields, they are less effective in dryer conditions and largely fail in woodlands. The introduction of infrared and later multispectral photography to aerial archaeology in the 1970s increased the visible range so that differences in soil moisture and vegetation growth could be used more effectively (Verhoeven 2008, 2012). However, inherent conceptual issues such as survey bias (Palmer 2005; Cowley 2016) could not be resolved through technological innovation.

In spite of these limitations, aerial archaeology has proven to be the single most effective method of archaeological site detection and mapping in Europe. The technique of taking oblique images with a handheld camera from a small aircraft has remained largely unchanged since the 1920s. Since then, systematic efforts such as English Heritage's National Mapping Programme (NMP; Horne 2009) and technological innovations such as digital cameras and positioning devices (e.g. GPS/INS) have increased the efficiency and effectiveness of the method (Leckebusch 2005; Doneus et al. 2016). As a result, today many European countries hold substantial collections of aerial images taken for the purpose of archaeological prospection. Important resources of archaeological information are also buried in the millions of vertical aerial photographs taken for purposes of military reconnaissance or cartography collected in major national archives (Cowley et al. 2010; Cowley and Stichelbaut 2012; Hanson and Oltean 2013). These historical images constitute a highly valuable resource, as many of them show sites and landscapes that have since been heavily altered, damaged or destroyed, e.g. through land consolidation, irrigation, urban sprawl, or armed conflict.

7.2.2 Earth Observation Remote Sensing

Like aerial archaeology, earth observation remote sensing grew out of military applications around the time of World War I. Systematic cartographic mapping based on aerial images began in the 1930s. Other early applications included land use studies, geology, hydrology and forestry (Lillesand et al. 2015). During World War II, millions of aerial photographs were taken for military reconnaissance, which brought about vastly improved methods of image capture, analysis and interpretation (Hanson and Oltean 2013). This time of conflict and the post-war years leading up to the Cold War also saw technological innovations such as colour and infrared photography that allowed new ways of studying land cover and vegetation. The basic remote sensing concept developed in those years, with aeroplanes serving as platforms for different sensors used for the systematic mapping of large areas, remains highly useful for earth observation until the present day. However, an important new branch developed in 1960 after the first satellites were launched into space. Photographs of the earth taken from manned spacecraft triggered the interest of the environmental sciences in spaceborne remote sensing, but the technological development was once again driven by military applications.

The first large-scale mapping of portions of the earth's surface from space was undertaken in the 1960s during the Cold War and for purposes of military espionage and reconnaissance. Consequently, the two main antagonists, the USA and UdSSR, mainly covered areas of geostrategic importance such as central Europe and the Near East (Fowler 2013). Images were captured by series of short-lived satellites, e.g. the American Corona and the Sowjet KOSMOS series, which initially produced black-and-white analogue images that were sent back to earth by parachute. Their recovery was complex and frequently failed (Day et al. 1998). The images had a spatial resolution of 1.2–12 m. In spite of great distortion due to their complex image geometry, they provide an invaluable data source for archaeology, for example, for the Near East prior to the time when mechanised agriculture, irrigation and urban sprawl destroyed many ancient sites and their surrounding landscapes (Goossens et al. 2006; Casana and Cothren 2008; Agapiou et al. 2016).

Beginning in the 1970s, government-run space agencies such as NASA initiated earth observation for scientific purposes. Landsat is NASA's most successful long-term programme with a series of seven satellites so far that capture multispectral images of large parts of the earth with a spatial resolution between 80 and 15 m (Lillesand et al. 2015). These images provide large-scale base data for applications, e.g. in geography, biology, climate and land use studies, urban planning, cartography, oceanography and numerous other disciplines. In spite of their low spatial resolution, they soon found first applications in archaeology (Sever 1990; Parcak 2009. For a recent overview of NASA's and ESA's activities related to archaeology, see Giardino 2011 and Stewart et al. 2015).

Technical developments in both platforms and sensors lead to a continuously increasing spatial resolution of spaceborne images. The best available spatial resolution from optical sensors mounted on earth observation satellites was 80 m in 1972, 30 m in 1982 and 5.8 m in 1995 (Lillesand et al. 2015). A paradigm shift occurred at the end of the last century when for the first time a commercial company launched a satellite into space with the sole purpose of selling the images to a wide range of clients (Ikonos 2, launched 1999). Consequently, commercial providers focused on very high spatial resolution (<1 m panchromatic) and up until today provide the highest available spatial resolution to private customers. However, recent satellites launched by government-run space agencies, while still featuring high spectral resolution, today reach spatial resolutions that come close to those of commercial satellites (Agapiou et al. 2015).

Table 7.1 lists selected satellites and sensors that have provided useful images for archaeological purposes in the past or have the potential of doing so in the future. This selection is necessarily subjective and incomplete. For additional data on satellites and sensors, see Remondino (2011) and Lillesand et al. (2015).

While spaceborne remote sensing blossomed, airborne remote sensing continued to be the workhorse for mapping and for environmental applications at smaller scales and has seen just as many technological innovations in recent years. One of them is the introduction of digital cameras, either following the traditional frame format or using linear array sensors (Lemmens 2011; Remondino 2011). Some digital cameras acquire oblique imagery, e.g. for urban mapping (Remondino and Gehrke 2015). Most of these new sensors cover also the near-infrared light, which makes them once again highly valuable for archaeological prospection using vegetation marks. At the same time, aeroplanes are common platforms for truly multispectral and hyperspectral sensors for environmental monitoring, as their lower altitude above ground allows higher spatial resolution than spaceborne platforms. These sensors are useful for a wide range of archaeological applications (Donoghue et al. 2006; Traviglia 2007; Beck 2011; Agapiou et al. 2014; Doneus et al. 2014).

7.3 Platforms, Sensors and Data

7.3.1 High to Low Altitude Platforms

Contrary to popular usage of the term, satellites themselves do not acquire images. Rather, they are platforms on which one or several sensors can be mounted which in turn capture images (Table 7.1). However, certain parameters of the platforms have an effect on image characteristics, among them orbit and altitude. For example, earth observation satellites carrying passive optical sensors, as well as many commercial satellites, circle the earth in sun-synchronous orbits roughly perpendicular to the equator to remain within the zone of sunlight at all times. Such a configuration entails that certain places on the surface of the earth are always visited at the same time of day. Another important parameter is altitude, which affects swath width and spatial resolution.

Table 7.1 Selected satellites and sensors ordered by ground resolution (pan, panchromatic; VIS, visible light; NIR, near infrared; MIR, mid infrared; TIR, thermal infrared)

Satellite	Sensor	Launched	Altitude (km)	Swath width (km)	Channels	Ground resolution (m)
Landsat 7	ETM+	1999	705	185	8 (pan, VIS, NIR, MIR, TIR)	15 (pan)
Landsat 8	OLI	2013	705	185	9 (pan, VIS, NIR, MIR)	15 (pan)
Terra	ASTER	1999	705	60	14 (VIS, NIR, MIR, TIR)	15 (VIS, NIR)
SPOT 5	HRG	2002	832	117	6 (pan, VIS, NIR, MIR)	5.0 (pan)
SPOT 6/7	NAOMI	2012/2014	694	60	5 (pan, VIS, NIR)	1.5 (pan)
Ikonos 2		1999	680	11	5 (pan, VIS, NIR)	1.0 (pan)
Quickbird 2		2001	450	16	5 (pan, VIS, NIR)	0.61 (pan)
Worldview 2		2009	770	16	9 (pan, VIS, NIR)	0.46 (pan)
Geoeye 1		2008	684	15	5 (pan, VIS, NIR)	0.41 (pan)
Worldview 3		2014	617	13	29 (pan, VIS, NIR, MIR)	0.31 (pan)
Worldview 4		2016	617	13	5 (pan, VIS, NIR)	0.31 (pan)

Typical orbits of satellites carrying optical sensors are 700–900 km above ground.

Aeroplanes operate in much lower ranges within the earth's troposphere, between several hundred metres (light aircraft) and 10–12 km (airliners). While earth observation remote sensing, depending on the purpose, may take advantage of this whole range, the flying height is usually closer to the lower end the higher the spectral resolution of the carried sensors is. This is to ensure a good trade-off between spectral and spatial resolution. While satellites operate on a global scale and follow a fixed schedule, aeroplanes operate on a regional scale and can be employed more flexible.

Low-altitude platforms on a local scale have seen a number of important innovations in recent years, especially in archaeological applications. While balloons, blimps and kites have been used for quite some time (Verhoeven 2009), unmanned aerial vehicles (UAVs) provide an unprecedented level of flexibility with regard to data acquisition ever since their first application in archaeology in 2004, especially with the introduction of autonomous navigation (Lambers et al. 2007; Gutiérrez and Searcy 2016; Campana 2017b. For more detail, see Chap. 10).

7.3.2 Active vs. Passive Sensors

There are two types of sensors mounted on airborne and spaceborne platforms: active and passive (Lillesand et al. 2015). Active sensors such as radar and lidar—not treated in this chapter but important to mention—use their own energy source to send a signal to the surface of the earth, from which it is partially reflected and then captured again by the sensor. Since the energy, a form of electromagnetic radiation, travels at the speed of light, the distance between the sensor and the surface can be calculated from the time interval between the emission and the return of the signal. This is called the time-of-flight (ToF) principle of range-based measurement. While radar uses radio/microwaves in different wavelengths, lidar uses visible or infrared light. Travelling between sensor and surface, the signal interacts with the atmosphere, with objects on the surface such as vegetation and with the surface itself in multiple and complex ways that need to be taken into account when reconstructing the surface geometry from the signal. An advantage of active sensing is its independence of sunlight and good weather: both methods work under cloudy/rainy conditions and by night. For accurate range measurements, the position and tilt or skew of the sensor needs to be determined with high accuracy, too. This is usually achieved with global navigation satellite system (GNSS) and inertial navigation system (INS) units.

Passive sensors do not have their own energy source but instead capture radiation emitted from or reflected by the earth's surface, the main source of which is the sun. They are often collectively called optical sensors, although many of them capture radiation outside the range visible to the human eye. Sensors for cartography, mapping and commercial purposes usually generate images with high spatial and limited spectral resolution, e.g. in the visible and near-infrared light (VNIR) range. On the other hand, sensors for earth observation often provide lower spatial yet higher spectral resolution, especially in the infrared range where atmospheric permeability is high and many relevant environmental parameters can be measured. Sensors that capture a limited number of—often disjoint—spectral bands produce multispectral images. Sensors that capture a high number of continuous narrow bands produce hyperspectral images. Since the total energy captured by a given sensor is limited, there is usually a trade-off between spectral and spatial resolution. Other relevant resolutions are the radiometric resolution, which expresses the range of digital numbers available to visualise an image (e.g. 8 bit: $2^8 = 256$ digital numbers), and the temporal resolution, which in the case of satellite images denotes the revisit time of the sensor over a given location on earth. As with radar and lidar data, passive sensing needs to take into account multiple atmospheric and other conditions that have an impact on image formation. Most importantly, passive imaging requires daylight and, in the case of satellite images, as little cloud cover as possible.

7.3.3 Analogue vs. Digital Images

While most optical images are today taken with digital sensors, there are huge archives from the era of analogue photography that contain a wealth of potentially useful information for archaeological purposes (Cowley et al. 2010; Cowley and Stichelbaut 2012; Hanson and Oltean 2013). Analogue photographs, be they negatives or positives, suffer from physical and chemical degradation and thus require measurements for their preservation and/or digitisation.

Aerial photographs for aerial archaeology are often taken with uncalibrated handheld cameras without registration of the exact position and tilt of the camera (Leckebusch 2005; Palmer 2005). Most of these images are oblique, showing the horizon (high) or not (low), in order to optimally capture crop, soil, shadow and other marks. Their acquisition depends on decisions of the operator. For all these reasons, their georeferencing is often difficult and depends on contextual information ideally provided by the operator.

In contrast, aerial photographs for cartography and military reconnaissance are usually taken with metric cameras from a near-vertical perspective in a systematic fashion that aims at the complete coverage of a given target area. Very often, there is a considerable overlap between consecutive images to enable stereoscopic analysis (Mikhail et al. 2001). Georeferencing is facilitated through positioning data collected along the flight path and ground control. In spite of these advantages, vertical aerial photographs are not always useful for archaeological purposes as they are often taken during unfavourable times of the year, e.g. in winter when there are no leafs on the trees, and of the day, e.g. around noon when shadows are minimal, when usual archaeological proxies such as crop marks do not show.

While analogue aerial cameras continue in use, most airborne and all spaceborne remote sensing today operates on digital sensors (Richards and Jia 2006 Lillesand et al. 2015).

Digital frame cameras, like analogue cameras, are pin-hole cameras that capture one individual scene at a time and produce rectangular images from it. Linear-array cameras, on the other hand capture many narrow strips of a scene corresponding to one line of pixels in the digital image, one after the other. The lines of pixels are then being combined into a continuous image.

Both frame and linear array cameras may produce multispectral images or stacks of images with each layer corresponding to one spectral channel, or band. Many airborne and spaceborne sensors are furthermore operated such that they produce stereoscopic images that can be used for 3D analysis. While overlap between consecutive image scenes ensures this for frame cameras, linear array cameras capture imagery in forward-, nadir- and backward-looking mode, such that multiple perspectives on each portion of the terrain are available from the captured imagery.

7.4 Archaeological Analysis of Remote Sensing Data

The best practice of analysing traditional remote sensing data such as aerial photographs for archaeological purposes has been described elsewhere (Brophy and Cowley 2005; Musson et al. 2013). We here focus on the archaeological analysis of digital remote sensing data using computational tools.

7.4.1 Recent Trends

In geodesy, cartography and earth observation, the complexity of digital remote sensing data has led to the development of a wide range of quantitative and computational tools for image processing and analysis since the 1970s (Richards and Jia 2006; Lasaponara and Masini 2012; Abrams and Comer 2013; Lillesand et al. 2015). While processing usually encompasses image correction, enhancement, transformation and registration, analysis often entails some level of classification of the image contents that assist in their interpretation. Furthermore, overlapping images may be analysed in 3D for the extraction of geometric information (Mikhail et al. 2001).

These techniques are usually systematically applied to whole images, or series of images, and, thus, to entire landscapes that they cover.

In contrast, archaeological image analysis originally focused on certain portions of images that were of archaeological interest, namely, traces of human activity in the landscape (Brophy and Cowley 2005). Consequently, the intensity of landscape coverage varied greatly. Site detection and mapping is still one of the most important goals in archaeological image analysis. However, in recent years, the theoretical turn towards landscapes as frames of reference for an archaeological enquiry has facilitated the adoption of full coverage remote sensing data originally not acquired and analytical tools originally not developed for archaeological purposes (Doneus 2013). This adoption of data and methods from earth observation remote sensing requires innovation and change in the practice of archaeological prospection (Cowley 2012; Verhoeven and Sevara 2016). For example, the thorough screening of individual aerial images by a human observer as in aerial archaeology is not scalable to the quantity and complexity of, e.g. multi-/hyperspectral images. On the other hand, existing analytical tools for object detection associated with digital remote sensing data, e.g. for road or building detection in cartography and mapping, usually fail when targeting faint, elusive archaeological traces.

Therefore, since the early 2000s, archaeologists, in close collaboration with experts from the earth sciences and computer science, have attempted to partly automate the archaeological analysis of remote sensing data, using digital image processing and analysis to detect and map archaeological traces (e.g. De Laet et al. 2007, 2009). Such attempts initially met with considerable scepticism (e.g. Hanson 2010) due to the unclear role that computer algorithms should play in the process of observing, analysing and interpreting archaeological traces in the landscape (Cowley 2012; Bennett et al. 2014). In the meantime, however, a number of projects have demonstrated the feasibility of such an approach (Traviglia et al. 2016). An interesting aspect here is that some automated approaches can be applied to both image data and range data. For example, algorithms have been developed to reliably detect burial mounds (Trier et al. 2009, 2015; Caspari et al. 2014; Sevara et al. 2016; Cerrillo-Cuenca 2017), stone tombs (Schuetter et al. 2013), charcoal kilns (Schneider et al. 2015), animal traps (Trier and Pilø 2012), trails (Vletter 2014) and tells (Menze and Ur 2012) in digital elevation models, high-resolution panchromatic images, or multispectral images. All of these handcrafted algorithms target well-known, clearly defined categories of recurrent, typical archaeological objects. They are thus designed to assist archaeological prospection and provide base data, not to replace fieldwork and archaeological interpretation. To illustrate this field of research more clearly, the following describes a case study from our own research.

7.4.2 Case Study: Archaeological Object Detection in the Silvretta Alps

The Silvretta Archaeological Project, directed by Thomas Reitmaier and conducted from 2009 to 2016 in the Silvretta mountains on the border between Switzerland and Austria, served as a case study to develop methods for a semi-automated archaeological analysis of optical remote sensing images. The main goal of the project was to investigate Holocene human-environment interaction and resource use in the alpine zone above the tree line, with a special focus on the prehistoric origins and further development of alpine pastoralism (Dietre et al. 2014, 2017; Kothieringer et al. 2015). An important category of archaeological sites relevant for this topic was ruins of livestock enclosures (LSEs) used for the management of sheep, goats and cattle during the annual grazing period in the short summer (Fig. 7.1). About 30 LSEs were registered during archaeological fieldwork from 2009 to 2016, dating from the Bronze and Iron Ages to the Modern Period. These known LSEs served as target objects for the development of an algorithm for archaeological object detection.

Fig. 7.1 Well-preserved livestock enclosure (LSE) in Val Urschai, Lower Engadine, Switzerland (photo: K. Lambers). About 30 LSEs were recorded in the Silvretta region, most of them older and less well preserved than this one

While the results have been described in detail elsewhere (Zingman 2016; Zingman et al. 2014, 2016), the following is a brief overview with a focus on the general idea behind the workflow.

Images captured in 2011 by the commercial satellite Geoeye 1 served as primary data source (cp. Table 7.1). They feature four bands in the VNIR spectrum and a spatial resolution of 0.5 m in the panchromatic band, downsampled from the original 45 cm due to legal restrictions. These images were chosen for two reasons. Firstly, they provided the only consistent, up-to-date data source for the entire study area of ca 500 km^2. Secondly, images of this type are a useful starting point for archaeological research in areas where other types of remote sensing data are not available or difficult to acquire or where access on the ground is difficult, as is the case in many remote or contested parts of the world. For the same reason, one aim of the project was to achieve as much as possible in terms of object detection based on the satellite images alone, without making use of contextual information, which in other cases might not be available. For reference, an orthoimage with 0.5 m resolution based on aerial images provided by SWISSTOPO was used as secondary data source.

The goal of the case study was to develop a workflow that would allow the quick and reliable detection of LSEs in optical remote sensing images of 0.5 m resolution. In order to be useful, the workflow needed to be robust to illumination changes and quick and enable a high detection rate combined with a manageable number of false detections. It should furthermore indicate the probability of the presence of a target object rather than yielding a binary yes/no classification, as the workflow was envisioned as an assisting step prior to archaeological fieldwork that would facilitate the ground truthing of its results.

The design of the workflow was determined by the nature of the target objects. While the LSEs show a wide variety of shapes, sizes, states of preservation and contexts (e.g. associated vegetation), they can all be described as roughly rectangular, though often incomplete objects. Furthermore, they are all located in open grassland. These features determined the detection approach.

Their location in open grassland meant that other portions of the study area—mainly forests, rocky/ice-covered areas and settlements—would have led to many false detections. Thus, in a first step, the images were segmented based on texture contrast such that open grasslands were distinguished from all other portions of the landscape. For this purpose, and based on mathematical morphology, two complementary operators were developed. The first, called morphological texture contrast (MTC), filters out high texture contrast regions. The second, called morphological feature contrast (MFC), highlights individual features in the remaining image portions, since those might be part of the target objects. The result is a binary image showing individual features in open areas. By filtering out irrelevant areas, computation time for all subsequent steps is reduced considerably.

The next, and crucial, step in the workflow was to determine which linear features in the binary images belonged to the target objects. Based on the geometric properties of LSEs, two constraints were defined: (1) a convexity constraint, requiring that linear features form a nearly convex hull, and (2) a rectangularity constraint, requiring that they meet at roughly right angles. First, candidate points in the images were determined which were surrounded by linear features in certain distances and configurations and were thus potential centre points of target objects. Second, from these candidates, the location and configuration of the linear features surrounding them were tested against the abovementioned constraints in a graph-based search. That way, most naturally occurring configurations such as random alignments of stones, streams or trails were rejected. Third, the remaining configurations were assessed based on how well they fulfilled the constraints, for which a rectangularity measure was introduced and assigned to the candidate points. Colour coding this quantitative measure results in a heat map in which red indicates a high probability of the presence of a target object, yellow a low probability and no colour a zero probability (Fig. 7.2). This heat map can serve as starting point for fieldwork, as it indicates the most likely locations where target objects can be detected.

Applying the above described workflow to large images results in a huge number of false detections. Therefore, the geometric properties of the known LSEs were used for filtering the results. Mapping their sizes against their rectangularity measures resulted in a clear distribution of the known LSEs towards one end of the overall distribution, such that a linear classifier could be defined that discarded the majority of the false detections. This classifier can be used in other contexts as well.

Testing the workflow on the original dataset showed that all known LSEs were reliably detected. In addition, a low number of hitherto unrecorded LSEs were detected, too. Applying the workflow to a similar dataset from the Bernese Alps also yielded promising results, which have yet to be validated in the field.

The Silvretta case study shows how computational tools can extract meaningful archaeological information from complex remote sensing data, thereby assisting archaeological fieldwork and enquiry. As in other disciplines, the combination of domain knowledge with methodological expertise from remote sensing and computer science is the key to tapping the full potential of remote sensing data.

7.5 A Look Ahead

For a long time, and in spite of their shared origins, aerial archaeology and earth observation remote sensing have followed own trajectories that overlapped only occasionally. In recent

Fig. 7.2 Geoeye 1 satellite image (panchromatic channel) of the Jam and Larein valleys above Galtür, Tyrol, Austria, with superimposed colour code indicating low (*yellow*) to high (*red*) probability of the presence of LSEs (image: I. Zingman, using copyrighted material of DigitalGlobe, Inc., All Rights Reserved). This heat map serves as starting point for ground-based archaeological survey

years, however, there is a clear and irreversible, highly promising trend towards convergence. Aerial archaeology has an enormous potential to adapt to new requirements (Verhoeven and Sevara 2016). At the same time, the continuously increasing use of data, tools and methods derived from earth observation remote sensing for archaeological purposes leads to exciting new opportunities and challenges. Airborne laser scanning, the first effective technique for large-scale archaeological prospection in woodlands, is just the most striking example (Crutchley and Crow 2009). At the same time, it is a good example of the data explosion (Bennett et al. 2014) or deluge (Bevan 2015) that archaeology now faces. Multidimensional, multi-resolution, multi-sensor remote sensing data are much more complex than traditional aerial photographs, and this requires new conceptual approaches to data processing, analysis and interpretation. While crowdsourcing is one way to address this problem (Casana 2014; Lin et al. 2014), computational approaches is another (Gattiglia 2015; Grosman 2016). Archaeology as a discipline has started to develop computational approaches in close collaboration with the earth and environmental sciences, engineering and computer science, as recent successful attempts towards automation in archaeological object detection show.

At the same time, these examples also reveal certain limitations. Handcrafted custom algorithms for object detection, while effective, have so far proven to be too specialised to be widely applied in cultural heritage management. They often target narrow object categories, sometimes require specific data, and are mostly not yet integrated into common working environments such as GIS. Clearly, more generic and user-friendly approaches are needed to make full use of computational power for the archaeological analysis of the rich content of remote sensing data. Currently, advanced machine learning techniques seem to offer the best solution for this problem. For example, deep learning based on convolutional neural networks has revolutionised computer vision in recent years, enabling considerable progress in such complex analytical problems as face recognition and image understanding. Whereas traditional methods of digital image analysis map and classify image contents, deep learning is capable of comprehensively analysing and describing them, e.g. in text (LeCun et al. 2015). This and related approaches thus seem to offer a great potential for a truly semantic analysis of remote sensing data for archaeological purposes. First archaeological case studies in this field (Zingman et al. 2016; Trier et al. 2017) show promising results.

References

Abrams MJ, Comer DC (2013) Multispectral and hyperspectral technology and archaeological applications. In: Comer DC, Harrower MJ (eds) Mapping archaeological landscapes from space. Springer, New York, pp 57–71. https://doi.org/10.1007/978-1-4614-6074-9_6

Agapiou A, Alexakis DD, Hadjimitsis DG (2014) Spectral sensitivity of ALOS, ASTER, IKONOS, LANDSAT and SPOT satellite imagery intended for the detection of archaeological crop marks. Int J Digital Earth 7:351–372. https://doi.org/10.1080/17538947.2012.674159

Agapiou A, Alexakis DD, Simon FX, Kalayci T, Papadopoulos N, Sarris A, Hadjimitsis D (2015) Extraction of archaeological information using high resolution FormoSAT-2 data. Int J Herit Digital Era 4:241–255. https://doi.org/10.1260/2047-4970.4.3-4.241

Agapiou A, Alexakis DD, Sarris A, Hadjimitsis DG (2016) Colour to greyscale pixels: re-seeing greyscale archived aerial photographs and declassified satellite CORONA images based on image fusion techniques. Archaeol Prospect 23:231–241. https://doi.org/10.1002/arp.1536

Beck A (2011) Archaeological applications of multi/hyper-spectral data – challenges and potential. In: Cowley DC (ed) Remote sensing for archaeological heritage management. Occasional Publication of the Aerial Archaeology Research Group No. 3. Europae Archaeologia Consilium, Brussels, pp 87–97

Bennett R, Cowley D, De Laet V (2014) The data explosion: tackling the taboo of automatic feature recognition in airborne survey data. Antiquity 88:896–905. https://doi.org/10.1017/S0003598X00050766

Bevan A (2015) The data deluge. Antiquity 89:1473–1484. https://doi.org/10.15184/aqy.2015.102.

Brophy K, Cowley D (eds) (2005) From the air: understanding aerial archaeology. Tempus, Stroud

Campana S (2017a) Remote sensing in archaeology. In: Gilbert AS (ed) Encyclopedia of geoarchaeology. Springer, Dordrecht, pp 703–725. https://doi.org/10.1007/978-1-4020-4409-0_122

Campana S (2017b) Drones in archaeology. State-of-the-art and future perspectives. Archaeol Prospect. doi: https://doi.org/10.1002/arp.1569

Casana J (2014) Regional-scale archaeological remote sensing in the age of Big Data. Adv Archaeol Pract 2:222–233. https://doi.org/10.7183/2326-3768.2.3.222

Casana J, Cothren J (2008) Stereo analysis, DEM extraction and orthorectification of CORONA satellite imagery: archaeological applications from the Near East. Antiquity 82:732–749. https://doi.org/10.1017/S0003598X00097349

Caspari G, Balz T, Gang L, Wang X, Liao M (2014) Application of Hough Forests for the detection of grave mounds in high-resolution satellite imagery. In: Proceedings of the 2014 IEEE geoscience and remote sensing symposium, pp 906–909. https://doi.org/10.1109/IGARSS.2014.6946572

Cerrillo-Cuenca E (2017) An approach to the automatic surveying of prehistoric barrows through LiDAR. Quat Int 435B:135–145. https://doi.org/10.1016/j.quaint.2015.12.099

Corsi C, Slapšak B, Vermeulen F (eds) (2013) Good practice in archaeological diagnostics: non-invasive survey of complex archaeological sites. Springer, Cham. https://doi.org/10.1007/978-3-319-01784-6

Cowley DC (2012) In with the new, out with the old? Auto-extraction for remote sensing archaeology. In: Bostater CR Jr, Mertikas SP, Neyt X, Nichol C, Cowley DC, Bruyant JP (eds) Remote sensing of the ocean, sea ice, coastal waters, and large water regions 853206. Proceedings of SPIE 8532. SPIE, Dresden. https://doi.org/10.1117/12.981758

Cowley DC (2016) What do the patterns mean? Archaeological distributions and bias in survey data. In: Forte M, Campana S (eds) Digital methods and remote sensing in archaeology. Springer, Cham, pp 147–170. https://doi.org/10.1007/978-3-319-40658-9_7

Cowley DC, Stichelbaut BB (2012) Historical aerial photographic archives for European archaeology. Eur J Archaeol 15:217–236. https://doi.org/10.1179/1461957112Y.0000000010

Cowley DC, Standring RA, Abicht MJ (eds) (2010) Landscapes through the lens: aerial photographs and historic environment. Occasional Publication of the Aerial Archaeology Research Group No. 2. Oxbow Books, Oxford

Crutchley S, Crow P (2009) The light fantastic: using airborne laser scanning in archaeological survey. English Heritage, Swindon

Day DA, Logsdon JM, Latell B (eds) (1998) Eye in the sky: the story of the CORONA spy satellites. Smithsonian Institution, Washington

De Laet V, Paulissen E, Waelkens M (2007) Methods for the extraction of archaeological features from very high-resolution Ikonos-2 remote sensing imagery, Hisar (southwest Turkey). J Archaeol Sci 34:830–841. https://doi.org/10.1016/j.jas.2006.09.013

De Laet V, Paulissen E, Meuleman K, Waelkens M (2009) Effects of image characteristics on the identification and extraction of archaeological features from Ikonos-2 and Quickbird-2 imagery: case study Sagalassos (southwest Turkey). Int J Remote Sens 30:5655–5668. https://doi.org/10.1080/01431160802705821

Dietre B, Walser C, Lambers K, Reitmaier T, Hajdas I, Haas JN (2014) Palaeoecological evidence for Mesolithic to Medieval climatic change and anthropogenic impact on the Alpine flora and vegetation of the Silvretta Massif (Switzerland/Austria). Quat Int 353:3–16. https://doi.org/10.1016/j.quaint.2014.05.001

Dietre B, Walser C, Kofler W, Kothieringer K, Hajdas I, Lambers K, Reitmaier T, Haas JN (2017) Neolithic to Bronze Age (4850–3450 cal. BP) fire management of the Alpine Lower Engadine landscape (Switzerland) to establish pastures and cereal fields. Holocene 27:181–196. https://doi.org/10.1177/0959683616658523

Doneus M (2013) Die hinterlassene Landschaft – Prospektion und Interpretation in der Landschaftsarchäologie. Mitteilungen der Prähistorischen Kommission 78. Austrian Academy of Sciences, Vienna

Doneus M, Verhoeven G, Atzberger C, Wess M, Rus M (2014) New ways to extract archaeological information from hyperspectral pixels. J Archaeol Sci 52:84–96. https://doi.org/10.1016/j.jas.2014.08.023

Doneus M, Wieser M, Verhoeven G, Karel W, Fera M, Pfeifer N (2016) Automated archiving of archaeological aerial mages. Remote Sens 8:209. https://doi.org/10.3390/rs8030209

Donoghue D, Beck A, Galiatsatos N, McManus K, Philips G (2006) The use of remote sensing data for visualising and interpreting archaeological landscapes. In: Baltsavias E, Gruen A, van Gool L, Pateraki M (eds) Recording, modeling and visualization of cultural heritage. Taylor & Francis, London, pp 317–326

Fowler MJF (2013) Declassified intelligence satellite photographs. In: Hanson WS, Oltean IA (eds) Archaeology from historical aerial and satellite archives. Springer, New York, pp 47–66. https://doi.org/10.1007/978-1-4614-4505-0_4

Gattiglia G (2015) Think big about data: archaeology and the Big Data challenge. Archäologische Informationen 38:113–124. https://doi.org/10.11588/ai.2015.1.26155

Giardino MJ (2011) A history of NASA remote sensing contributions to archaeology. J Archaeol Sci 38:2003–2009. https://doi.org/10.1016/j.jas.2010.09.017

Goossens R, De Wulf A, Bourgeois J, Gheyle W, Willems T (2006) Satellite imagery and archaeology: the example of CORONA in the Altai mountains. J Archaeol Sci 33:745–755. https://doi.org/10.1016/j.jas.2005.10.010

Grosman L (2016) Reaching the point of no return: the computational revolution in archaeology. Annu Rev Anthropol 45:129–145. https://doi.org/10.1146/annurev-anthro-102215-095946

Gutiérrez G, Searcy MT (eds) (2016) Special issue: Drones in archaeology. SAA Archaeol Rec 16

Hanson WS (2010) The future of aerial archaeology in Europe. Photo Interprétation. Eur J Appl Remote Sens 46:3–11

Hanson WS, Oltean IA (eds) (2013) Archaeology from historical aerial and satellite archives. Springer, New York. https://doi.org/10.1007/978-1-4614-4505-0

Horne PD (2009) A strategy for the national mapping programme. English Heritage, Swindon

Johnson JK (ed) (2006) Remote sensing in archaeology: an explicitly North American perspective. University of Alabama Press, Tuscaloosa

Kothieringer K, Walser C, Dietre B, Reitmaier T, Haas JN, Lambers K (2015) High impact: early pastoralism and environmental change during the Neolithic and Bronze Age in the Silvretta Alps (Switzerland/Austria) as evidenced by archaeological, palaeoecological and pedological proxies. Z Geomorphol 59(2):177–198. https://doi.org/10.1127/zfg_suppl/2015/S-59210

Lambers K, Eisenbeiss H, Sauerbier M, Kupferschmid D, Gaisecker T, Sotoodeh S, Hanusch T (2007) Combining photogrammetry and laser scanning for the recording and modelling of the Late Intermediate Period site of Pinchango Alto, Palpa, Peru. J Archaeol Sci 34:1702–1712. https://doi.org/10.1016/j.jas.2006.12.008

Lasaponara R, Masini N (eds) (2012) Satellite remote sensing: a new tool for archaeology. Springer, Dordrecht. https://doi.org/10.1007/978-90-481-8801-7

Leckebusch J (2005) Aerial archaeology: a full digital workflow for aerial photography. Archaeol Prospect 12:235–244. https://doi.org/10.1002/arp.260

LeCun Y, Bengio Y, Hinton G (2015) Deep learning. Nature 521:436–444. https://doi.org/10.1038/nature14539

Lemmens M (2011) Digital aerial cameras. GIM Int 25 (4):18–25

Lillesand TM, Kiefer RW, Chipman JW (2015) Remote sensing and image interpretation, 7th edn. Wiley, New York

Lin AYM, Huynh A, Lanckriet G, Barrington L (2014) Crowdsourcing the unknown: the satellite search for Genghis Khan. PLoS One 9(12):e114046. https://doi.org/10.1371/journal.pone.0114046

Menze BH, Ur JA (2012) Mapping patterns of long-term settlement in Northern Mesopotamia at a large scale. PNAS 109(14):E778–E787. https://doi.org/10.1073/pnas.1115472109

Mikhail EM, Bethel JS, McClone JC (2001) Introduction to modern photogrammetry. Wiley, New York

Musson C, Palmer R, Campana S (2013) Flights into the past: aerial photography, photo interpretation and mapping for archaeology. Occasional Publication No. 4 of the Aerial Archaeology Research Group. http://archiv.ub.uni-heidelberg.de/propylaeumdok/volltexte/2013/2009. Accessed 3 Feb 2017.

Palmer R (2005) If they used their own photographs they wouldn't take them like that. In: Brophy K, Cowley DC (eds) From the air: understanding aerial archaeology. Tempus, Stroud, pp 94–116

Parcak SH (2009) Satellite remote sensing for archaeology. Routledge, London

Remondino F (2011) Heritage recording and 3D modeling with photogrammetry and 3D scanning. Remote Sens 3:1104–1138. https://doi.org/10.3390/rs3061104

Remondino F, Gehrke M (2015) Oblique aerial imagery – a review. In: Fritsch D (ed) Photogrammetric week '15. Wichmann, Berlin, pp 75–84

Richards JA, Jia X (2006) Remote sensing digital image analysis: an introduction, 4th edn. Springer, Berlin

Schneider A, Takla M, Nicolay A, Raab A, Raab T (2015) A template-matching approach combining morphometric variables for automated mapping of charcoal kiln sites. Archaeol Prospect 22:45–62. https://doi.org/10.1002/arp.1497

Schuetter J, Goel P, McCorriston J, Park J, Senn M, Harrower M (2013) Autodetection of ancient Arabian tombs in high-resolution satellite imagery. Int J Remote Sens 34:6611–6635. https://doi.org/10.1080/01431161.2013.802054

Sevara C, Pregesbauer M, Doneus M, Verhoeven G, Trinks I (2016) Pixel versus object – a comparison of strategies for the semi-automated mapping of archaeological features using airborne laser scanning data. J Archaeol Sci Rep 5:485–498. https://doi.org/10.1016/j.jasrep.2015.12.023

Sever TL (1990) Remote sensing applications in archeological research: tracing prehistoric human impact upon the environment. PhD dissertation, University of Colorado. University Microfilms, Ann Arbor

Stewart C, Rast M, Sarti F, Arino O (2015) ESA activities in earth observation for cultural heritage applications. Int J Herit Digital Era 4:325–338. https://doi.org/10.1260/2047-4970.4.3-4.325

Traviglia A (2007) MIVIS hyperspectral sensors for the detection and GIS supported interpretation of subsoil archaeological sites. In: Clark JT, Hagemeister EM (eds) Digital discovery: exploring new frontiers in human heritage. CAA 2006 Fargo. Archaeolingua, Budapest, pp 287–299

Traviglia A, Cowley DC, Lambers K (2016) Finding common ground: human and computer vision in archaeological prospection. AARGnews 53:11–24. http://hdl.handle.net/1887/43751. Accessed 3 Feb 2017

Trier ØD, Pilø LH (2012) Automatic detection of pit structures in airborne laser scanning data. Archaeol Prospect 19:103–121. https://doi.org/10.1002/arp.1421

Trier ØD, Larsen SØ, Solberg R (2009) Automatic detection of circular structures in high-resolution satellite images of agricultural land. Archaeol Prospect 16:1–15. https://doi.org/10.1002/arp.339

Trier ØD, Zortea M, Tonning C (2015) Automatic detection of mound structures in airborne laser scanning data. J Archaeol Sci Rep 2:69–79. https://doi.org/10.1016/j.jasrep.2015.01.005

Trier ØD, Salberg AB, Pilø LH (2017) Semi-automatic mapping of charcoal kilns from airborne laser scanning data using deep learning. In: Uleberg E, Matsumoto M (eds) Proceedings of the 44th annual conference on computer applications and quantitative methods in archaeology (CAA 2016)

Trümpler C (2005) Aerial photography in archaeology and its pioneers. In: Trümpler C (ed) The past from above: photographs by Georg Gerster. Frances Lincoln, London, pp 9–23

Verhoeven G (2008) Imaging the invisible using modified digital still cameras for straightforward and low-cost archaeological near-infrared photography. J Archaeol Sci 35:3087–3100. https://doi.org/10.1016/j.jas.2008.06.012

Verhoeven G (2009) Providing an archaeological bird's eye view – an overall picture of ground-based means to execute low-altitude aerial photography (LAAP) in archaeology. Archaeol Prospect 16:233–249. https://doi.org/10.1002/arp.354

Verhoeven G (2012) Near-infrared crop mark archaeology: from its historical use to current digital implementations. J Archaeol Method Theory 19:132–160. https://doi.org/10.1007/s10816-011-9104-5

Verhoeven G, Sevara C (2016) Trying to break new ground in aerial archaeology. Remote Sens 8:918. https://doi.org/10.3390/rs8110918

Verhoeven G, Sevara C, Karel W, Ressl C, Doneus M, Briese C (2013) Undistorting the past: new techniques for orthorectification of archaeological aerial frame imagery. In: Corsi C, Slapšak B, Vermeulen F (eds) Good practice in archaeological diagnostics: non-invasive survey of complex archaeological sites. Springer, Cham, pp 31–67. https://doi.org/10.1007/978-3-319-01784-6_3

Vletter WF (2014) (Semi-)automatic extraction from Airborne Laser Scan data of roads and paths in forested areas. In: Hadjimitsis DG, Themistocleous K, Michaelides S, Papadavid G (eds) Second international conference on remote sensing and geoinformation of the environment (RSCy2014), Proceedings of SPIE 9229 SPIE, Paphos 92291D. https://doi.org/10.1117/12.2069709

Wiseman J, El-Baz F (eds) (2007) Remote sensing in archaeology. Springer, New York. https://doi.org/10.1007/0-387-44455-6

Zingman I (2016) Semi-automated detection of fragmented rectangular structures in high resolution remote sensing images with application in archaeology. PhD dissertation, University of Konstanz. http://nbn-resolving.de/urn:nbn:de:bsz:352-0-386546 Accessed 3 Feb 2017

Zingman I, Saupe D, Lambers K (2014) A morphological approach for distinguishing texture and individual features in images. Pattern Recogn Lett 47:129–138. https://doi.org/10.1016/j.patrec.2014.03.019

Zingman I, Saupe D, Penatti OAB, Lambers K (2016) Detection of fragmented rectangular enclosures in very high resolution remote sensing images. IEEE Trans Geosci Remote Sens 54:4580–4593. https://doi.org/10.1109/TGRS.2016.2545919

Paleoenvironmental Research in the Semiarid Lake Manyara Area, Northern Tanzania: A Synopsis

8

Felix Bachofer, Geraldine Quénéhervé, Christine Hertler, Liane Giemsch, Volker Hochschild, and Michael Maerker

Abstract

The Lake Manyara area is the focus of several paleo-archeological investigations. The Manyara basin is located approximately 70 km east of Olduvai Gorge, where important paleoanthropological artifacts are traced back to *Homo habilis*. In the Manyara basin itself, two hominin-bearing sites (0.78–0.633 Ma) and plenty of vertebrate bones and teeth as well as stone artifacts from different periods were discovered, especially close to the Makuyuni River. Different methodological approaches with a main emphasis on remote sensing were utilized to contribute to the understanding of the paleo-landscape development. In order to investigate the morphotectonic evolution of the study area, lineaments were detected from Synthetic Aperture Radar satellite scenes. The complex lacustrine development of the Lake Manyara and its paleo-stages was investigated by delineating the extent of paleo-lake sediments (older than 0.633 Ma) with multispectral ASTER data. In addition, lake terraces and shorelines on different levels (up to 80 m above today's lake level) and an outlet to the neighboring Engaruka basin were detected by analyzing the backscattered intensity of TerraSAR-X data. The distribution of topsoils, identified from multisensory remote sensing datasets, indicates soil formation as well as erosional and depositional processes. The fossils and artifacts were then

F. Bachofer (✉) • G. Quénéhervé • V. Hochschild
Institute of Geography, University of Tuebingen, Ruemelinstr. 19-23, 72070 Tuebingen, Germany
e-mail: felix.bachofer@uni-tuebingen.de; geraldine.queneherve@uni-tuebingen.de; volker.hochschild@uni-tuebingen.de

C. Hertler
Heidelberg Academy of Sciences and Humanities, Ruemelinstr. 19-23, 72070 Tuebingen, Germany

Senckenberg Forschungsinstitut, Senckenberganlage 25, 60325 Frankfurt am Main, Germany
e-mail: christine.hertler@senckenberg.de

L. Giemsch
Archäologisches Museum Frankfurt, Karmelitergasse 1, 60311 Frankfurt am Main, Germany
e-mail: liane.giemsch@stadt-frankfurt.de

M. Maerker
Heidelberg Academy of Sciences and Humanities, Ruemelinstr. 19-23, 72070 Tuebingen, Germany
e-mail: michael.maerker@uni-tuebingen.de

© The Author(s) 2018
C. Siart et al. (eds.), *Digital Geoarchaeology*, Natural Science in Archaeology,
DOI 10.1007/978-3-319-25316-9_8

characterized, and their distribution probabilities were determined using a statistical model. The proposed methods contribute to a better understanding of the paleoenvironmental interrelations within the study area.

Keywords
Paleolithic • Vertebrate fossils and fish bones • Landscape reconstruction • Site prediction • Stochastic modeling • SAR • Remote sensing • EARS

8.1 Introduction and Study Area

The East African Rift System (EARS) (Fig. 8.1) is perceived as a migration corridor linking eastern and southern Africa and providing potential ways "out of Africa" for early humans. In this context, we focus on the Northern Tanzanian section of the eastern branch of the EARS, the so-called Gregory Rift. As already stated by Leakey (1979) and numerous other authors, Tanzania is well known for prehistoric specimens of early hominids found, for example, in Olduvai Gorge, at Lake Masek or at Laetoli (Fig. 8.2). For the evolution of hominids and especially hominins, the EARS seems to be a crucial region. The geographic center of this study lies in the vicinity of Olduvai, yet within the half-graben of the Gregory Rift around the eastern margins of Lake Manyara and along the Makuyuni River. Lake Manyara (954 m a.s.l.) is located in an endorheic basin (Fig. 8.[illegible]) and is a shallow soda lake that periodically dries out completely (Deus et al. 2013). The primary structural unit of the Manyara basin is an asymmetrically shaped rift. To the west extends the 250–900 m high escarpment. A west-dipping monocline and the volcano Essimingor prevail in the east. The water supply originates from springs at the base of the rift escarpment and from several seasonal drainages (Schwartz et al. 2012).

Paleo-lake evidences, like paleo-shorelines, sediments, and fossils, were found by Jaeger (1913) and Reck (1921) to the east of Lake Manyara already in the beginning of the twentieth century. Pleistocene sequences in the valley of the Makuyuni River were discovered early on by Louis and Mary Leakey and were later examined by Kent in 1935 (Kent 1941, 1942). Keller and colleagues collected Pleistocene faunal material and Middle Stone Age (MSA) lithics and published several stratigraphic sections (Keller et al. 1975). Recent studies investigated the geology, paleontology, and archeology of the Lake Manyara Beds, e.g., Kaiser et al. (1995, 2010), Kaiser (2000), Ring et al. (2005a), Frost et al. (2012), Schwartz et al. (2012), Mana et al. (2012), and Giemsch (2015).

Previous studies have shown that two fossil-bearing layers of different ages exist in the area, namely, the Lower and Upper Manyara Beds (UMB and LMB, respectively). Correlations with the sequence in Olduvai seem to indicate Lower and Middle Pleistocene ages for the Manyara Beds (Ring et al. 2005a; Frost et al. 2012; Schlüter 1987). Kaiser et al. (2010) show the corresponding stratigraphic succession. Upper terrestrial beds and lower lacustrine beds are separated by a clear boundary characterized by a color change from grayish (lacustrine sediments) to reddish (terrestrial deposits). The entire stratigraphy is interspersed with presumably reworked tuff deposits. Moreover, UMBs and LMBs bear fossil layers.

The large number of (fossil) bearing sites, including specimens of fossil vertebrates and artifacts, detected during various field campaigns of the authors from 2007 to 2014, was the motivation for a more extensive analysis of the spatial distribution of these sites, in relation to environmental characteristics and processes. Different methodological approaches, with a major emphasis on remote sensing, were utilized to contribute to the understanding of the paleo-landscape development and their relations in order to find assemblages.

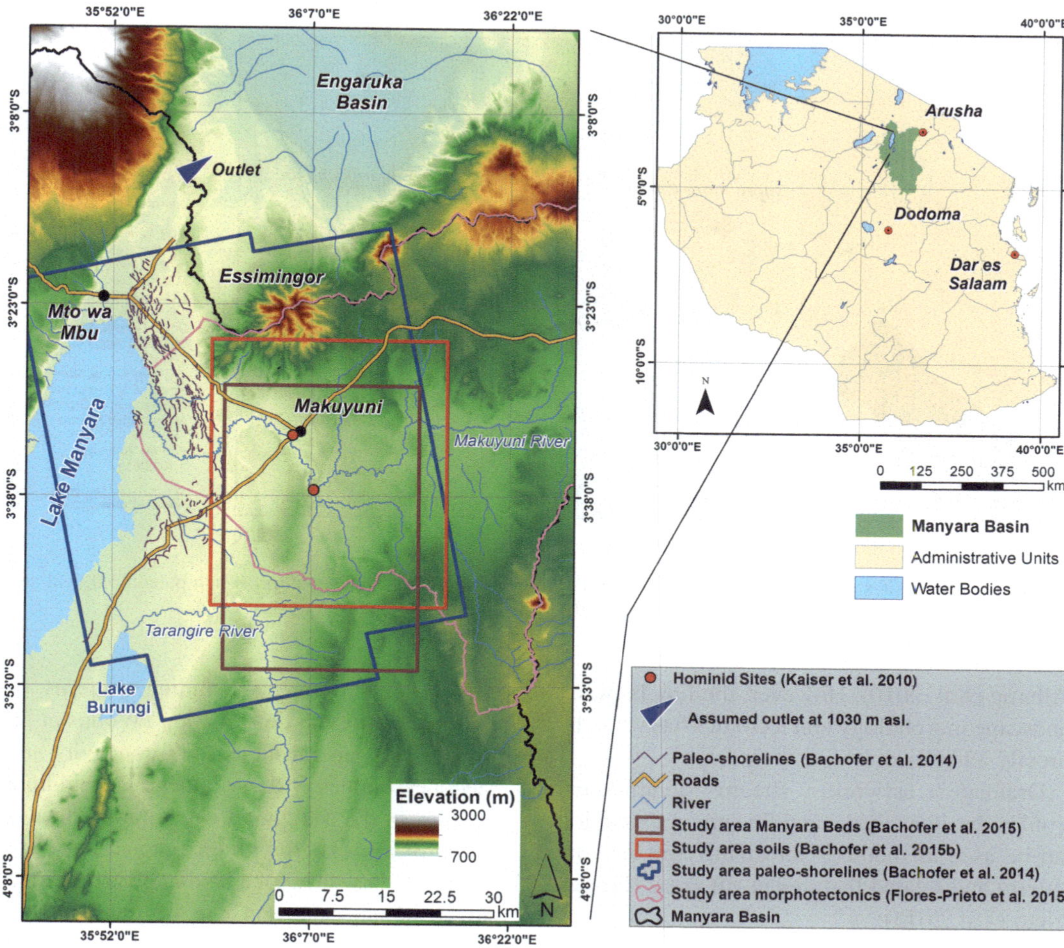

Fig. 8.1 The locations and extent of the study areas in the Lake Manyara basin in Northern Tanzania

We investigate the morphotectonic evolution of the study area, the complex lacustrine dynamics of the Lake Manyara, and its paleo-stages as well as the related lake terraces and shorelines at different levels (up to 80 m above today's lake level). Additionally, we study the distribution of topsoils and outcropping substrates to decipher soil formation processes as well as erosional and depositional processes. Finally, the fossils and artifacts were characterized, and the probability distributions were determined using a stochastic approach. The proposed methods contribute to a better understanding of the paleoenvironmental interrelations within the study area.

8.2 Morphotectonics and Their Interpretation

In the EARS, landforms are caused by effects of the active continental extension zone within the graben systems as well as by volcanic, erosional, and depositional processes. Tectonic processes formed the recent drainage systems and landforms. The tectonics of this region were previously studied in terms of (1) kinematic and structural geology (Ring et al. 2005b), (2) faulting, (3) Neogene tectonics and volcanism (Dawson 1992, 2008; Le Gall et al. 2008; Albaric et al. 2009), (4) and structural geology

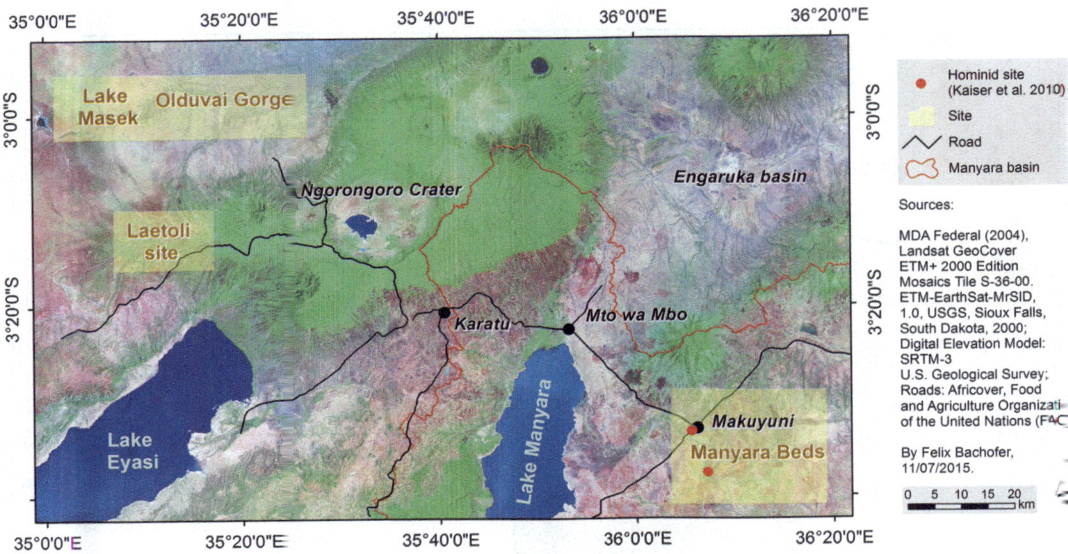

Fig. 8.2 Outlined sites of the Olduvai, Masek, Laetoli, and Manyara Beds, which comprise rich paleontological and archeological findings [with permission of Bachofer (2015)]

(Albaric et al. 2010). However, the link between landscape morphology and tectonics has yet to be directly addressed.

Drainage network, stream longitudinal profiles, basin analysis, and lineament extraction can be used as methods for identifying tectonic activity and related features in rift areas (Flores-Prieto et al. 2015):

1. Relief and landscape patterns of the Lake Manyara basin reflect the evolution of a complex part of the rift systems that underwent combined effects of tectonic factors inherent to its geological settings and Quaternary geomorphological processes. In the Makuyuni catchment, lineaments were extracted from a Synthetic Aperture Radar (SAR) image (ENVISAT ASAR, 2011-08-02) (Fig. 8.3). They show an N–S orientation and reveal tectonic activity characterized by deviations and knickpoints of streambeds and the occurrence of erosion processes.
2. Concavity and steepness values of drainage network segments depend on basin morphology, underlying rock strengths, and hydraulic geometry (Snyder et al. 2000). These values are especially used for evaluating river system responses to different landscape forming and modifying processes, including patterns of tectonic uplift and deformation (Wobus et al. 2006). The general increase of concavity and steepness values within drainage longitudinal profiles in the Makuyuni catchment indicates an uplift associated with tectonic activity.
3. A distinct base level of a landscape is related to an erosional stage and can be considered a product of erosional tectonic events (Golts and Rosenthal 1993). The development and migration of knickpoints due to changes in the base level can be related to lineament structures.
4. The morphology of the stream longitudinal profiles (Shahzad and Gloaguen 2011), basin tilting (Hare and Gardner 1985) and basin hypsometry (Pérez-Peña et al. 2009) suggest the importance of tectonic activity in governing Quaternary geomorphological processes, such as river incision and soil erosion, and hence, the landscape evolution of this region.

This analysis has allowed the differentiation of the Makuyuni catchment into two distinctive areas: (a) tectonic activity influences the hydrological controlled northern section, and (b) a more stable southern section, where the Precambrian lithology and less intense tectonics prove dominant.

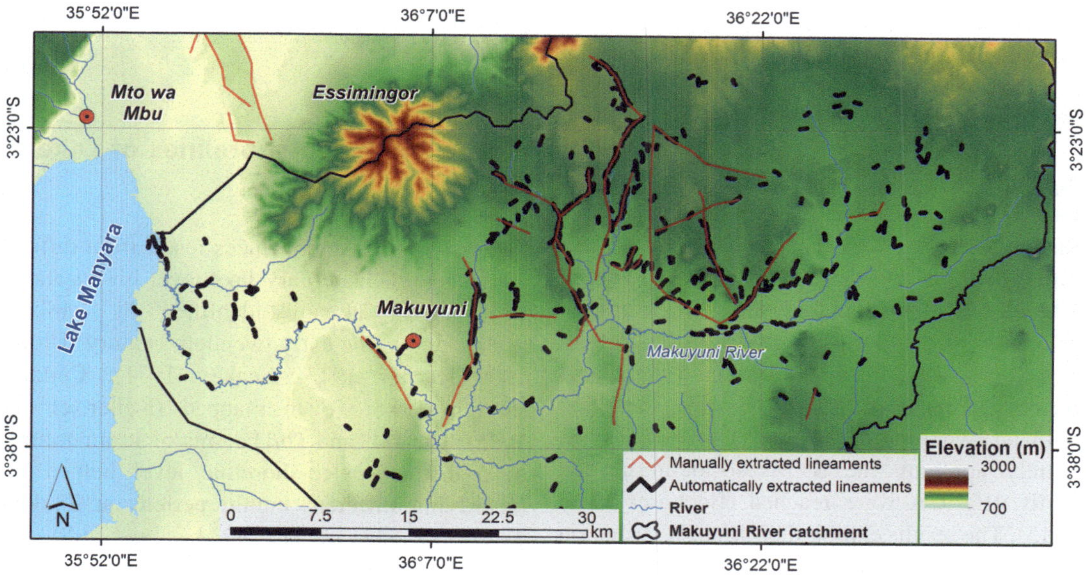

Fig. 8.3 Automatically and manually extracted lineaments

8.3 Delineation of the Manyara Beds

The stratigraphy of the Manyara Beds includes two fossil-bearing beds, namely, the LMBs and the UMBs (Schwartz et al. 2012). The UMB is composed of terrestrial deposits of up to 13 m mudstones, siltstones, conglomerates, and breccias. The deposition took place between 0.633 Ma and 0.44 (0.27) Ma (Ring et al. 2005a; Schwartz et al. 2012). A distinct layer of tephra is situated at the transition of the UMB to the LMB, which was $^{40}Ar/^{39}Ar$ dated to 0.633 Ma (Ring et al. 2005a; Schwartz et al. 2012). Schwartz et al. (2012) conclude, from assumed sedimentation rates, the beginning of the deposition of the LMBs between 1.3 Ma and 0.98 Ma. The Manyara Beds are underlain Precambrian intermediate gneisses in the south of the study area and by a sequence of nephelinitic conglomerates, nephelinitic lavas, and rocks in the vicinity of Essimingor, which have a minimum age of 1.4 Ma (Bagdasaryan et al. 1973). The LMB represents lacustrine deposits, grayish in color, and consist mainly of tuff, marls, siltstones, mudstones, and diatomite. The LMBs indicate the maximum extent and highest level of Lake Manyara.

An ASTER multispectral satellite image (2006-08-23), and topographic information derived of a SRTM-X digital elevation model (DEM), was used as input datasets for the delineation of the Manyara Beds. The ASTER system is composed of a sensor for visible and near-infrared wavelengths (VNIR: 0.52–0.86 μm; 15 m ground resolution), a shortwave sensor (SWIR: 1.6–2.43 μm; 30 m ground resolution), and a thermal infrared sensor (TIR: 8.125–11.65 μm; 90 m ground resolution) (Fujisada 1995). For the bands of the VNIR and SWIR spectral region, 35 spectral indices were calculated (Bachofer et al. 2015). Spectral indices are ratios of distinct spectral bands emphasizing the presence or absence of vegetation and different mineral compositions. Twenty-five additional topographic indices were calculated from a SRTM-X DEM (Bachofer et al. 2015).

For the delineation of the specific stratigraphic layers of the Manyara Beds, we applied a support vector machine (SVM) classifier based on the library of support vector machines (LIBSVM) (Chang and Lin 2011). SVM is a supervised classification approach, which maximizes the margin for the input features (spectral and topographic indices), describing the intended target classes (UMB and LMB). For the optimization of margins between several input features, it is necessary to transfer the data into an n-dimensional feature space. Kernel functions are used to separate the independent input features by the target classes (Vapnik 1995, 1998). Four hundred ninety-eight ground reference points, which were collected in field campaigns, were used to train and validate the datasets. The UMB and LMB were identified with an overall accuracy (OA) of 80%, when merely applying the SVMs to delineate the LMB 91% OA were reached (Bachofer et al. 2015). The results exhibit the surface distribution of the LMB and UMB (Fig. 8.4). In the eastern part of the study area, lacustrine sediments were identified, which are similar to the LMB and bear also vertebrate fossils. Based on indications of previous research, due to the topographic position of these sediments, we propose the existence of a cascaded lake system parallel to an extended paleo-lake Manyara following the main faults toward the east.

8.4 The Paleo-shorelines of Lake Manyara

Paleo-shorelines and terraces result from different stages of quaternary lake-level high stands, which were already identified by various researchers in the early twentieth century (Uhlig 1909; Jaeger 1913; Leakey 1931). Certain shorelines were even mapped (Keller et al. 1975). The 14C and Th/U dating of stromatolites of a distinct paleo-shoreline level helped in identifying different humid periods at 22 ka, 27–23 ka, 35–32 ka, 90 ka, and 140 ka (Hillaire-Marcel et al. 1986; Casanova and Hillaire-Marcel 1992; Casanova 1986). The ages were supplemented by diatom analysis of two drilling cores in the Lake Manyara basin, which supported humid periods between 27.5–26 ka and 12.7–8 ka (Holdship 1976; Barker 1990; Goetz 1990) (Fig. 8.5). An

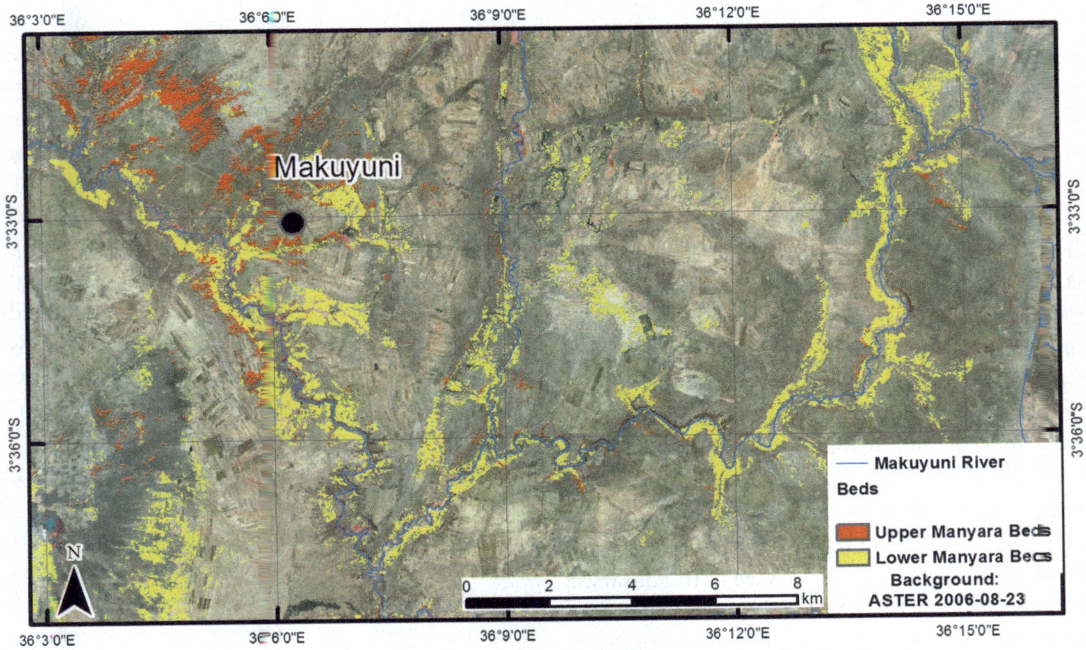

Fig. 8.4 Section with the classification result of UMB and LMB (based on ASTER and SRTM-X data)

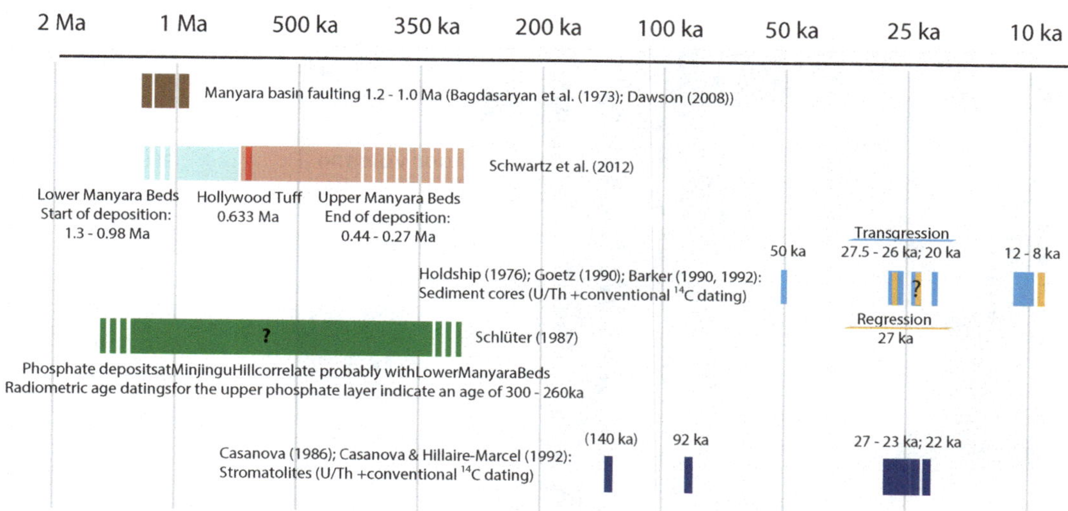

Fig. 8.5 Evidence for high paleo-lake levels in the Lake Manyara basin and basin faulting [with permission of Bachofer (2015)]

in-depth spatial consideration of the shorelines had thus far been missing.

Paleo-shorelines appear as small-scale landforms, which disappear in optical remote sensing images because of the uniform surface cover compared to their surroundings. A spectral analysis of the landforms proved to be difficult and led to unsatisfying results. Therefore, it was made use of the capabilities of SAR images. Most SAR systems are active remote sensing systems, which operate in the microwave electromagnetic spectrum. SAR systems transmit energy pulses and record the backscattered energy of objects (Lillesand et al. 2004). The backscattered signal is influenced mainly by the physical properties of a surface. In the case of paleo-shorelines and terraces, which occur as elongated steps and ridges, their geometry and the roughness of their coarse covering rocks led to an intense backscattered signal. The images from 2011 to 2013 were acquired by the TerraSAR-X satellite system (Bachofer et al. 2014). The scenes were radiometrically calibrated, and the paleo-shorelines were delineated using a Canny edge operator (Canny 1986). For the edge analysis, the elevation was extracted of a SRTM-X digital elevation model. Several prominent paleo-lake levels were identified and validated with reference data from field surveys (Fig. 8.6) (Bachofer et al. 2014). The maximum lake level identified resembles an identical elevation to the lowest possible outlet of the endorheic Manyara basin (Fig. 8.6A). The position of the outlet indicates an overspill into the neighboring Engaruka and Natron-Magadi basins (Fig. 8.7).

8.5 Lithosphere and Surface Soil Mapping

The distribution of soils and rocks yields valuable information for the interpretation of landscape evolution. To acquire information regarding the surface characteristics, especially in areas with scarce vegetation cover, remote sensing techniques prove highly suitable.

During fieldwork and through laboratory analysis of surface soil samples, nine soil and lithological target classes were identified. As input datasets for the classification served (1) a high resolution WorldView-2 scene, (2) the backscattered intensity information of TerraSAR-X and ENVISAT ASAR SAR images, (3) medium resolution ASTER spectral bands and corresponding spectral indices, as well as

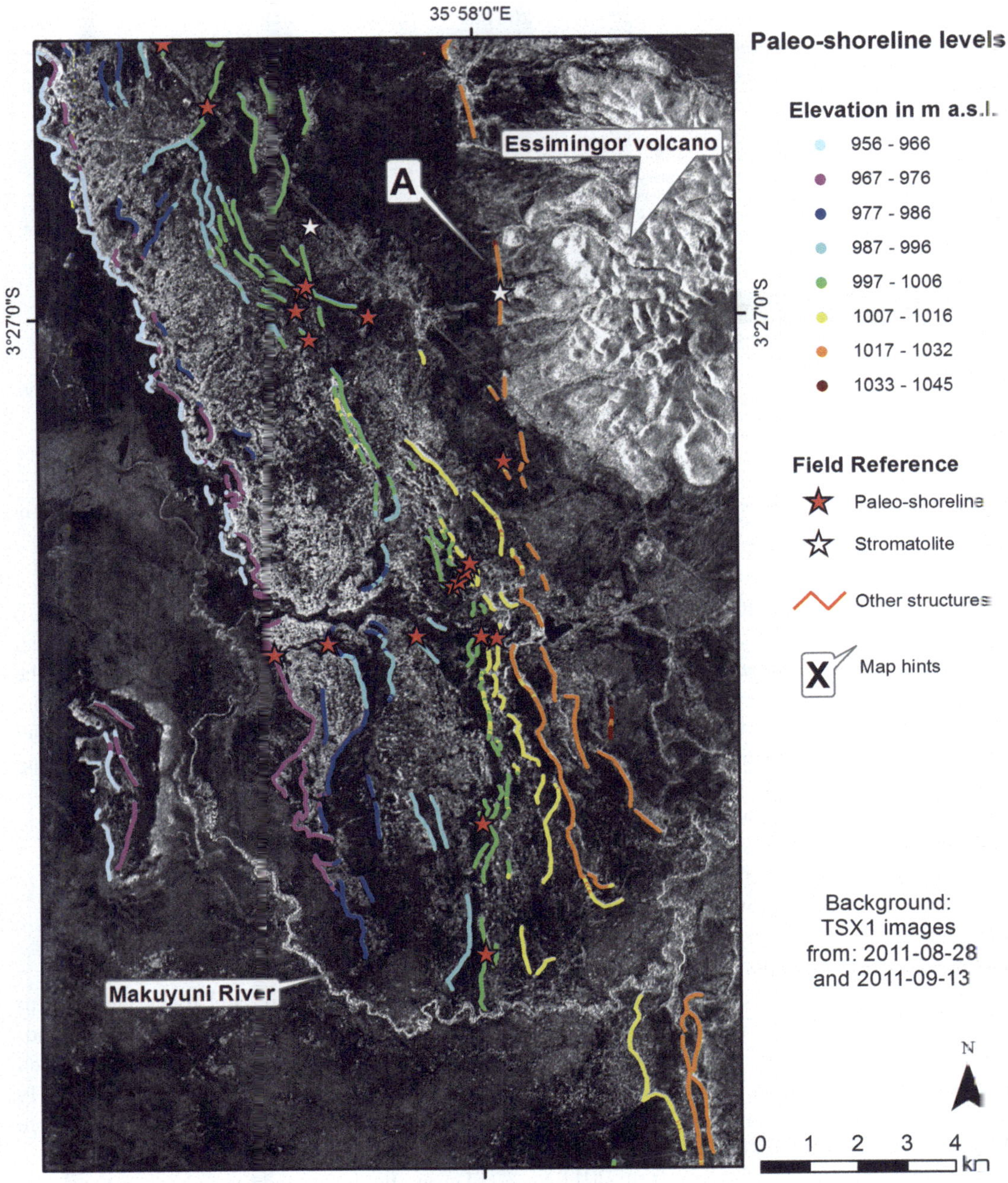

Fig. 8.6 Identified levels of paleo-shorelines after (Bachofer et al. 2014). The *colors* indicate different elevations of the paleo-shorelines. "*A*" shows the highest shoreline with more than 80 m above the present lake level. Stromatolites were found on this level

(4) SRTM-X-derived topographic indices. Because of the size of the input dataset, an image object segmentation was conducted to derive relative homogenous objects with the average value information of the underlying raster cells (Baatz and Schäpe 2000). Vegetated areas were excluded from further analysis by applying a normalized difference vegetation index (NDVI) threshold

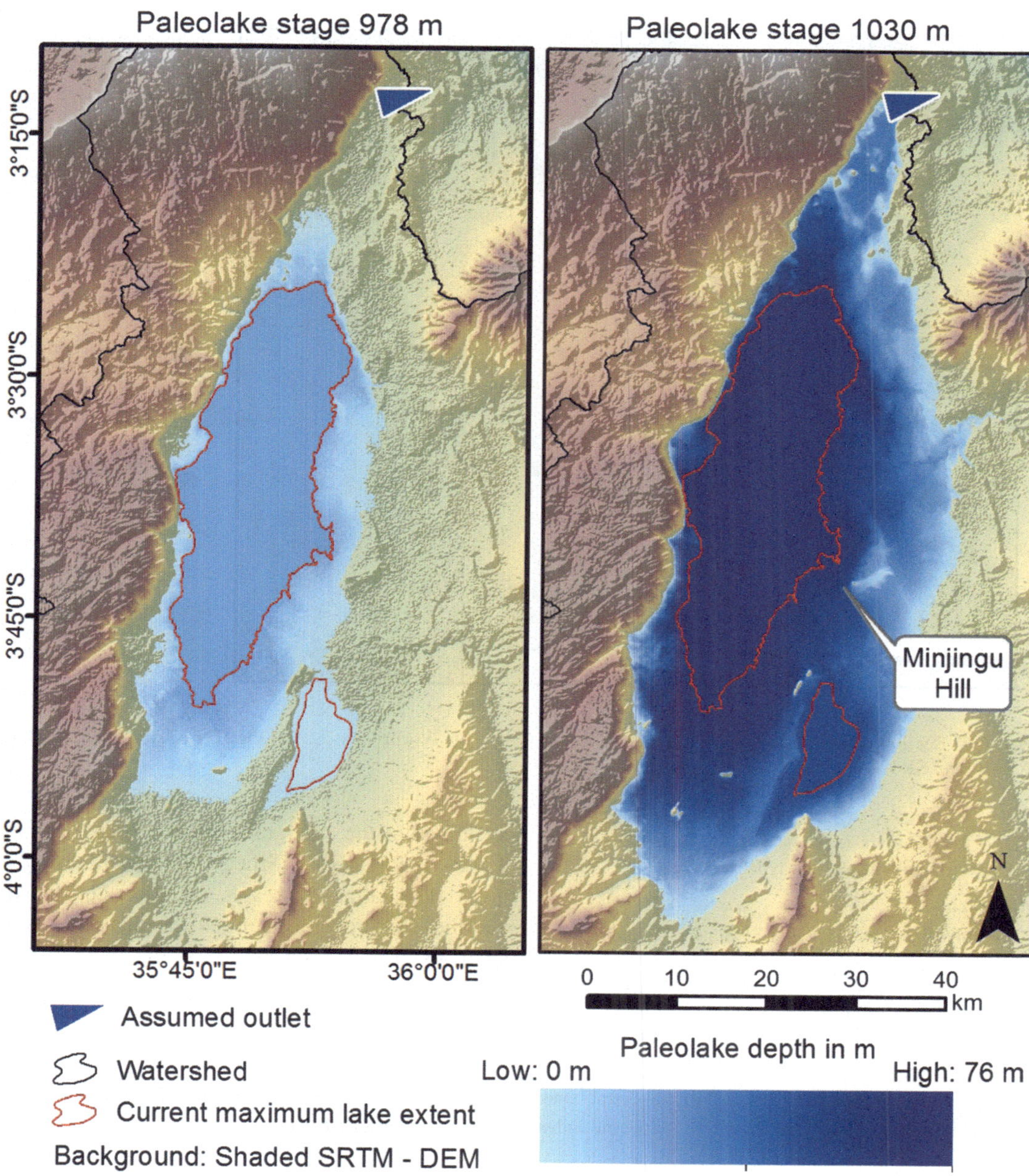

Fig. 8.7 The paleo-lake levels for the Manyara basin at 978 m and 1030 m

(Rouse et al. 1974). A nonlinear SVM approach was applied to classify the input dataset. Based on the spectral, physical, and topographical properties of the surface material, the nine target classes were identified with an overall accuracy of 71.9% (Fig. 8.8). By merging similar target classes such as "carbonate-rich sediments" and "calcaric topsoil," the accuracy could be increased significantly.

By means of recursive feature elimination (RFE), the features with the highest impact on the classification are identified (Guyon and Elisseeff 2003). The six most important input features are presented in Table 8.1.

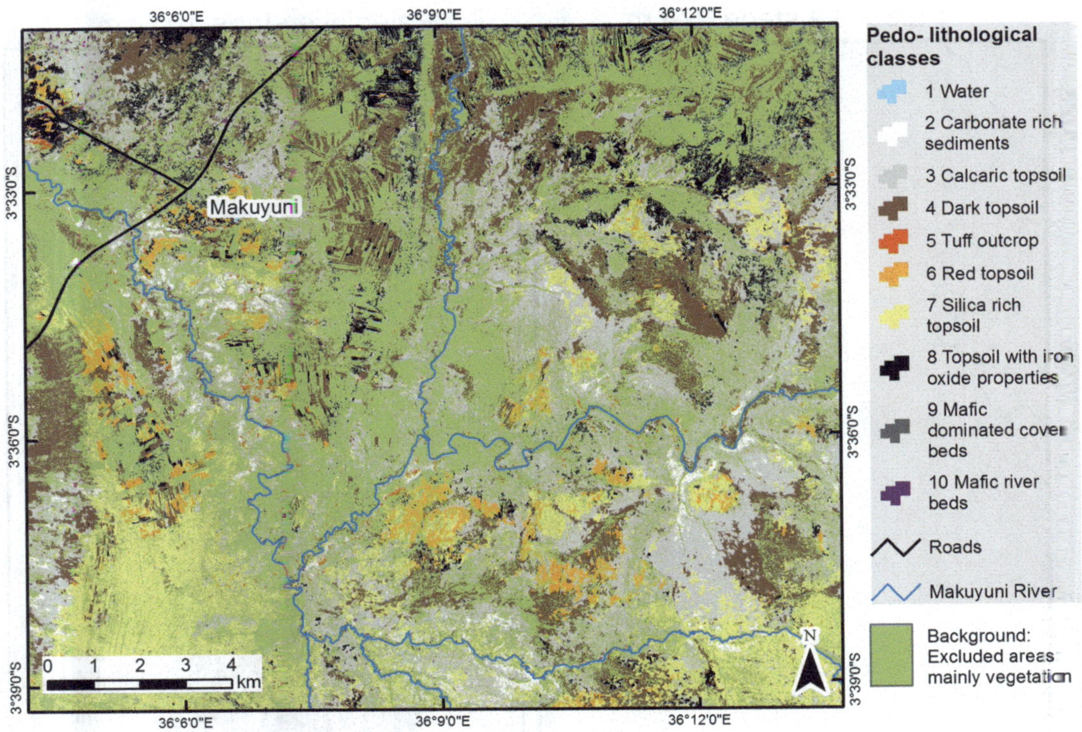

Fig. 8.8 Section of the pedo-lithological classification (based on WorldView-2, ASTER, TerraSAR-X, ENVISAT ASAR, and SRTM-X data)

Table 8.1 The six features with the highest impact on the classification identified by RFE

Input feature	Type
Geomorphons (Jasiewicz and Stepinski 2013)	Topographic index
Multiresolution index of ridge top flatness (Gallant and Dowling 2003)	Topographic index
WorldView-2 spectral band 3 (green)	Spectral band
Ferric iron index (Mitchell et al. 2013)	Spectral index
WorldView-2 spectral band 1 (coastal)	Spectral band
Calcite index (Pour and Hashim 2011)	Spectral index

The results show that topsoils with iron oxide properties are predominant when associated with mafic outcrops of the slopes of the Essimingor volcano. "Carbonate-rich sediments" represent chiefly the LMBs, while "calcaric topsoil" indicates soil development processes upon the LMBs or on soils influenced by secondary carbonates. The thin outcrops of tuffs were identified at the stratigraphic border between the LMBs and UMBs. "Silica-rich topsoil" results from denudation processes on soils developed on the Proterozoic intermediate quartzite and gneisses of the Masai Plateau in the south of the study area.

8.6 Archeological Settings and Site Prediction

The large number of sites with specimens of fossil vertebrates and artifacts detected during different field campaigns was the motivation for a more thorough analysis of the spatial distribution of these sites in relation to present-day environmental characteristics and processes. Therefore, we developed an integrative spatial modeling concept using GIS, remote sensing, and sophisticated statistical methodologies based on statistical mechanics, e.g., MaxEnt (Jaynes 1957; Phillips et al. 2006). This method is able to handle presence-only datasets such as the locations proper.

The predictor variables consist of 50 continuous raster datasets applied in the modeling. The topographic indices characterize erosion transport and deposition processes as well as climatic and geologic variations in the landscape. They therefore not only reflect the immediate vicinity of a specific spot but additionally refer to a wider territorial context. Furthermore, we exploited spectral satellite data from the ASTER platform from which we calculated derivatives (spectral indices and band ratios) as predictor variables. Though most indices are designed for specific materials, they may also be used for a relative distinction between various minerals, depending on the abundant materials in the study area.

The target (also called response or dependent) variables consist in sites with paleontological evidence such as fossils or artifacts. In total, 102 sites were identified with differential GPS and transformed into point vector objects. Subsequently, we distinguished between archeological findings (stone artifacts only) from Early Stone Age (ESA) ($n_A = 45$) and from Middle as well as Late Stone Age (MSA/LSA) ($n_B = 14$) (Giemsch 2015) and for paleontological locations ($n_C = 43$; fossils only).

The MaxEnt model predicts potential areas in which further fossil and/or artifact sites (paleontological sites) may be located. Figure 8.9 shows the areas with high potential in red to yellow and areas with low potential in dark blue. The research method considers not only site-specific characteristics but implicitly also the related pedogenetic and morphogenetic processes. The areas with high potential are mainly related to the paleo-drainage network terraces and paleo-lake deposits.

In order to evaluate the model's predictive performance beyond classification matrices, we calculated the receiver operating characteristics (ROC) curve integral, also known as area under curve (AUC). Model performance is given by AUC values of 0.95 for artifact sites and 0.90 for fossil sites.

The relative variable importance charts illustrate the overall importance of the salinity index (Al-Khaier 2003). The index relies on the ASTER bands four and five. This parameter may describe terrain elements with high salinity related to the former lake shores where salt concentration was high due to high salt contents of the lake itself and high evaporation rates of arid to semiarid paleoclimatic phases. The same bands can also be used to delineate soils with high iron content (Bierwirth 2002; Cudahy 2012). Other important ASTER-derived parameters are the Burn Ratio (Hudak et al. 2004) as well as the Laterite index (Bierwirth 2002). The Burn Ratio is designed to highlight burned areas and however in our case correlates well with dark vertisols with sparse grass vegetation, which cover the lower river terraces and the LMB/UMB transition. The Laterite index shows high iron contents present in the terrestrial tephra deposits. The DEM gives hints about the absolute elevation, in this case correlating with the former lake levels. The aspect, which is derived from the DEM, indicates the geological structure with main fault systems oriented from northwest to southeast.

Generally, the analysis shows that the paleontological sites are correlated mainly with the LMB/UMB transition and the paleo-river network. The results indicate the dependence of early hominids on water resources and food resources such as game, which also concentrates in these ecological units. However, the major difference between the paleontological sites and the ESA/MSA/LSA sites is related to general higher elevations where fossils occur. Moreover, the fossils seem to be related to stable paleo-landscape features indicated by the quartz-rich rocks index (Rowan et al. 2005) and the Laterite index. In fact, the sites are mainly associated with cape rock material forming on lateritic paleo-soil horizons and outcrops of the felsic basement. The LS factor, a derivative of the DEM, points to a transport of the material by surface runoff (Moore et al. 1991).

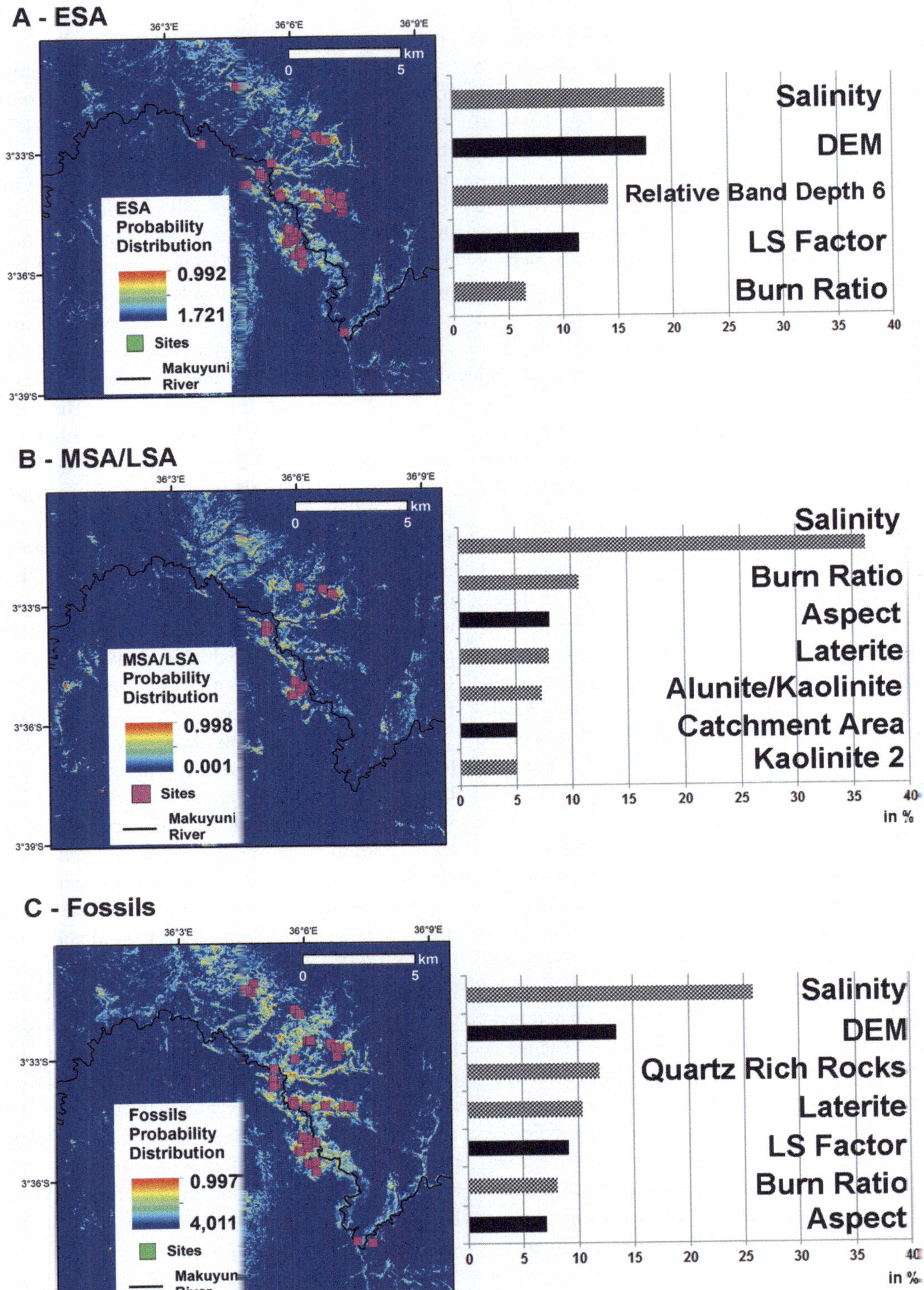

Fig. 8.9 Modeled probabilities and their variable importance in percentage for (**a**) Early Stone Age (ESA) artifacts, (**b**) Middle and Late Stone Age (MSA/LSA) artifacts, and (**c**) fossil sites

8.7 Conclusions

The presented results from different digital geospatial methods draw a comprehensive picture of the landscape development in the Lake Manyara basin. Remote sensing data enabled area-wide analysis and interpretations. Attention should be paid on the individual contributions of remote sensing sensors. The application of SAR data and different optical sensors with high spatial and/or spectral resolution allowed a detailed analysis of present-day and paleo-landscape features. Stochastic- and machine-based learning algorithms provided insights into the driving factors concerning the spatial distribution of paleo-landscape features and find locations. In detail, we obtained the following results:

1. The analyses of morphotectonics of the Makuyuni catchment coupled with comprehensive research on topography, drainage networks, stream longitudinal profiles, and lineaments revealed a morphostructural control with an N–S trend for the uplifted Masai Block, as well as tectonic deformation in the NE of the study area.
2. The Manyara Beds were identified, and evidence of lacustrine sediments further east of the study area has been detected. Besides a probable direct connection to the paleo-lake Manyara, it is also possible that parallel endorheic lacustrine and/or palustrine systems existed further east, which were subsequently cut and drained by the Makuyuni River.
3. The Lake Manyara underwent different transgression and regression phases during the abovementioned humid periods. The distinct morphological features of the paleo-shorelines prove their extent on different elevations in the Lake Manyara basin. The most elevated shoreline proves an overspill into the Engaruka/Natron basins.
4. By analyzing a combination of remote sensing-derived surface characteristics and terrain features, the spatial distribution of topsoils and related soil types was derived. The results indicate a complex and heterogeneous evolution of the study area and of the Manyara basin as a whole. We identified strong relations between silica-rich topsoils and the Precambrian Masai Block, topsoils with iron oxide properties, and andosols on volcanic layers, as well as between carbonate-rich substrates and the Lower Manyara Beds.
5. We assessed the paleontological as well as archeological sites using stochastic models allowing also a prognosis of potential find locations. As shown, the highest potential is related to paleo-river network terraces and paleo-lake deposits. There are distinct differences between the fossil sites and the artifact sites. The latter seems to be closer to the paleo-shorelines and the paleo-river network, whereas the fossil sites seem to be related to cape rock formations in the higher parts of the fluvial terrace systems.

These above-summarized results have demonstrated the enormous potential of innovative remote sensing methods and statistical, machine-based learning algorithms to derive valuable information on paleo-landscape features on larger spatial scales. However, detailed fieldwork activity is required to properly calibrate and validate the proposed analyses.

Acknowledgments This study was funded by the Heidelberg Academy of Sciences and Humanities' research center "The Role of Culture in Early Expansions of Humans (ROCEEH)." The WorldView-2 scene is courtesy of the DigitalGlobe Foundation. We would like to thank the DLR and the German Remote Sensing Data Center (DFS) for providing the TerraSAR-X and the SRTM/X-SAR data. The ASTER L1B data were obtained through the online data pool at the NASA Land Processes Distributed Active Archive Center (LP DAAC), USGS/Earth Resources Observation, and Science (EROS) Center, Sioux Falls, South Dakota, USA.

References

Albaric J, Déverchère J, Petit C, Perrot J, Le Gall B (2009) Crustal rheology and depth distribution of earthquakes: insights from the central and southern East African Rift System. Tectonophysics 468(1–4):28–41. https://doi.org/10.1016/j.tecto.2008.05.021

Albaric J, Perrot J, Déverchère J, Deschamps A, Le Gall B, Ferdinand RW, Petit C, Tiberi C, Sue C, Songo M (2010) Contrasted seismogenic and rheological behaviours from shallow and deep earthquake sequences in the North Tanzanian Divergence, East Africa. J Afr Earth Sci 58(5):799–811. https://doi.org/10.1016/j.jafrearsci.2009.09.005

Al-Khaier F (2003) Soil salinity detection using satellite remote sensing. M.Sc. thesis, ITC Enschede, Enschede, Netherlands

Baatz M, Schäpe A (2000) Multiresolution segmentation: an optimization approach for high quality multi-scale image segmentation. In: Paper presented at the Angewandte Geographische Informationsverarbeitung XII. Beiträge zum AGIT-Symposium Salzburg

Bachofer F (2015) Assessment of paleo-landscape features using advanced remote sensing techniques, modelling and GIS methods in the Lake Manyara Basin, Northern Tanzania. Dissertation, University of Tuebingen, Tuebingen, Germany

Bachofer F, Quénéhervé G, Märker M (2014) The delineation of paleo-shorelines in the Lake Manyara Basin using TerraSAR-X Data. Remote Sens 6(3):2195–2212. https://doi.org/10.3390/rs6032195

Bachofer F, Quénéhervé G, Märker M, Hochschild V (2015) Comparison of SVM and boosted regression trees for the delineation of lacustrine sediments using multispectral ASTER Data and topographic indices in the Lake Manyara Basin. Photogrammetrie, Fernerkundung, Geoinformation (PFG) 1/2015:81–94. doi:10.1127/pfg/2015/0251

Bagdasaryan G, Gerasimovskiy V, Polyakov A, Gukasyan R, Vernadskiy V (1973) Age of volcanic rocks in the rift zones of East Africa. Geochem Int 10:66–71

Barker PA (1990) Diatoms as palaeolimnological indicators: a reconstruction of Late Quaternary environments in two East African salt lakes. Dissertation, Loughborough University of Technology, Loughborough, UK

Bierwirth P (2002) Evaluation of ASTER satellite data for geological applications. Consultancy Report to Geoscience Australia

Canny J (1986) A computational approach to edge detection. IEEE Trans Pattern Anal Mach Intell 8(6):679–698. https://doi.org/10.1109/tpami.1986.4767851

Casanova J (1986) Les stromatolites continentaux: paléoécologie, paléohydrologie, paléoclimatologie. Application au Rift Gregory. Dissertation, Université Aix-Marseille - II

Casanova J, Hillaire-Marcel C (1992) Chronology and paleohydrology of late quaternary high lake levels in the Manyara basin (Tanzania) from isotopic data (18O, 13C, 14C, ThU) on fossil stromatolites. Quat Res 38(2):205–226. https://doi.org/10.1016/0033-5894(92)90057-p

Chang C-C, Lin C-J (2011) LIBSVM: a library for support vector machines. ACM Trans Intell Syst Technol 2(3):1–27. https://doi.org/10.1145/1961189.1961199

Cudahy T (2012) Satellite ASTER geoscience product notes for Australia. CSIRO, Australia

Dawson JB (1992) Neogene tectonics and volcanicity in the North Tanzania sector of the Gregory Rift Valley: contrasts with the Kenya sector. Tectonophysics 206(1–2):81–92. https://doi.org/10.1016/0040-1951(92)90271-7

Dawson JB (2008) The Gregory rift valley and Neogene-recent volcanoes of Northern Tanzania. Geological Society Memoir 33. Geological Society, London. doi:https://doi.org/10.1144/M33

Deus D, Gloaguen R, Krause P (2013) Water balance modeling in a semi-arid environment with limited in situ data using remote sensing in Lake Manyara, East African Rift, Tanzania. Remote Sens 5(4):1651–1680

Flores-Prieto E, Quénéhervé G, Bachofer F, Shahzad F, Maerker M (2015) Morphotectonic Interpretation of the Makuyuni Catchment in Northern Tanzania using DEM and SAR data. Geomorphology 248:427–439. https://doi.org/10.1016/j.geomorph.2015.07.049

Frost SR, Schwartz HL, Giemsch L, Morgan LE, Renne PR, Wildgoose M, Saanane C, Schrenk F, Harvati K (2012) Refined age estimates and Paleoanthropological investigation of the Manyara Beds, Tanzania. J Anthropol Sci 90:151–161. https://doi.org/10.4436/jass.90001

Fujisada H (1995) Design and performance of ASTER instrument. In: Paper presented at the Proceedings of SPIE: advanced and next-generation satellites, Paris, France, 25 Sept 1995

Gallant JC, Dowling TI (2003) A multiresolution index of valley bottom flatness for mapping depositional areas. Water Resour Res 39(12):1347. https://doi.org/10.1029/2002WR001426

Giemsch L (2015) Makuyuni; Fundstellen des Acheuléen am Lake Manyara, Tansania. Ein Beitrag zur Erforschung der mittelpleistozänen Kultur in Ostafrika, vol 7. Tübinger Arbeiten zur Urgeschichte. Verlag Marie Leidorf GmbH, Rahden, Germany

Goetz C (1990) Traçage isotopique et chronologie des processus d'altération et de sédimentation par l'étude des déséquilibres U & Th: application aux systèmes lacustres de Magadi (Kenya) et Manyara (Tanzanie). Dissertation, Universite Aix-Marseille II [S.l.]

Golts S, Rosenthal E (1993) A morphotectonic map of the northern Arava in Israel, derived from isobase lines. Geomorphology 7(4):305–315. https://doi.org/10.1016/0169-555X(93)90060-F

Guyon I, Elisseeff A (2003) An introduction to variable and feature selection. J Mach Learn Res 3:1157–1182

Hare PW, Gardner T (1985) Geomorphic indicators of vertical neotectonism along converging plate margins, Nicoya Peninsula, Costa Rica. In: Morisawa M, Hack J (eds) Tectonic geomorphology. Allen and Unwin, Boston, MA, pp 75–104

Hillaire-Marcel C, Carro O, Casanova J (1986) 14C and ThU dating of Pleistocene and Holocene stromatolites from East African paleolakes. Quat Res 25 (3):312–329. https://doi.org/10.1016/0033-5894(86)90004-9

Holdship SA (1976) The paleolimnology of Lake Manyara, Tanzania: a diatom analysis of a 56 meter sediment core: a diatom analysis of a 56 meter sediment core. Dissertation, Duke University

Hudak AT, Robichaud P, Evans JS, Clark J, Lannom K, Morgan P Stone C (2004) Field validation of Burned Area Reflectance Classification (BARC) products for post fire assessment. In: Greer JD (ed) Remote sensing for field users; proceedings of the tenth Forest Service remote sensing applications conference, Salt Lake City, UT, April 5–9 2004. vol Paper 220. American Society of Photogrammetry and Remote Sensing, p 12

Jaeger F (1913) Das Hochland der Riesenkrater und die umliegenden Hochländer Deutsch-Ostafrikas; 2 Länderkundliche Beschreibung. Mitteilungen aus den deutschen Schutzgebieten/Ergänzungshefte; 8. Mittler, Berlin

Jasiewicz J, Stepinski TF (2013) Geomorphons – a pattern recognition approach to classification and mapping of landforms. Geomorphology 182:147–156. https://doi.org/10.1016/j.geomorph.2012.11.005

Jaynes ET (1957) Information theory and statistical mechanics. Phys Rev 106(4):620–630

Kaiser TM (2000) Die Taphonomie plio-pleistozäner Hominidenfundstellen Ostafrikas mit besonderer Berücksichtigung der Säugetierfaunen des Laetoli- und Lake Manyara-Gebietes in Nordtansania. Archäologische Informationen 23(1):139–142

Kaiser T, Bromage TG, Schrenk F (1995) Hominid Corridor Research Project update: New Pliocene fossil localities at Lake Manyara and putative oldest Early Stone Age occurrences at Laetoli (Upper Ndolanya Beds), Northern Tanzania. J Hum Evol 28 (1):117–120. https://doi.org/10.1006/jhev.1995.1010

Kaiser TM, Seiffert C, Hertler C, Fiedler L, Schwartz HL, Frost SR, Giemsch L, Bernor RL, Wolf D, Semprebon G, Nelson SV, Schrenk F, Harvati K, Bromage TG, Saanane C (2010) Makuyuni, a new Lower Palaeolithic Hominid Site in Tanzania. Mitteilungen Hamburgisches Zoologisches Museum und Institut 106:69–110

Keller CM, Hansen C, Alexander CS (1975) Archaeology and paleoenvironments in the Manyara and Engaruka Basins, Northern Tanzania. Geogr Rev 65(3):364–376

Kent PE (1941) The recent history and pleistocene deposits of the plateau north of Lake Eyasi, Tanganyika. Geol Mag 78(03):173–184. https://doi.org/10.1017/S0016756800071867

Kent PE (1942) A note on pleistocene deposits near Lake Manyara, Tanganyika. Geol Mag 79(01):72–77. https://doi.org/10.1017/S0016756800073532

Le Gall B, Nonnotte P, Rolet J, Benoit M, Guillou H, Mousseau-Nonnotte M, Albaric J, Deverchère J (2008) Rift propagation at craton margin. Tectonophysics 448(1-4):1–19. https://doi.org/10.1016/j.tecto.2007.11.005

Leakey LSB (1931) East African Lakes. Geogr J 77 (6):497–508. https://doi.org/10.2307/1785041

Leakey MD (1979) Olduvai Gorge: my search for early man. Collins, London

Lillesand TM, Kiefer RW, Chipman JW (2004) Remote sensing and image interpretation, 5th edn. Wiley, New York

Mana S, Furman T, Carr MJ, Mollel GF, Mortlock RA, Feigenson MD, Turrin BD, Swisher CC (2012) Geochronology and geochemistry of the Essimingor volcano: melting of metasomatized lithospheric mantle beneath the North Tanzanian Divergence zone (East African Rift). Lithos 155:310–325. https://doi.org/10.1016/j.lithos.2012.09.009

Mitchell J, Shrestha R, Moore-Ellison C, Glenn N (2013) Single and multi-date Landsat classifications of basalt to support soil survey efforts. Remote Sens 5 (10):4857–4876

Moore ID, Grayson RB, Ladson AR (1991) Digital terrain modelling: a review of hydrological, geomorphological, and biological applications. Hydrol Process 5 (1):3–30. https://doi.org/10.1002/hyp.3360050103

Pérez-Peña JV, Azañón JM, Azor A (2009) CalHypso: an ArcGIS extension to calculate hypsometric curves and their statistical moments. Applications to drainage basin analysis in SE Spain. Comput Geosci 35(6):1214–1223. https://doi.org/10.1016/j.cageo.2008.06.006

Phillips SJ, Anderson RP, Schapire RE (2006) Maximum entropy modeling of species geographic distributions. Ecol Model 190(3-4):231–259. https://doi.org/10.1016/j.ecolmodel.2005.03.026

Pour AB, Hashim M (2011) Application of advanced spaceborne thermal emission and reflection radiometer (ASTER) data in geological mapping. Int J Phys Sci 6(33):7657–7668

Reck H (1921) Eine neue diluviale Säugetierfundstelle am Minjonjo in Deutsch-Ostafrika. Sitzungsberichte der Gesellschaft Naturforschender Freunde zu Berlin 1–3:25–36

Ring U, Schwartz HL, Bromage TG, Sanaane C (2005a) Kinematic and sedimentological evolution of the Manyara Rift in Northern Tanzania, East Africa. Geol Mag 142(4):355–368. https://doi.org/10.1017/s0016756805000841

Ring UWE, Schwartz HL, Bromage TG, Sanaane C (2005b) Kinematic and sedimentological evolution of the Manyara Rift in Northern Tanzania, East Africa. Geol Mag 142(04):355. https://doi.org/10.1017/s0016756805000841

Rouse JW, RH Haas, JA Shell, DW Deering and JC Harlan (1974) Monitoring the vernal advancement of retrogradation of natural vegetation. Final report, type III. NASA/GSFC, Greenbelt, MD

Rowan LC, Mars JC, Simpson CJ (2005) Lithologic mapping of the Mordor, NT, Australia ultramafic complex by using the Advanced Spaceborne Thermal Emission and Reflection Radiometer (ASTER).

Remote Sens Environ 99(1–2):105–126. https://doi.org/10.1016/j.rse.2004.11.021

Schlüter T (1987) Paleoenvironment of lacustrine phosphate deposits at Minjingu, Northern Tanzania, as indicated by their fossil record. In: Matheis G, Schandelmeier H (eds) Current research in African earth sciences. Berlin, pp 223–226

Schwartz H, Renne PR, Morgan LE, Wildgoose MM, Lippert PC, Frost SR, Harvati K, Schrenk F, Saanane C (2012) Geochronology of the Manyara Beds, Northern Tanzania: new tephrostratigraphy, magnetostratigraphy and 40Ar/39Ar ages. Quat Geochronol 7:48–66. https://doi.org/10.1016/j.quageo.2011.09.002

Shahzad F, Gloaguen R (2011) TecDEM: A MATLAB based toolbox for tectonic geomorphology. Part 1: Drainage network preprocessing and stream profile analysis. Comput Geosci 37(2):250–260. https://doi.org/10.1016/j.cageo.2010.06.008

Snyder NP, Whipple KX, Tucker GE, Merritts DJ (2000) Landscape response to tectonic forcing: digital elevation model analysis of stream profiles in the Mendocino triple junction region, northern California. Geol Soc Am Bull 112(8):1250–1263 https://doi.org/10.1130/0016-7606(2000)112<1250:lrttfd>2.0.co;2

Uhlig C (1909) Die Ostafrikanische Bruchstufe und die angrenzenden Gebiete zwischen den Seen Magad und Lawa ja Mweri sowie dem Westfuß des Meru. Mitteilungen aus den deutschen Schutzgebieten/ Ergänzungshefte; 2. Mittler, Leipzig

Vapnik VN (1995) The nature of statistical learning theory. Springer, New York

Vapnik VN (1998) Statistical learning theory. Adaptive and learning systems for signal processing, communications, and control. Wiley, New York

Wobus C, Whipple KX, Kirby E, Snyder N, Johnson J, Spyropolou K, Crosby B, Sheehan D (2006) Tectonics from topography: procedures, promise, and pitfalls. Geol Soc Am Spec Pap 398:55–74. https://doi.org/10.1130/2006.2398(04)

Open Access This chapter is licensed under the terms of the Creative Commons Attribution 4.0 International License (http://creativecommons.org/licenses/by/4.0/), which permits use, sharing, adaptation, distribution and reproduction in any medium or format, as long as you give appropriate credit to the original author(s) and the source, provide a link to the Creative Commons license and indicate if changes were made.

The images or other third party material in this chapter are included in the chapter's Creative Commons license, unless indicated otherwise in a credit line to the material. If material is not included in the chapter's Creative Commons license and your intended use is not permitted by statutory regulation or exceeds the permitted use, you will need to obtain permission directly from the copyright holder.

In Search of the Optimal Path to Cross the Desert: Geoarchaeology Traces Old Trans-Saharan Routes

9

Olaf Bubenzer, Andreas Bolten, and Heiko Riemer

Abstract

Due to today's broad and often free availability of detailed satellite data, it became possible to examine desert areas on a large scale and, ideally, down to tiny details for old trans-Saharan route systems, which are abandoned since centuries or even thousands of years ago. Additionally, digital elevation models (DEMs) can be used to evaluate the geomorphological situation. In conjunction with historical sources and ground-truth data, these data allow us to reconstruct the position of the desert routes with reasonable accuracy, here exemplified for the Western Desert of Egypt. On the central limestone plateau, where stony ground (hamada, serir) prevails, most of the routes can precisely been seen from space because of their specific natural preservation conditions. In contrast, sandy surfaces usually do not allow recognition of routes in the satellite image. Additionally, the quite narrow donkey tracks of the pharaonic routes are mostly invisible from space. On the other hand, DEMs (ASTER, SRTM) allow calculating ideal routes by means of geographical information systems (GIS), usually applied in cases were traditional routes have not survived. In the present study, we confront both methods using the example of the Darb el-Tawil to test the hypothesis that these old roads largely follow the ideal route. This 250 km long road has been one of the primary arteries between the Nile Valley and the Western Desert oases during the past 4000 years. Due to modern demographic and economic development, it has been selected for the construction of a new paved highway, which will irrevocably destroy the still existing tracks and archaeological objects.

O. Bubenzer (✉)
Institute of Geography, Physical Geography, Heidelberg University, Heidelberg, Germany
e-mail: olaf.bubenzer@uni-heidelberg.de

A. Bolten
Institute of Geography, University of Cologne, Cologne, Germany
e-mail: andreas.bolten@uni-koeln.de

H. Riemer
Institute of Prehistoric Archaeology, African Archaeology, University of Cologne, Cologne, Germany
e-mail: heiko.riemer@uni-koeln.de

© Springer International Publishing AG 2018

C. Siart et al. (eds.), *Digital Geoarchaeology*, Natural Science in Archaeology,
DOI 10.1007/978-3-319-25316-9_9

Keywords
Desert road archaeology • Linear structures • Satellite images • Least cost paths • DEM • ASTER • SRTM • GIS

9.1 Introduction

Geoarchaeological investigation on ancient desert roads as linear structures connecting, for example, the Egyptian oases with the Nile Valley is a comparably new field of research. Riemer and Förster (2013) stated in the introduction of their ground-basing book that "Desert Road Archaeology" (Förster and Riemer 2013) has to be multi- and interdisciplinary, which combines archaeological, geographical, ethnological, logistical, etc., aspects. In the Eastern Sahara, two ancient caravan routes are well known: the Darb el-Arbain and the Darb el-Tawil (Harding King 1912). However, numerous other routes and tracks were reported (e.g. Riemer and Förster 2013). They were in use at least since pharaonic times (Förster 2015) and supplied sub-Saharan Africa with salt, cloth, weapons etc., and eventually European traders with gold, slaves, ivory, aromatic substances and other commodities (e.g. Austen 2010).

Ancient caravan routes had to conquer the inhospitable desert. To find the shortest route between the starting point and the destination was the most important aspect for road construction (Bubenzer and Bolten 2013). However, the shortest way is not consequently the easiest one. The primary constraints in desert landscapes are topography and surface cover (Förster et al. 2010). Due to today's broad and often free availability of detailed satellite data, it became possible to examine desert areas on a large scale and, ideally, down to tiny details. Satellite remote sensing data are, in general, progressively in use for archaeological questions in deserts, mainly with regard to archaeological sites (e.g. Parcak 2009; Lasaponara and Masini 2012). Recently, Tapete and Cigna (2016) reported trends of space-borne Synthetic Aperture Radar (SAR) remote sensing for archaeological landscape and cultural heritage applications. However, studies applying remote sensing data with regard to the analysis of ancient linear road structures in deserts are rare (Förster et al. 2010 Bubenzer and Bolten 2013; Riemer and Förster 2013; De Laet et al. 2015). This applies in particular for the application of digital elevation models. Bubenzer and Bolten (2013) used the Darb el-Tawil as an example to show that the visible detection of caravan routes depends on the surface conditions. Where stony ground (hamada, serir) prevails that was not overprinted by modern activities such as road construction or mining, most of the ancient camel routes can precisely be seen from space because of their specific natural preservation conditions. In contrast, narrow donkey tracks and, on sandy surfaces, even wide bundles of camel tracks, usually do not allow a clear recognition of routes in the satellite image. In addition, after a comparison of satellite images, ground-truth data and a relief profile of the Darb el-Tawil caravan route, they hypothesised that these old roads largely follow the ideal topographical route. In order to prove this hypothesis, for the study at hand, we use two different free worldwide available elevation models, ASTER GDEM (NASA JPL 2017a) and SRTM 1 (NASA JPL 2017b). Both models are based on different sources; however, they show similar resolution and accuracy (e.g. Table 9.1).

9.2 The Egyptian Limestone Plateau and the Darb-el Tawil Route

9.2.1 Geological and Geomorphological Conditions

The Darb el-Tawil crosses the Egyptian Limestone Plateau (also named Abu Muhariq Plateau), which is situated between the semi-circle of the oases of the hyperarid Western Desert

Table 9.1 Basic data about the used elevation models [after Japan Space Systems (2017) and Tachikawa et al. (2017)]

	ASTER GDEM V2	JPL SRTM-1 V3
Data source	ASTER	Shuttle Radar Topo. Mission 2000
Release year	2011	2014
Resolution	1 arcsec	1 arcsec
Absolute vert. acc.	7–14 m	10 m
RMSE	<12 m	<12 m

and the Nile Valley (Fig. 9.1). The plateau elevates from around 500 m above sea level (a.s.l.) in the south (Abu Tartur subplateau) to 200 m a.s.l. in the north and towards the Nile Valley. The villages of Balat and Teneida in the most south-eastern part of the Dakhla Oasis have altitudes of around 100 m a.s.l. The oasis floor consists mainly of Cretaceous sedimentary rocks (limestone, shale, sandstone), and on the plateau surface Palaeocene and Eocene limestones prevail (Klitzsch et al. 1987). The oasis is bordered to the east by a steep escarpment, which overlooks the oasis floor by up to 350 m. On the limestone plateau, the relief documents a (paleo-)karstic landscape of plains or rounded hill tops, which are intercepted by minor escarpments (e.g. Abu Gerara) and several flat depressions. Flat draining channels are the result of former wetter climate phases (Kindermann et al. 2006). However, the recent hyperaridity results in strong wind abrasion with typical stony desert surfaces (serir and hamada), some depressions covered by Aeolian sands and the Abu Muhariq dune belt (Bubenzer and Bolten 2013).

9.2.2 History and Course of the Darb el-Tawil Route

The Darb el-Tawil (Arabic "the long road") was the primary corridor leading from the Nile Valley, the heartland of the ancient Egyptian culture, to Dakhla Oasis, one of the large groundwater-fed depressions in the Western Desert. Remote sensing and initial field investigations have revealed that the Darb el-Tawil is not a single route; rather, it can be defined as a road system with a primary trunk road, some major branch routes and several shorter auxiliary routes. At the ends of the road trunk, several auxiliary tracks branch of to connect to other small towns or minor locations. Major branch routes exist especially on the Dakhla side. At Abu Gerara, the Darb el-Tawil forks into two major branches, of which one leads to the most eastern part of Dakhla at the town of Balat, while another one heads for the towns in the oasis' western part, some 30 to 50 km away from Balat. The main trunk of the Darb el-Tawil connects the eastern part of the Dakhla Oasis (Balat and Teneida) with Manfalut and Beni Adi and Beni Adi in the Nile Valley. On the whole route, evidence of wells or other freshwater sources is lacking. Unfavourable sections for caravans with profound loose sand are located along the escarpment edges and the area of the Abu Muhariq traverse. Here, the route does not follow the direct course (Bubenzer and Bolten 2013).

Up to the early twentieth century, camel caravans along this road provided the only direct communication with this oasis, especially for trading dates from the oasis (Giddy 1987). Absolute walking times without breaks of 62.5 hours (Beadnell 1909) and 63 hours (Edmonstone 1822) are reported. Cailliaud (1822) reports a total travel time of 5 days, which is equal to a daily distance of around 53 km. The camel is famous as the "ship of the desert" because it has a high load-carrying capacity and is well adapted to withstand long periods without any external source of water, even under high temperatures. A good pack camel is capable of going for up to 10 days without water, so that a journey of 5–6 days with a walking speed of 4 km/h along the Darb el-Tawil is in the usual range of camel caravans. Long-lasting and repeated traffic along this route is also evident

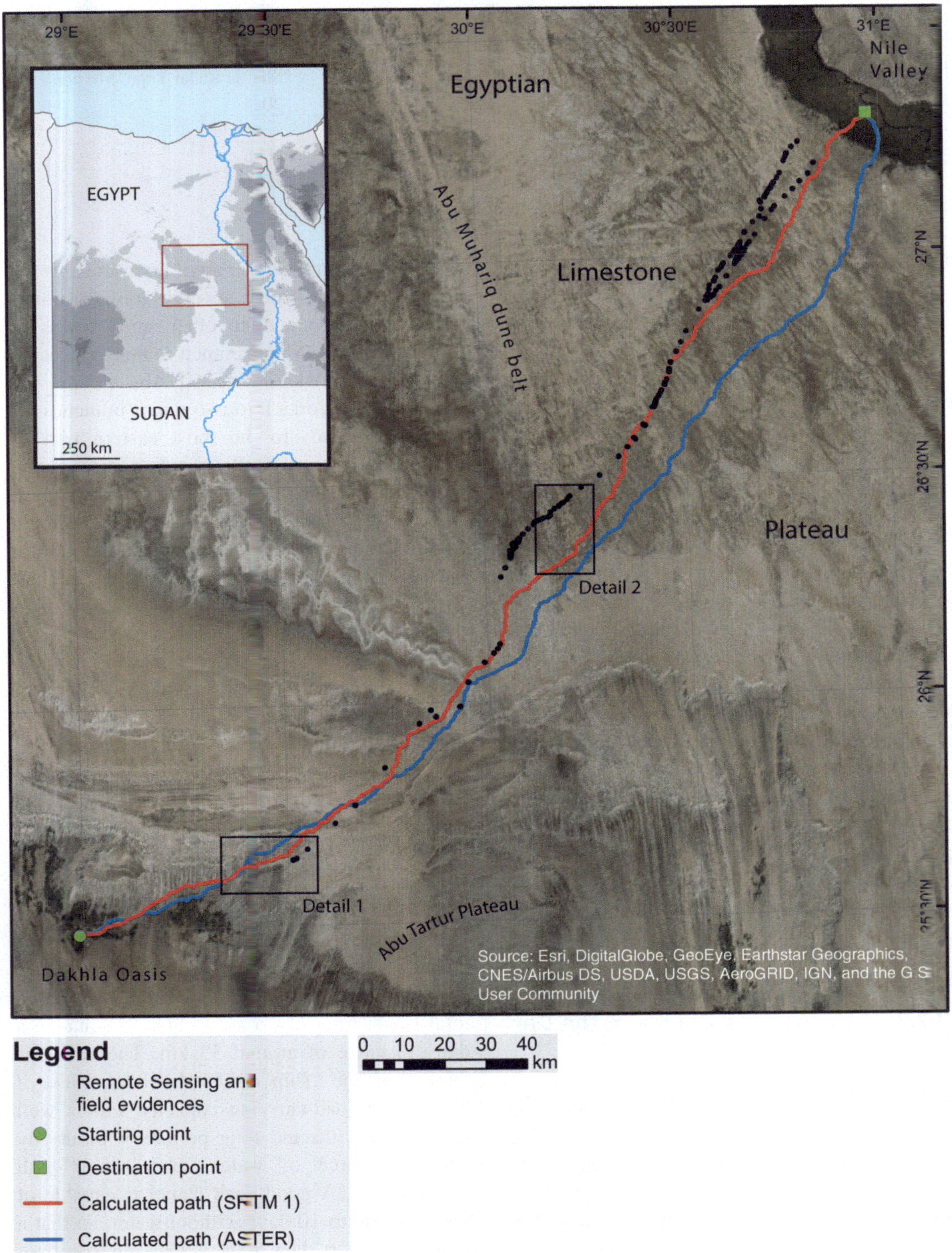

Fig. 9.1 Least cost paths based on different elevation models (ASTER and SRTM 1) in contrast to remote sensing and field data

in broken pottery that was lost during the journeys (Eichhorn et al. 2005). The attribution of pottery to known ceramic productions in the Nile Valley and the oases indicates that this route was heavily trafficked in Ptolemaic and Roman times when the oases were known for their production of grain and wine. However, there is also some indication that the history of the Darb el-Tawil goes back to much earlier times of the pharaonic state, when donkeys were the beasts of burden. As early as the Old Kingdom (around 2500 BC), pharaonic Egyptians had colonised Dakhla Oasis where they established agriculture, settlements and the palace of the provincial governor. Given the need for regular communication and exchange with the Nile Valley, and especially with the then capital of the pharaonic state in Memphis (south of Cairo), the Darb el-Tawil can be suggested as the main artery to and from this oasis during this period.

With the appearance of early motor transport, camels were gradually replaced and traffic shifted to other routes easier to traverse by car. When the first asphalt road connected the Nile Valley via Kharga to Dakhla, the Darb el-Tawil became abandoned. Nevertheless, its historical importance as the prime route to Dakhla is still visible in the bundles of animal tracks beaten in the desert surface (Bubenzer and Bolten 2013). Its route is waymarked by hundreds of simple cairns (arabic "alamat", sing. "alam"), which functioned as a guiding system for the caravan leaders (Förster et al. 2010).

9.3 Methods and Data

Within the interdisciplinary subproject A1 "Climatic Change and Human Settlement between the Nile Valley and the Central Sahara" as part of the Cologne Collaborative Research Centre 389 ACACIA, a wide range of geoscientific, (archaeo-)botanical and archaeological methods were used all over the funding period (1995–2007) (Bubenzer et al. 2007; Bubenzer and Bolten 2013). Alongside the analysis of satellite data, point information was recorded in field-work expeditions (Riemer and Förster 2013). Therefore, several indications about anthropogenic changes of the landscape are visible in the field. Linear structures (Fig. 9.2) and historic landmarks were recorded by GPS measurement and documented in logbooks. These ground-truth data represent an unique archive and merit a direct countercheck of the satellite data interpretation. The impression in the field, in particular the linear structures, provides an indication to get a closer look to high-resolution satellite images.

Within the subproject E1 of the CRC 389 "Arid Climate, Adaptation and Cultural Innovation in Africa" (ACACIA), for scientific purposes, since 2000 there has been the possibility to examine the Western Desert of Egypt area-wide and in sufficient detail for a reconstruction of visible routes. Additionally, at that time the new stereoscopic ASTER data allowed us to calculate first digital elevation models and therefore to evaluate the geomorphological situation in the surrounding area of these routes. Today, based on the ASTER data, a worldwide elevation model is available for free.[1] In conjunction with freely available satellite images, historical sources and ground-truth information (e.g. GPS points of route markers, documented in the logbooks of the ACACIA subproject A1), these data have made it possible, for the first time, to document and reconstruct the position of caravan routes in the Western Desert almost completely and with reasonable accuracy.

9.3.1 Remote Sensing

Due to the good availability of high-resolution satellite images even for the less accessible regions of arid Africa, it is possible with the use of virtual globes (e.g. Google Earth or Bing maps) to inspect abroad regions about the same anthropogenic linear structures seen in the field.

New commercial satellites reach resolutions of better than 0.3 m for each pixel, such as the World-View 4 imager.[2] Therefore, the resolution reaches the scale of aerial photos and enables this

[1] https://asterweb.jpl.nasa.gov/gdem.asp

[2] http://www.digitalglobe.com/

Fig. 9.2 Camel tracks marking the Darb el-Tawil caravan route east of Dakhla Oasis. View to the west (Photo: H. Riemer)

good quality to be used for inaccessible regions as well. However, at the same time, the amount of storage space quadrupled for each bisection of the pixel size. Consequently, more and more data storage and processing power are needed to handle the data quantity. In addition, the visual inspection of the data takes more and more time, so the current research tries to use automatic object filtering to enhance and accelerate the use of high-resolution satellite data (De Jong and Van der Meer 2006; Lillesand et al. 2015). However, this highest resolution enables to test verified areas with different algorithm, in order to identify scale independent methods and to identify a threshold scale value. Earlier studies showed us, for example, the strong dependence of the visibility of linear structures on the surface setting and sediment colour (Bubenzer and Bolten 2013).

9.3.2 Digital Elevation Data

Digital elevation models represent the mean elevation of the bare earth in a distinct area, which are processed to use with a computer system (Campbell and Waynne 2011; Lillesand et al. 2015). In these models, the resolution of an area part (pixel), the absolute accuracy and the relative accuracy play an important role (cf. Table 9.1). With the use of digital elevation models, several geomorphometric tools are available to calculate information about the relief position of a pixel or a part of the elevation model (Wilson and Gallant 2000; Pike and Dikau 1995; Dikau 1993). The simplest derivation is the slope or the exposition. However, the georelief classification is still an object of research (Evans 2012; Jasiewicz and Stepinski 2013).

9.3.3 GIS and Cost Path Analysis

For the reconstruction of possible caravan routes, the analysis and use of a cost raster is a powerful tool. Based on several assumptions and existing data sets, it is possible to define a least cost path. The general cost distance formula is the following (esri 2017):

$$\begin{aligned} Cost = & \, Cost\ of\ travel\ over\ surface^* \\ & Characteristics\ of\ the\ mover^* \\ & Movement\ characteristics\ on\ the\ surface \end{aligned}$$

One of the simplest influences of the cost raster is the uphill or downhill movement, which can be allocated by the slope of a surface. Several other properties of the landscape force humans to take a distinct route independent from the Euclidean distance. In the presented example, the roughness of the surface or the smoothness of the sediment could influence the caravan to choose a non-Euclidean route. Some of the factors can be derived by remote sensing data. The texture of a high-resolution satellite image could be a proxy for the surface character, or a high-resolution elevation model could be a base for the roughness calculation.

In general, a cost raster C is a function of all the different force factors F with different weight l and the walking distance D plus a not defined uncertainty U:

$$C = U + \sum l_i F_i D$$

For this reason, in the field between exterior driving factors, there are always difficult calculable human factors, which influence the decision for a selected route.

In the presented example, the path analysis consists of the defined start and destination point and the assumption that the caravan uses less slope values preferably. Using up-to-date Geographic Information Systems, a cost analysis tool is implemented to derive a least cost path. Using the destination point, at first an Euclidean distance raster is calculated, which represents the cost only for the movement to the destination. Using a slope raster and the starting point, a cost raster is calculated which accumulates more cost weather more absolute slope values used on the route. The combination of distance raster and the cost raster gives the least cost path. Connecting further information coming from remote sensing or elevation model analysis, the cost raster could be combined by several drivers with different weights.

9.4 Results

Figure 9.1 shows the calculated routes derived by least cost analysis based on the two different elevation models. First of all, it is noticeable that both routes course not the direct direction from the start to the destination point. In addition, both routes match in parts the remote sensing and field evidences. However, there are two positions where one or both models differ to the observations.

The first differentiating region is the climb to the escarpment (cf. Fig. 9.3). Here, the algorithm, summarising low slope values as possible, induces to select sand ramp positions to climb up the escarpment. Expert knowledge of the escarpment characteristic give a completely different position. Field investigation findings see complex routes up to the escarpment (cf. Fig. 9.3). The further calculation, based on the SRTM 1 data, matches very well the observed field and remote sensing data. The second region with a different course is presented in Fig. 9.4. Again, this passage is dominated more by expert knowledge than external driving factors. The models choose a more southern bypass around the Abu Muhariq dune belt, where the barchans distribution is smaller. The observed route is divided into a distinct passage through the several single dunes at a more northern position and a calculated route alongside a dissolved part of the dune belt.

Overall, the SRTM 1 dataset, in combination only with the slope factor, in the least cost analyses performs better than the ASTER GDEM model. This could be due to the higher accuracy of the model and/or due to the origin of the dataset. To enhance the calculation of the

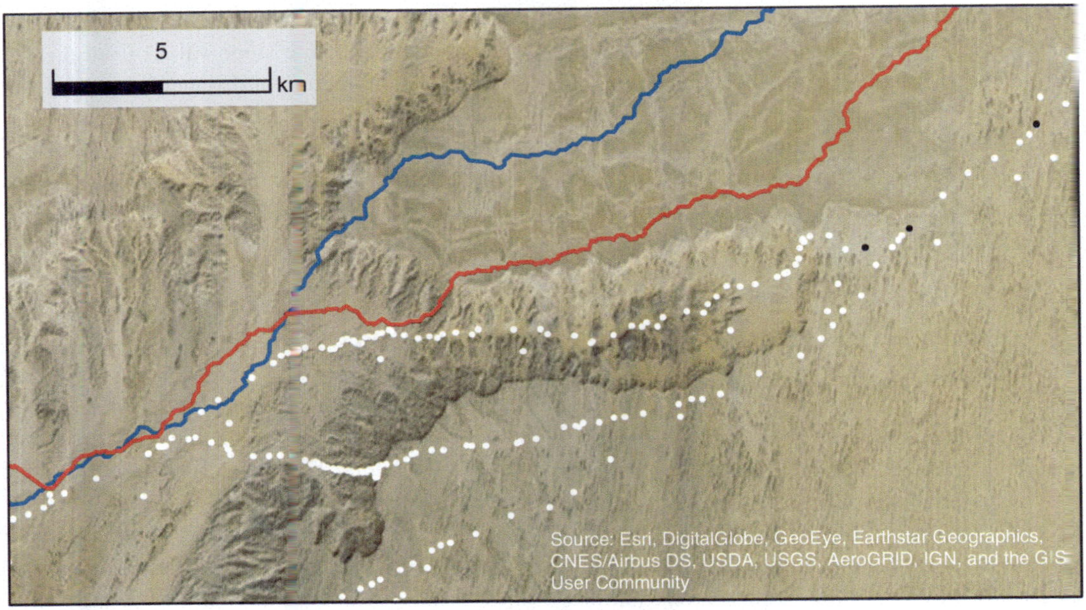

Fig. 9.3 Detail 1 (cf. Fig. 9.1). The two calculated routes (*blue* and *red*) prefer steep ramp-like ascents through the escarpment northeast of Dakhla Oasis. Three routes proven by remote sensing and field survey (*white dots*) indicate both steep ramp-like routes through the eastern escarpment and a route along the valley bottom with moderate slope values. Legend, see Fig. 9.1

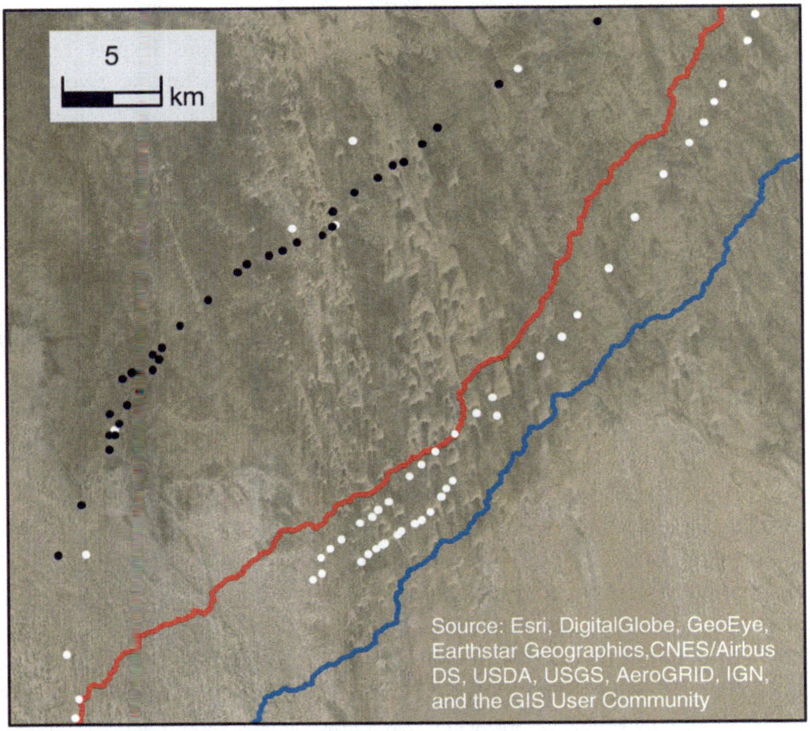

Fig. 9.4 Detail 2 (cf. Fig. 9.1). Two routes are proven by remote sensing and partly by field evidence. A northern main route zigzags across the almost impassable dune belt (*black dots*), while a southern route made an easier crossing through a dissolved part of the dune belt (*white dots*). Legend, see Fig. 9.1

route further cost information like the petrography, the surface roughness or the smoothness of the sediment could be implemented. For this, several studies about the position of field data in context to remote sensing information and the weight of each parameter are necessary in order to calibrate a possible cost raster.

9.5 Discussion and Conclusions

In combination with the ground-truth data and the knowledge about the preservation conditions of tracks, the satellite images provide a huge increase in information with the result that it was possible to correct and extend the known network of caravan routes. Connections between known parts of caravan routes could be completed by the use of least cost operations to give a hint for a possible course and to narrow possible ground investigations. These results are transferable to other deserts in general. The fact that significantly more caravan routes were not found on the Egyptian Limestone Plateau in the surrounding of the Darb el-Tawil may be because the existing routes, which had been known for centuries, were the fastest and easiest links between the oases and the Nile Valley. The caravans had to cross long distances on routes through the desert with neither water nor vegetation. This was an extremely arduous enterprise. Thus, the lengths of the existing caravan routes calculated within the framework of this research suggest that due to the distances and the consequently greater duration of journeys, alternative regional routes were probably not considered. The indicators revealed in the satellite image do not allow any clear conclusions to be drawn about the frequency of use or the age of individual caravan routes. Beyond the area of the limestone plateau, e.g. in the sand cover Western Desert of Egypt, it was, with a few exceptions, not possible to locate caravan routes in the satellite image. The Nubian Sandstone Formation with the associated layer of sand on the surface south of the Dakhla–Kharga line has, quite obviously, a negative effect on the natural preservation and the visibility of caravan routes in the satellite image. Here, the presented technique may also help to find remains of formerly existing routes.

Due to modern demographic and economic development, the course of the famous Darb el-Tawil has been selected as course for the construction of a new paved highway. Therefore, most of the today still existing ancient tracks and archaeological objects will be irrevocably destroyed. For the sake of protection and with regard to cultural heritage management, it is urgent and essential to aim at a thorough documentation of ancient trade routes by combining advanced remote sensing techniques and fieldwork as an example for further interdisciplinary desert road archaeology.

Acknowledgements The study at hand is part of the research project "The Darb el-Tawil: A caravan route across the Egyptian Western Desert and its potential for methods and concepts in desert road archaeology", financed by the German Research Foundation (DFG).

References

Austen RA (2010) Trans-Saharan Africa in World History. The New Oxford World History. Oxford University Press, Oxford

Beadnell HJL (1909) An Egyptian oasis: an account of the oasis of Kharga in the Libyan desert. John Murray, London

Bubenzer O, Bolten A (2013) Top down: new satellite data and ground truth data as base for a reconstruction of ancient caravan routes – examples from the western desert of Egypt. In: Förster F, Riemer H (eds) Desert road archaeology. Africa Praehistorica, vol 27. Heinrich-Barth-Institut, Köln, pp 61–75

Bubenzer O, Bolten A, Darius F (eds) (2007) Atlas of cultural and environmental change in arid Africa. Africa Praehistorica, vol 21. Heinrich-Barth-Institut, Köln

Cailliaud F (1822) Travels in the Oasis of Thebes, and in the deserts situated East and West of the Thebaid; in the years 1815, 1816, 1817, and 1818. Shackell and Arrowsmith, London

Campbell JB, Waynne RH (2011) Introduction to remote sensing. Guilford Press, London

De Jong SM, Van der Meer FD (2006) Remote sensing image analysis. Springer, Dordrecht

De Laet V, van Loon G, Van der Perre A, Deliever I, Willems H (2015) Integrated remote sensing investigations of ancient quarries and road systems in

the Greater Dayr al-Barsha Region, Middle Egypt: a study of logistics. J Archaeol Sci 55:286–300
Dikau R (1993) Geographical information systems as tools in geomorphology. In: Z Geomorph N.F. Suppl 92:231–239
Eichhorn B, Hendrickx S, Riemer H, Stern B (2005) Desert roads and transport vessels from Late Roman-Coptic times in the Eastern Sahara. J Afric Archaeol 3 (2):213–229
Edmonstone A (1822) Journey to two of the oases of Upper Egypt. John Murray, London
esri (2017) ArcGIS 10.5 Desktop Help. Redlands
Evans IS (2012) Geomorphometry and landform mapping: what is a landform? Geomorphology 137:94–106
Förster F (2015) Der Abu Ballas-Weg. Eine pharaonische Karawanenroute durch die Libysche Wüste. Africa Praehistorica, vol 28. Heinrich-Barth-Institut, Köln
Förster F, Riemer H (eds) (2013) Desert road archaeology, Africa Praehistorica, vol 28. Heinrich-Barth-Institut, Köln
Förster F, Riemer H, Bolten A, Bubenzer O, Hendrickx S, Darius F (2010) Tracing linear structures: remote sensing, landscape classification and the archaeology of desert roads in the Eastern Sahara. In: Möhlig W, Bubenzer O, Menz G (eds) Towards interdisciplinarity. Experiences of the long-term ACACIA project. Topics in interdisciplinary African studies, vol 16. Cologne, pp 49–75
Giddy LL (1987) Egyptian oases: Bahariya, Dakhla, Farafra and Kharga during Pharaonic times. Aris & Phillips, Warminster
Harding King WJ (1912) Travels in the Libyan Desert. Geogr J 39:133–137;192
Jasiewicz J, Stepinski TF (2013) Geomorphons – a pattern recognition approach to classification and mapping of landforms. Geomorphology 182:147–156
Japan Space Systems (2017) ASTER GDEM—comparison with Other DEMs, 2012. http://www.jspacesystems.or.jp/ersdac/GDEM/E/2.html (Accessed: 21 Feb 2017)
Kindermann K, Bubenzer O, Darius F, Nussbaum S, Pöllath N, Riemer H, Smettan U (2006) Palaeoenvironment and Holocene land use of Djara, Western Desert of Egypt. Quat Sci Rev 25:1619–1637
Klitzsch E, List FK, Pöhlmann G, Handley R, Hermina M, Meissner B (1987) Geological map of Egypt 1:500,000, NG 35 NE. Farafra, Conoco, Egypt
Lasaponara R, Masini N (eds) (2012) Satellite remote sensing – a new tool for archaeology. Springer, New Delhi. https://doi.org/10.1007/978-90-481-8801-7
Lillesand T, Kiefer RW, Chipman J (2015) Remote sensing and image interpretation. Wiley, New York
NASA JPL (2017a) ASTER Global Digital Elevation Model, 2009. Available online, doi:https://doi.org/10.5067/ASTER/ASTGTM.002
NASA JPL (2017b) NASA Shuttle Radar Topography Mission Global 1 arc second. NASA LP DAAC. 2013. Available online, doi:https://doi.org/10.5067/MEaSUREs/SRTM/SRTMGL1.003
Pike R, Dikau R (eds) (1995) Advances in geomorphometry. Z Geomorph N.F, Suppl 101, Stuttgart
Parcak SH (2009) Satellite remote sensing for archaeology. Routledge, London
Riemer H, Förster F (2013) Ancient desert roads: towards establishing a new field of archaeological research. In: Förster F, Riemer H (eds) Desert road archaeology, Africa praehistorica 28. Heinrich-Barth-Institut, Köln, pp 19–58
Tachikawa T, Kaku M, Iwasaki A, Gesch D, Oimoen M, Zhang Z, Danielson J, Krieger T, Curtis B, Haase J et al. (2017) ASTER Global Digital Elevation Model Version 2—summary of validation results; technical report; available online: http://www.jspacesystems.or.jp/ersdac/GDEM/ver2Validation/Summary_GDEM2_validation_report_final.pdf. Accessed 21 Feb 2017
Tapete D, Cigna F (2016) Trends and perspectives of space-borne SAR remote sensing for archaeological landscape and cultural heritage applications. J Archaeo Sci Rep. https://doi.org/10.1016/j.jasrep.2016.07.017
Wilson JP, Gallant JC (eds) (2000) Terrain analysis principles and applications. Wiley, New York

10 Combined Aerial and Ground-Based Structure from Motion for Cultural Heritage Documentation

Christian Seitz

Abstract

Structure from motion (SFM) and unmanned aerial systems (UAS) are two technologies which are well suited for archaeological documentation purposes and complementing each other very well. In the following, we would like to give an overview of the methods, systems, and software available as well as two short practical applications for archaeology and geography. UAS are usually small drones for aerial imaging and video. The UAS section will give an overview on the systems available, restrictions, and challenges, but the focus will be on the capabilities of drones for different purposes. SFM is a subset of the wide field of photogrammetry, using photos to calculate 3D data of the objects pictured. A selection of both free-to-use and commercial software will be introduced, while the main part of the SFM section will be discussing the basic work flow. It covers taking the photos, restoration of the camera positions in 3D space, and creating a surface from the computed point cloud. The combination of both techniques to create 3D models of places or parts of buildings barely reachable is addressed then. Examples from excavations and the documentation of the *King's Hall* at the UNESCO World Heritage Site *Lorsch Abbey* will follow. The last part will show the future possibilities with new technologies, new software, and some challenges for the future.

Keywords

Structure for motion (SFM) • Unmanned aerial systems (UAS) • Drones • Photogrammetry • Heritage documentation • 3D models

10.1 Introduction

Mapping and surveying as common tasks in archaeology and geography are challenging, especially when dealing with larger or very complex areas. Archaeological documentation often lacks

C. Seitz, M.A. (✉)
Research Group "Optimization, Robotics, and Biomechanics", Institute for Computer Engineering, Heidelberg University, Berliner Str. 45, Heidelberg 69120, Germany
e-mail: christian.seitz@iwr.uni-heidelberg.de

© Springer International Publishing AG 2018
C. Siart et al. (eds.), *Digital Geoarchaeology*, Natural Science in Archaeology,
DOI 10.1007/978-3-319-25316-9_10

aerial views, which would offer an overview of the excavated areas and the feature correlations. Documentation in three dimensions still is only infrequently done. In geographic applications terrestrial laser scanning (TLS) is a great method, but it sometimes is difficult with objects not entirely visible from the scan positions, while some other necessary viewpoints might not be accessible at all.

The approach using an UAS and SFM is well suited as a stand-alone application or in combination with other methods to fill gaps. Along with a general overview on each of the single topics, some use cases of different applications will be presented.

The two techniques presented gained importance in the last years driven by powerful computing hardware and even more by improved algorithms. Different approaches and systems appeared along with this process, both open source and commercial.

This article intends to give you an overview on state-of-the-art unmanned aerial systems (UAS) for the use in archaeology and geography and the different approaches of the photogrammetric SFM. Some use cases of both elements and their combination will show the practical use out in the fields.

10.2 Unmanned Aerial Systems

The UAS or, more commonly speaking, *drones* are a wide field of aircraft systems without a pilot. Among them are large planes for military purposes like the General Atomics MQ-1 *Predator* with a wingspan of 16.8 m[1] as well as palm-sized multirotor toys as recently used in drone races. Especially multicopters could be developed because of the progress in microchip technology, particularly the ARM-series, which is nowadays often used in a wide range of hobby projects. They are inexpensive and powerful, which are necessary for the fast control of the multicopter's motors to keep the attitude of the aircraft.

The most interesting type of drones for applications in archaeology and geography can be divided into two main groups, large drones and so-called micro aerial vehicles.

Large drones are usually helicopters with a rotor diameter of up to or even over two meters for heavy payload and with distinguished flight control systems. They can carry large equipment loads like laser scanners or sensor arrays. A good starting point to get information on this topic still is Eisenbeiß (2009, 34 ff.). Since there are major disadvantages in comparison to smaller systems, like their high price and strict legal restrictions on flights, they will not be addressed in this article.

The second group emerged from remote-controlled models having a long tradition all over the world. There are several projects which had developed UAS as well as there are many active ones working on flight control systems. There are also various companies selling complete ready-to-fly devices of this type for diverse purposes and in different price classes.

The two most important classes of UAS are the *rotary wing drones* and *fixed-wing systems*. The rotary wing aircrafts are using a propeller to create a lift, like helicopters or multicopters. Fixed-wing aircrafts are basically remote-controlled airplanes in varied body types and wingspans, where the wings' lift is keeping the aircraft up. They need a specific minimal speed to fly; therefore they are not suited for every application. There are different basic setups, like shoulder-wing planes and flying wings, for different purposes.

10.2.1 Hardware Overview

There are many different autopilot systems on the market, as shown in the following. They usually can not only control the aircraft itself but also adjust the camera, trigger it on certain waypoints, and even do more complex tasks like keeping the camera pointed to a certain point. The user has to find the system which fits his tasks best. For numerous or repetitive small area documentations, a small multicopter is sufficient.

[1] http://www.af.mil/AboutUs/FactSheets/Display/tabid/224/Article/104469/mq-1b-predator.aspx

Table 10.1 Aircraft overview and suitable applications (+ suited/ • possible/ – not working)

Drone type / Application	Small multicopter	Large multicopter	Fixed-wing plane
Mapping of objects like buildings and monuments	+	+	–
Mapping of small excavation sites, fields, etc.	+	+	+
Mapping of large areas like cities or landscapes	–	•	+

as maintenance both in time and costs have to be considered. For more sophisticated aerial photography, a larger multirotor UAS is useful, having from 4 to 12 vertically aligned rotors and the ability to hover like a helicopter. For large-scale documentations, from several hundred square meters to some square kilometers, fixed-wing aircrafts should be preferred, since they do not have to lift their weight by propellers; thus the flight duration increases to more than 30 min.

The following scheme and some guiding questions might be considered useful in the process of finding a convenient UAS (Table 10.1).

- **Is it useful to get a pre-built system?**
  This has to be recommended if time and usability are main factors, because the maintenance is easy or has to be done by the companies. These systems are easy to use, and the responsible persons and pilots usually get a professional training on the UAS. Of course, all this will result in a stiff initial price and also will cause higher running expenses.
- **What does it take to build a custom aircraft?**
  The systems offering different setups will need persons experienced in remote-controlled models as well as in computers and electronics. They usually will have to find an adequate system of autopilot, frame type, and the correct drive setup. The person will have to install and fit a lot of electronic components, adjust the software to match the used aircraft, and so on. This needs a lot of time over a longer period and might cause some mistakes and crashes.
- **Which size should the aircraft have?**
  The aircrafts range from small flying wings for compact cameras to systems carrying a full DSLR camera. This results in wingspans from less than one to three meters and more. There are many different aircrafts available, and it is also necessary to choose them with respect to the autopilot system used.
- **Are there any legal aspects?**
  In many countries it is not allowed to fly without legal permission. Thus the pilot has to obey different rules. In some cases the permissions are connected to specific obligations like a license of the pilot and a marked UAS. A special RC insurance is always recommended and in most cases a prerequisite for such a permission.

10.2.2 Commercial Systems

The commercial systems are giving the user a sophisticated and reliable aircraft. They usually include training classes for the designated pilots. The autopilot system can support the pilot with different flight modes. They reach from fully autonomous flight, including the automated landing, over assisted flying to full manual control.

Four UAS built by different vendors will be presented here:

- The *eBee* from the Swiss company *senseFly*, a lightweight flying wing—www.sensefly.com
- *MAVinci Sirius*, a fixed wing aircraft—www.mavinci.de
- *DJI*'s *Phantom*, a very popular quadrotor system—www.dji.com
- *Mikrokopter Okto XL*, an octocopter with a reliable autopilot system—www.mikrokopter.de

Fig. 10.1 *senseFly*'s *eBee* just before start, fieldwork at Troezen, Greece (Photo: Author)

This list is far from complete and is intended to give an overview of different drone types.[2]

10.2.2.1 senseFly's eBee

The Swiss company *senseFly* has been developing UAS for several years now, making literally waves with their *Swinglet CAM*'s flight mode, as they are shooting photos with deactivated engine. The actual generation with several improvements is called *eBee* (Fig. 10.1) (Sensefly 2016), coming with a complete set of flight planning and processing software. This UAS can fly with one battery load for about 40 min covering up to 12 km.[2] The UAS is started by just throwing it up in the air, and it needs only a small patch of grass to land. The drone can carry a Sony compact camera with maximal 18 MPx for the mapping, a multi-spectrum camera, or a thermal camera.

The flight planning and the control are done via laptop attached to transmitter box. The flight can be live monitored and adjusted in real time, for example, to set the landing vector or update the mission plan.

The *eBee* has a good balance of weight and power; it is suited to map large areas of interest in very short time. The resulting digital surface model (DSM) is fairly precise. For archaeological use this system is only of limited suitability, because at low altitudes and under windy conditions some of the images tend to get blurry. Furthermore the resolution is too low to map archaeological remains adequately.

On the other hand, one big advantage is the possibility to replace the standard RGB camera with a multi-spectral camera, capturing near-infrared (NIR) as well as red edge and other

[2] The selection of the systems is based on the author's experience with the different manufacturers. While in the process of writing this contribution, the author started to work at MAVinci GmbH but tries to be as impartial as possible.

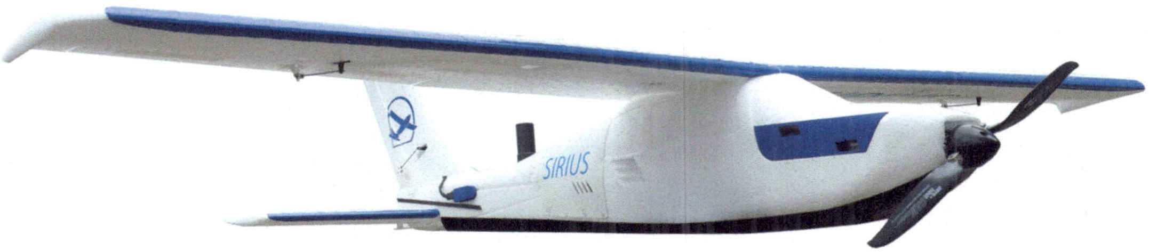

Fig. 10.2 The *MAVinci* Sirius (Photo: MAVinci)

spectrum areas. Using this technology, it might be possible to detect crop marks not visible for the human eye.

10.2.2.2 MAVinci Sirius

The German UAS manufacturer *MAVinci*, recently bought by *Intel*, is offering a complete system for survey tasks with different positioning systems. The Sirius Basic is without Real Time Kinematics (RTK), but the standard DGPS is sufficient for basic mapping tasks. For improved accuracy the Sirius Pro uses a high precision DGPS with RTK, resulting in an accuracy of up to 1.9 cm. They also offer a NIR camera and different additional systems. The Sirius aircraft (MAVinci 2016b) has a wingspan of 163 cm and a maximum takeoff weight of 2.7 kg carrying a payload of 550 g for 50 min (Fig. 10.2).

MAVinci is developing their own route planning software, an own autopilot system, and also a lot of other tools like an air traffic monitoring system; thus it is closed source. The processing of the images is usually done with *Agisoft PhotoScan* with a special plugin to simplify the postprocessing. A helpful feature is the import of previously mapped data, for example, digital surface models. These can be used for a terrain adopted flight plan, delivering constant ground sampling distances. This is especially important in rough mountainous terrains to both keep a steady ground sampling distance and avoid collisions.

Common applications for the UAS are surveying of large mines and other projects requiring large-scale data collection, especially when law regulations permit flights in greater altitude and beyond the line of sight.

The application in archaeology is yet to be tested, but one example can be shown here: the documentation of the UNESCO World Heritage Site Lorsch Abbey (see the examples section).

10.2.2.3 DJI

Phantom The *Phantom*, currently in version four (DJI 2016b), is one of the most popular drones at the moment (Fig. 10.3). It is easy to fly and can carry a fixed Ricoh GR system for up to 20 min or a GoPro camera with a gimbal[3] for 30 min. The drone is set up via mobile phone, and the comparatively low price makes this device very interesting for different purposes in both archaeology and geography. Additionally it has a convenient packing size and can be transported in a small case.

Naza and WooKong *Naza* (DJI 2016a) and *WooKong* (DJI 2016c) are a closed-source autopilot systems developed for different multicopter frames. The Naza is aimed at hobby pilots with the option to expand the functions. WooKong is for professional flying with GPS and waypoint support and thus a system for the use in geo-archaeology—and of course more expensive. Both platforms offer different software for the setup and route planning, all closed source.

Other DJI Systems DJI is offering more aircrafts and autopilot systems which are interesting for geo-archaeological purposes, too. They offer different cameras for their systems;

[3] A system to adjust and stabilize the camera to compensate UAS movements

Fig. 10.3 The *DJI Phantom*, Version II. (Photo: DJI technologies https://commons.wikimedia.org/w/index.php?curid=38812304)

one is a special combination of system camera and gimbal, offering a good image quality.

10.2.2.4 Mikrokopter Okto XL

One of the oldest multirotor autopilot projects and still further developed is the Mikrokopter hardware, produced by the corresponding company; see Mikrokopter (2016). The hardware is sophisticated and also fail-safe. The software is Windows only and has some bugs, but the main features are working properly. The maps in software are functioning offline but somehow tricky to use.

They are not only selling the autopilot system but also complete UAS aircrafts. Their Okto XL (Fig. 10.4) is a pre-built octocopter and needs only some additional hardware. It can carry up to 4 kg for about 30 min, depending on the wind. This system is adjustable; the user can choose from different setups to fit the law regulations and the camera system for his purposes.

If the customer buys the autopilot system only, he can choose his other UAS parts independently.

This system is marking the step to the open-source autopilots, since parts of the software implementation are open to the community. Many projects are using this hardware in either configuration; see, for example, Haubeck and Prinz (2013) and Rinaudo et al. (2012).

10.2.3 Open-Source Systems

In this section, the systems are more or less the autopilot hardware, not complete systems. The user usually has to choose an autopilot system fitting his purposes and also find his own type of aircraft, as well as the propulsion system. This provides high flexibility but the user needs patience and a lot of skills. The lack of it may cause mistakes as well as setbacks. A lot of knowledge and time are also necessary for maintenance.

10.2.3.1 Paparazzi Project

When started, this project was focusing on fixed-wing aircrafts only, but due to high community activities, this system can now control most aircrafts. For low-cost projects, preferably with fixed-wing airplanes, this might be an option to give a chance. The documentation is very good. A more detailed overview is giving Hattenberger et al. (2014).

Fig. 10.4 The *Mikrokopter* Okto XL (Photo: Mikrokopter)

10.2.3.2 ArduPilot

The *ArduPilot* project started in 2009 and, as the name already shows, is based on the *Arduino* microcontroller development board. The project got a large community due to low costs and soon developed its own hardware platform, the *APM*. It is, like the Arduino, based on ARM processors for the flight software. This system is still in development and the software has many contributors. For projects with small budgets also worth a try, see ArduPilot (2016a).

10.2.3.3 Pixhawk

From the *ArduCopter* project expanded a branch of new hardware with a more powerful board and more possibilities of additional sensor integration, beginning in 2012 with the *PX4* and since 2013 the *Pixhawk*. Main development is done by Lorenz Meier of *ETH Zürich* (Meier et al. 2012). Through its origin it offers the possibility to use both autopilot software, the Pixhawk suite as well as the ArduPilot. It offers some very distinctive sensor features like an airspeed sensor, an optical flow measurement, and ultrasound distance meters. Even a kind of laser distance scanner is available. Additionally the source code is very well documented and described, and there is really good route planning software available.

As complex as the features are, this system is quite simple to build and to fly. Of course, some experience with electronics and RC models is helpful, but not mandatory.

10.3 Structure from Motion

SFM is a photogrammetric approach to compute 3D models from a series of photos. They have to be taken from different viewpoints regularly spread around the object in order to compute a model. Since photographic documentation is long established in archaeology, the necessary equipment is available for this effort, and therefore its emergence in the last few years is no surprise. Additionally, the object's size does not matter except for the selection of a matching lens. It is possible to record a small cylinder seal as well as complete strongholds, cities, or even mountains.

There is a lot of software available, both open source and commercial. Lots of it are free to use or at least offered to a reduced price for academic purposes.

10.3.1 Free-to-Use or Open-Source Software

The most used free-to-use software is C. Wu's *VisualSFM* (VSFM) (Wu 2016; Wu et al. 2011), since it is very easy to install and to use as well as delivering really nice results in combination with

Y. Furukawa's *CMVS/PMVS* (Furukawa and Ponce 2010). Open-source ones are *Bundler* (also in combination with CMVS/PMVS to compute the dense point cloud), indeed one of the oldest software and quite slow, but it is still available. Newer ones with better parallelization and GPU support are, for example, *Theia-SFM*. Both are far less comfortable than VisualSFM, but they might be an option for difficult reconstructions or benchmarking.

10.3.2 Commercial Software

Agisoft *PhotoScan* (Agisoft 2016) is widely used not only in archaeology and has various functions and options for the computation of 3D models. The user interface is intuitive and offers many tools for setting masks and many more. *Pix4D* is focusing on UAS applications (Pix4D 2016) and brings several different tools for a lot of tasks. Many companies are also working on their own reconstruction pipelines; quite new is *3DF Zephyr* (3DFlow 2016). The new algorithms appearing at the moment are promising faster computations with better details, like the one from Slovakian company *Capturing Reality* (CapturingReality 2016). This has yet to be proven.

10.3.3 Method

The structure-from-motion method, no matter which software, is basically the same if the positions of the camera are not known.

Find Points of Interest This very first step is analyzing each photo for distinctive points or features. They have to have different characteristics; the points should not change by rotation or scaling. There are different approaches on how to find interests. VisualSFM, for example, uses the SIFT algorithm (Lowe 2004), originally developed to create multiline panoramas. There are plenty other algorithms available, only to name a few: SURF, BRISK, Freak, and more. An overview offers (Schaeffer 2013).

Feature Matching The interests are now compared to find similar objects shown in the images. To do this, all interests of each photo have to be compared with all other photos. Therefore the amount of computation steps is very high, while the size of a single interest is small. Thus, they can be computed with the support of the graphic cards, which accelerates the computation a lot (Wu 2007). Another approach is using statistics to sort the similarities at first and then only compute useful image matches. A short overview gives Hartmann et al. (2016).

Camera Position Computation The position where each photo was taken from can now be reconstructed in 3D space, based on the informations coming from the matching. Each image is referred as camera from now on. This step is called *bundle adjustment*. The approach is to take two photos with high similarities and use the interests to find their positions by using epipolar geometry (Luhmann 2010, 274ff). After the two camera positions are known, a third camera is added and the positioning is repeated. This is done until either no more cameras are left or the errors are too large. With each step the camera calibration improves, resulting in a higher precision of all camera positions (Wu et al. 2011). During this position reconstruction, the interest points are also reconstructed in 3D space, resulting in a sparse point cloud of the object. There are more approaches than the presented ones, but for this one is easy to explain, it stands as an example. Other approaches are optimizing the camera positions globally or do a lot more computing to improve the results.

If the camera position is known, for example, with information coming from a GPS receiver, this step is reduced to just optimize the cameras.

Dense Point Cloud Generation Now that the camera positions are available, they can be used to reconstruct each point visible on at least two photos by an inverse step of the position reconstruction described above. In recent approaches, depth maps are computed for each image pair, from which the so-called dense point cloud can be derived. There is different software available to solve this; VisualSFM is using CMVS/PMVS but can also feed programs like *CMP-MVS* (Jancosek and Pajdla 2011) or *SURE* (Rothermel et al. 2012). *SURE* is a very powerful tool, resulting in highly detailed and accurate dense point clouds, while it is at the same time very resource friendly. Problems occur with surfaces of homogeneous color, no structure, glossiness, or even slight transparency.

Surface Mesh Generation The final step of the creation is to compute a surface over the dense point cloud. This is a step with only a few algorithms; the most common is the Poisson Surface Reconstruction, as it is implemented in open-source postprocessing tools like *CloudCompare* (Girardeau-Montaut 2016) or *MeshLab* (Meshlab 2016).

Postprocessing and Analysis It is important to realize that the resulting data is not scaled, since no distances are known without any surveyed information. But most tools offer the possibility to use ground-based markers for georeferencing or to set a scale. After that the analysis can start, which can be measurements, area and volume estimations, slope computation, and many more. Also the surface can be analyzed and the visibility of specific features can be enhanced. Applications of this kind are shown in the contributions of H. Mara and B. Rieck in this book.

10.3.4 Applications

The application spectrum of SFM is, as mentioned before, vast. Objects made from stone or other nontransparent materials are usually easy to capture, as their surface is not too homogeneous. When they are worked in addition, the results are very accurate and do not need too many photos. Sometimes mica can cause small holes in the model, but they can easily be closed by hole-fill algorithms. Some examples will be shown below, especially the ornaments of Lorsch Abbey which are a good example for this type of objects. Different behavior is detectable when the stone has a glossy surface, like polished steatite as used for seals. A solution is then to take more photos to correct the reflections. If the objects are transparent, shiny, or having contrast-less, homogeneous surfaces, some photographic experience is necessary. But even then, some objects are just impossible to capture or have to be treated specially. Objects made of clay, especially pots or potsherds, but also the many other object types of clay found in archaeology, are usually straightforward to document. Sometimes the surface lacks contrast, but this is rather unusual. More effort needs some high glossy ware, white and also black areas.

Especially when it comes to documenting small objects, some special effort is necessary. The longer the focal length of the lens, the smaller is the focus range. In combination the autofocus can fail more often. The first effect makes it necessary to take more photos of different views, while the latter causes more work for the photographer, as he has to change the focus measurement point quite often or it is even necessary to take the photos in manual focus mode. In comparison, larger objects are usually easier to capture. With SFM it is also possible to create models of excavation areas, historical buildings, or even whole landscapes—of course with imagery coming from UAS, manned aircrafts, or even satellites.

10.4 Combined Approach

Mapping larger areas, for example, an archaeological excavation, makes it necessary to use aerial imagery. Ground-based photos are usually not sufficient to reconstruct plain surfaces because the cameras' angle of view to the surface is too flat. There are other possibilities, for example, large poles or tripods with heights of 10 m or more are used, but this is strenuous and slow.[4]

[4] For information on this topic, a good starting point is to search for *pole aerial photography*.

Some of the previous mentioned UAS systems are offering flight planning tools which include also a planner for grid-like routes to cover whole areas. In Mission Planner (for Pixhawk and ArduPilot), the user is able to choose his camera, lens, and a flight altitude as well as the overlap settings for the pictures. This is what we basically realized for the software of the ArchEye project a few years ago (Seitz and Altenbach 2011). But this is only suitable for flat areas, not for uprising objects even though some more recent approaches are supporting the use of landscape data. Therefore it is necessary to have video preview of the photos the UAS is taking to get them from the required positions (ArduPilot 2016b). MAVinci's flight planner is offering a different solution, allowing the user to import terrain data from previous flights. This very detailed surface model is then used to get a continuous ground sampling distance as well as improved flight patterns for steep areas (MAVinci 2016a).

It is also possible to combine SFM data with different 3D methods like TLS. This method, further explained in the LiDAR section in this book, is resulting in point clouds which can be filled up with SFM data. The TLS data with their high absolute accuracy can be used to verify the SFM data when the overlap between both datasets is large enough.

10.5 Examples

The examples are chosen to show the manifold applications of SFM and the use of UAS. They also focus on the presentation of useful topic-related information; therefore the historic facts are kept to a minimum.

10.5.1 Lorsch Abbey, Germany

The UNESCO World Heritage Site *Lorsch Abbey* is a Carolingian foundation, situated roughly between Heidelberg and Frankfurt (Main). The preserved compound consists today of the abbey's wall in the eastern and southern region, a remain of the church, and the world famous *King's Hall* in the northwest (Untermann 2011) (Fig. 10.5).[5]

The first example is one of the eight capitals of the *King's Hall*. This example is ground based only. It shows how the digital model can support the documentation by showing parts of the capital, which cannot be easily drawn while inserted in the wall because parts are concealed. Each capital was photographed in 20 min, while the processing took 10 h at that time.

The second example is an impost of the *church fragment*, showing all the details of the ornaments, like the little upward-looking monster face (Fig. 10.6). The visibility is enhanced using the *Multiscale Integral Invariant* algorithm (Fig. 10.7) in H. Mara's software GigaMesh (Mara et al. 2010); see also his contribution in this book.

The third example is the *King's Hall* as a whole (Fig. 10.8). The current model consists of 1306 photos; roughly half of them were taken by ArchEye's oktocopter. The resulting 3D model consists of 33.2 million points and derived from those are two point clouds with 15 and 5 million faces. All important ornaments are visible, although not completely modeled in this quality setup. The model could be better if a lot more computation time would be used.

Recently a flight with the *MAVinci Sirius Pro* took place over the abbey and the city of Lorsch. The result is a 3D model of the whole area and, by using a special lens, even with details on the facades, which usually are not mapped with standard methods. This dataset is very large, as it consists of 2430 photos, each georeferenced by RTK-GPS. The flight with the MAVinci Sirius Pro was with a 8 mm wide-angle lens to get more information than with the standard lens. The ground resolution was 3.71 $^{cm}/_{px}$ over an area of 85.7 ha. The point cloud has over 127 million points. Processing took 6 days. The result is a

[5] The author would like to thank Dr. K. Papajanni (TU Munich, Staatliche Schlösser und Gärten Hessen) and Dr. H. Schefers (Site manager, World Heritage Site Lorsch Abbey) for the possibility and support of testing.

Fig. 10.5 Capital of the King's Hall, eastern side, *left*. *Left* shows the colored model, *right* the uncolored model enhanced with MeshLab's radiance scaling

Fig. 10.6 The impost of the church fragment at Lorsch Abbey, colorized model

complete 3D model of the abbey area and part of the city along with a *digital elevation model* (DEM) and a high-resolution orthographic photo (Fig. 10.9).

10.5.2 Ancient City of Troezen, Greece

The antique Greek city of *Troezen* is situated in the east part of the Peloponnese, Greece. Despite

Fig. 10.7 The impost of the church fragment at Lorsch Abbey. Visualization on the *right* using the *MSII-Algorithm* in *GigaMesh*

Fig. 10.8 The model of the King's Hall, Lorsch Abbey. *Left side view*, right northern facade

being mystically loaded, the acropolis and the lower city were only roughly surveyed in the 1930s. In recent years, Professor Reinhard Stupperich from the Institute for Classical Archaeology, Heidelberg University, started with the Australian Troizen Archaeology Project to update the archaeological informations of the remains. In that effort, a 3D model of the so-called Frankish Tower was created. In 2012 a ground-based model was generated, while in 2015 a complete model with ArchEye's oktocopter was created. The first model already showed something that could not easily be observed by naked eye, a third building phase (Fig. 10.10).

During the 2015 remote sensing campaign, we mapped a large area of the ancient city and the sanctuary area with an eBee, provided by senseFly for the master's thesis of Michael Ebner, who operated the UAS. The result is an orthographic photo of the whole area, shown in Fig. 10.11.

10.6 Outlook

In the future, drones will more likely support documentation tasks, whether in archaeology or geography. Methods for live 3D data creation will increase and become more efficient, while SFM will be the tool to compute detailed and scientifically relevant data. The technical possibilities and developments in both fields seem not to have reached their peak yet.

For the practical application, more automatic systems, easier controls, new algorithms, and

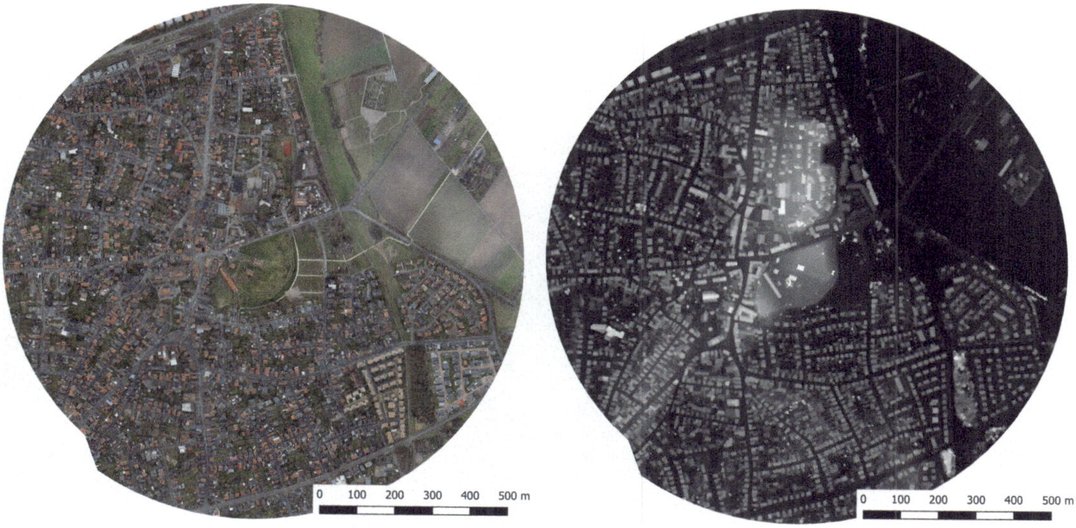

Fig. 10.9 The results of the circular flights at Lorsch, *left* the orthographic photo, *right* the digital surface model (Data and Images: MAVinci)

Fig. 10.10 Detail view of the *Frankish Tower* model, northern wall, showing the three building phases, the lower two probably from Hellenistic times, while the topmost is from the Frankish time

improved sensors can be expected. We will probably soon see UAS doing their many different tasks fully autonomously on a daily basis.

The legislative situation at the moment is tense, as the low prices lead to a high number of UAS sales in the recent years. In Germany, at least, many UAS operators do not know the regulations for their flights or do not respect the personal rights.[6] Some of the operators are even careless by hazarding people when flying over crowds or close to airports. This leads to stricter

[6] A short overview for Germany, Austria, and Switzerland is available at Mikrokopter.de: http://wiki.mikrokopter.de/rechtlicheGrundlagen

Fig. 10.11 The orthographic photo of the city area of Troezen with a detail on the temple of Hippolytus, Greece, April 2015 (Data: Michael Ebner, Map: Author)

regulations, though affecting mainly commercial and scientific UAS applications.

UAS will doubtlessly be another important tool in the tool box of the scientist, maybe equivalent to a photo camera—no matter if archaeologist or geographer. Therefore the need to teach about these tools will have to be included in the curriculum as soon as possible.

References

3DFlow (2016) Zephyr. http://www.3dflow.net/3df-zephyr-pro-3d-models-from-photos/

Agisoft (2016) Photoscan. http://www.agisoft.com/

ArduPilot (2016a) http://ardupilot.org

ArduPilot (2016b) Planning a mission with waypoints and events. http://ardupilot.org/planner/docs/common-planning-a-mission-with-waypoints-and-events.html

CapturingReality (2016) www.capturingreality.com

DJI (2016a) Naza-m. http://www.dji.com/de/product/naza-m-v2

DJI (2016b) Phantom. http://www.dji.com/de/product/phantom-4

DJI (2016c) Wookong. http://www.dji.com/de/product/wookong-m

Eisenbeiß H (2009) UAV photogrammetry. ETH Zurich, Switzerland

Furukawa Y, Ponce J (2010) Accurate and dense robust multi-view stereopsis. IEEE Trans Pattern Anal Mach Intell 32(8):1362–1376

Girardeau-Montaut D (2016) Cloudcompare. http://www.danielgm.net/cc/

Hartmann W, Havlena M, Schindler K (2016) Recent developments in large-scale tie-point matching. ISPRS J Photogramm Remote Sens 115:47–62

Hattenberger G, Bronz M, Gorraz M (2014) Using the paparazzi UAV system for scientific research. In: IMAV 2014, International micro air vehicle conference and competition 2014, pp 247–252

Haubeck K, Prinz T (2013) A UAV-based low-cost stereo camera system for archaeological surveys – experiences from Doliche (Turkey). Int Arch Photogramm Remote Sens Spat Inf Sci XL-1/W2:195–200

Jancosek M, Pajdla T (2011) Multi-view reconstruction preserving weakly-supported surfaces. In: IEEE conference on computer vision and pattern recognition (CVPR), 2011, IEEE, pp 3121–3128

Lowe DG (2004) Distinctive image features from scale-invariant keypoints. Int J Comput Vis 60(2):91–110. https://doi.org/10.1023/B:VISI.0000029664.99615.94

Luhmann T (2010) Nahbereichsphotogrammetrie. Grundlagen, Methoden und Anwendungen, vol 3. Wichmann VDE-Verlag, Berlin

Mara H, Krömker S, Jakob S, Breuckmann B (2010) GigaMesh and gilgamesh – 3D multiscale integral invariant cuneiform character extraction. In: Artusi A, Joly M, Lucet G, Pitzalis D, Ribes A (eds) Proceedings of the 11th international conference on virtual reality, archaeology and cultural heritage, Eurographics Association, Paris, pp 131–138. http://diglib.eg.org/EG/DL/WS/VAST/VAST10/131-138.pdf

MAVinci (2016a) Mavinci flight planning software. http://www.mavinci.de/mavinci-system/mission-control-software/

MAVinci (2016b) Sirius. http://www.mavinci.de/mavinci-system/uav/

Meier L, Tanskanen P, Heng L, Lee GH, Fraundorfer F, Pollefeys M (2012) Pixhawk: a micro aerial vehicle design for autonomous flight using onboard computer vision. Auton Robot 33(1–2):21–39

Meshlab (2016) http://meshlab.sourceforge.net/

Mikrokopter (2016) http://www.mikrokopter.de

Pix4D (2016) https://pix4d.com/

Rinaudo F, Chiabrando F, Lingua AM, Spanò AT (2012) Archaeological site monitoring: UAV photogrammetry can be an answer. Int Arch Photogramm Remote Sens Spat Inf Sci 39(B5):583–588

Rothermel M, Wenzel K, Fritsch D, Haala N (2012) Sure: photogrammetric surface reconstruction from imagery. In: Proceedings LC3D workshop, Berlin, vol 8

Schaeffer C (2013) A comparison of keypoint descriptors in the context of pedestrian detection: Freak vs. surf vs. brisk. Cité en p 12

Seitz C, Altenbach H (2011) Project archeye–the quadrocopter as the archaeologist's eye. ISPRS Int Arch Photogramm Remote Sens Spat Inf Sci 38(1): C22

Sensefly (2016) ebee. https://www.sensefly.com/drones/ebee.html

Untermann M (2011) Die Torhalle. In: Hessisches Landesmuseum Darmstadt, Verwaltung der Staatlichen Schlösser und Gärten Hessen (eds) Kloster Lorsch. Vom Reichskloster Karls des Großen zum Weltkulturerbe der Menschheit, Imhof, Petersberg, pp 194–214

Wu C (2007) SiftGPU: a GPU implementation of scale invariant feature transform (SIFT). http://cs.unc.edu/~ccwu/siftgpu

Wu C (2016) Visualsfm. http://ccwu.me/vsfm/

Wu C, Agarwal S, Curless B, Seitz S (2011) Multicore bundle adjustment. IEEE transactions on computer vision and pattern recognition (CVPR), pp 3057–3064

Part III

Laser Scanning Applications

Introduction to LiDAR in Geoarchaeology from a Technological Perspective

11

Martin Hämmerle and Bernhard Höfle

Abstract

LiDAR is a remote sensing method established in the geosciences for capturing highly accurate three-dimensional geodata. It is increasingly used to support geoarchaeological research due to a range of advantages, including survey-grade data quality, real 3D geodata, nonselective coverage of scenes with high measurement density, on-demand data capturing, and comprehensive filtering options based on geometric and radiometric information.

Different LiDAR measurement principles are used to derive 3D geodata, e.g., time-of-flight, phase shift, and structured light. The captured datasets include real 3D XYZ coordinates (point clouds). Depending on the sensor system, also radiometric information is gathered for each point, e.g., RGB, strength, and full waveform of the backscattered signal.

The processing of point clouds in general follows a similar workflow. After data acquisition, the point clouds are registered. Depending on available radiometric information and research question, the data is radiometrically calibrated. After removing outliers, the point cloud is filtered according to the requirements of the study, and derivative products (e.g., DTMs) are generated (Chaps. 12 and 13). Finally, geospatial analyses are conducted directly in the final point cloud or derived elevation models (e.g., raster datasets, TINs), and the data can be used for visualization purposes.

With LiDAR, a wide range of spatial scales can be captured: (1) Object scale: single objects and their surfaces can be captured in high detail, allowing for studies, e.g., of engravings. (2) On-site scale: specific areas or

M. Hämmerle (✉)
GIScience & 3D Spatial Data Processing Research Group, Institute of Geography, Heidelberg University, Im Neuenheimer Feld 368, 69120 Heidelberg, Germany
e-mail: haemmerle@uni-heidelberg.de

B. Höfle
GIScience & 3D Spatial Data Processing Research Group, Institute of Geography, Heidelberg University, Im Neuenheimer Feld 368, 69120 Heidelberg, Germany

Heidelberg Center for the Environment, Im Neuenheimer Feld 229, Heidelberg, Germany
e-mail: hoefle@uni-heidelberg.de

© Springer International Publishing AG 2018

C. Siart et al. (eds.), *Digital Geoarchaeology*, Natural Science in Archaeology,
DOI 10.1007/978-3-319-25316-9_11

objects like excavation sites or building complexes are surveyed and documented with static or dynamic scanners applied on-site (TLS, ULS, etc.). Such datasets covering sites of tens or several hundreds of meters are used in studies which examine the local spatial relations of objects and their immediate surroundings (see also Chap. 13). (3) Off-site scale: on a spatial scale covering whole regions, off-site scanning, mainly from an airborne platform, is applied.

However, the LiDAR method has some shortcomings, e.g., high costs for equipment, trained personnel, and processing the large amounts of data. Here, low-cost approaches can offer complementary data sources.

The described workflow leads to important implications for DGA regarding the filtering and calculation of derivatives. There is no universal filter but only tailored filters, depending on the available data and the aim of the study. Furthermore, the original, raw point cloud should always be available to enable subsequent analyses with new methods or different aims and objects of interest.

Emerging fields offering new possibilities for LiDAR in DGA are, e.g., multi-platform and multi-sensor approaches, the combination of spatial scales, multi-wavelength devices, multi-temporal datasets, and refined filtering based on radiometric information, e.g., full waveform.

Keywords
LiDAR • Laser scanning • 3D spatial data processing • 3D models • Geometry • Radiometry • Active remote sensing

11.1 Introduction

LiDAR is an active remote sensing method for capturing 3D geodata in high detail and with high accuracy. Archaeological studies working with LiDAR data are well situated within the discipline of geoarchaeology by establishing a methodological bridge between geosciences and archaeology (Renfrew 1976; Ghilardi and Desruelles 2009; Rapp and Hill 2009; Diskin et al. 2013).

This introduction chapter gives an overview of LiDAR principles and applications with a focus on geoarchaeology. First, LiDAR is introduced regarding technical aspects. The method's advantages and drawbacks are described with respect to the applicability in geoarchaeology, and a general workflow of LiDAR data acquisition and processing is outlined. Selected case studies covering a wide range of applications are presented before concluding with notes on research gaps and future trends.

11.2 How It Works: Principles of Capturing 3D Geodata with LiDAR

The acronym LiDAR stands for "Light Detection And Ranging." The method is also referred to as laser scanning. Both expressions hint at the underlying measurement principle: LiDAR instruments emit laser shots and derive the range between the device and an object surface hit by the shot.

The most common range measurement techniques are time-of-flight, phase shift, and structured light (Petrie and Toth 2009a; Beraldin et al. 2010). In case of time-of-flight measurements, the time between single emitted laser shots and the respective backscattered echo recording is used to derive the travelled distance of the laser shot. Phase-shift laser scanners emit a continuous laser beam. The range can then be derived from the phase-angle shift of the emitted and the received signal. Structured light scanners derive 3D coordinates from the discrepancies between a reference light pattern and the pattern's distortions when projected onto an object or a scene.

To cover an entire scene with measurements, i.e., to "scan" an entire scene, multiple laser shots have to be distributed over the area of interest, such as by deflecting the shots via a rotating mirror or even by rotating the entire scanner head (Petrie and Toth 2009b, c). The scanning device can be mounted on either a static (terrestrial laser scanning, TLS) or a dynamic platform. Especially the set of dynamic platforms is growing continuously and includes:

- Airborne laser scanning with the LiDAR device mounted on a plane or a helicopter (ALS, Fernandez-Diaz et al. 2014)
- Ground-based mobile laser scanning, for example, from cars or trains (MLS, Liang et al. 2014a; Elberink and Khoshelham 2015)
- UAS-borne laser scanning with unmanned aerial systems carrying the scanner (ULS, Wallace et al. 2012; Pfennigbauer et al. 2014)
- Scanners mounted on boats (Boeder et al. 2011)
- Personal laser scanning in which scanners are mounted like backpacks or carried by hand (PLS, Liang et al. 2014b; Zlot et al. 2014)

The most suitable platform for your application has to be carefully selected, for example, regarding accuracy and precision of the measurements, and terrain accessibility. Also, the size of the object and the spatial extent of the area of interest influence the choice of platform: in case an entire landscape needs to be covered, laser scanning from an off-site dynamic airborne platform is the standard method of choice. If a single small object, an exhibit, needs to be captured in high resolution, accuracy, and submillimeter detail, a static structured light scanner will produce appropriate datasets. Furthermore, if measurements have to be taken multiple times, the temporal resolution, i.e., the frequency of repeated scanning campaigns, will limit the use of certain approaches. Monitoring an excavation, for example, will be done most efficiently with a static terrestrial laser scanner placed on-site. An overview on the coverage of spatial and temporal scales by different platforms and device types is given in Fig. 11.1.

Additionally to the range measurements, the horizontal and vertical emission angles for each laser shot are recorded. The range and angle measurements are subsequently transformed into a set of XYZ coordinates called point cloud (Heritage and Large 2009). Point clouds are the basic dataset for all subsequent analyses.

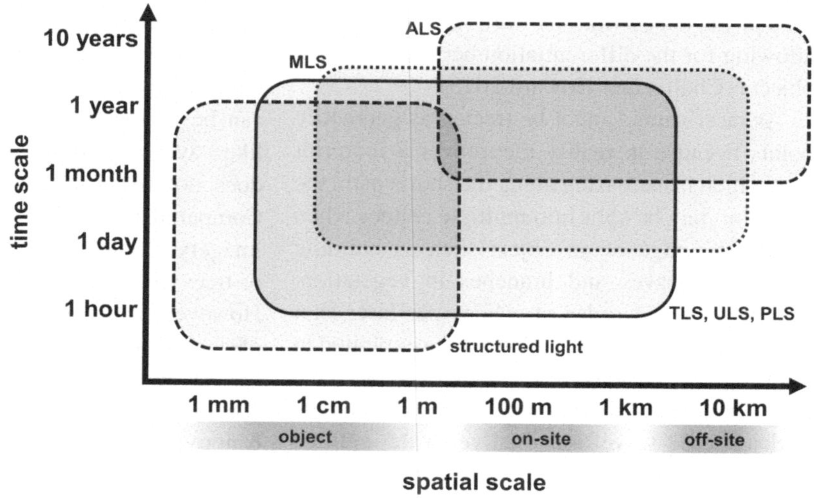

Fig. 11.1 Coverage of spatial scale (size of examined object, spatial extent of area of interest) and time scale (frequency of repeated LiDAR campaigns) by different laser scanning approaches (after Böhler and Heinz 1999; Heritage and Hetherington 2007)

Fig. 11.2 (**a**) RGB-colored point cloud of the gunpowder magazine of Heidelberg Castle and (**b**) same scene colored according to backscatter signal strength (*gray gradient*, differing surface backscatter characteristics; *green*, very low backscatter strength, e.g., on vegetation)

In addition to the core information of a point cloud, i.e., the XYZ coordinates, further features can be recorded for each point. Very common is the assignment of RGB values to each single point (Fig. 11.2a) which facilitates navigating through the data, allows for analyses based on the point colors, and helps to achieve realistic visualizations. Furthermore, radiometric information can be captured for each point. For example, the strength of the recorded signal depends on the backscatter characteristics of the scanned object surface. Differently scattering surfaces, thus, result in different backscatter signal strength (Fig. 11.2b; see Höfle and Pfeifer 2007 for a physical description of LiDAR backscatter), allowing for the differentiation between scanned objects (Challis and Howard 2013).

As laser shots cannot be treated as geometric points because in reality they cover a footprint area which is increasing along the shot's path, the footprint may be split into multiple echoes when hitting the edge of an object such as building corners or leaves and branches in vegetation. The number and order of echoes per laser shot provide another feature which can be exploited in point clouds. Increasingly examined is also the full waveform of a laser shot. Deviations of the backscattered waveform relative to the emitted signal as well as the full waveform of echoes along the laser shot's propagation path provide further valuable information about captured surfaces and objects (Stilla and Jutzi 2009; Doneus et al. 2010; Fernandez-Diaz et al. 2014).

11.3 Advantages and Drawbacks of LiDAR

Being a remote sensing method is a general advantage of LiDAR, especially regarding the application in geoarchaeology. No destructive sampling is necessary for the measurements, and the object or scene of interest can be captured without changing its properties. Another straightforward benefit of LiDAR systems is the active emission of laser shots. LiDAR is, thus, not depending on lighting conditions so that scans can be performed in mediocre-lit environments like caves or rooms. The time of the day also does not restrict LiDAR scanning campaigns. Compared to passive remote sensing based on imagery, shadows cast by objects like buildings or trees do not influence LiDAR measurements. However, as laser shots cannot penetrate solid obstacles, occlusion occurs when conducting LiDAR measurements. Nevertheless, laser shots can "look" through small gaps in the vegetation canopy. The terrain can, thus, be extracted from a point cloud of an area covered by trees and shrubs—another major advantage of LiDAR

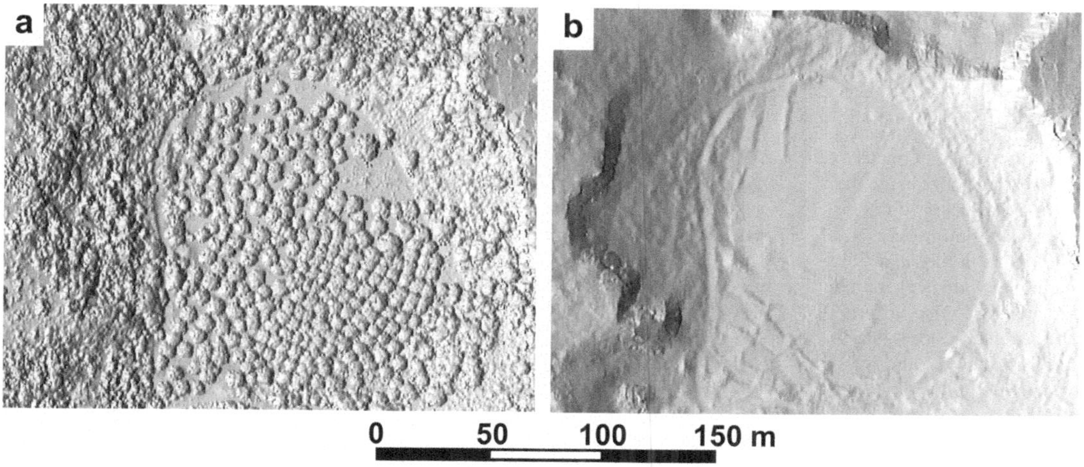

Fig. 11.3 Bird's eye view on a doline near Kroustas, Crete, Greece (cf. Siart et al. 2013). Shading of a digital surface model (DSM) including vegetation, (**b**) digital terrain model (DTM) representing the bare earth void of vegetation

(Pfeifer and Mandlburger 2009; Briese 2010; Risbøl 2013; Fernandez-Diaz et al. 2014; Johnson and Oumiet 2014; Fig. 11.3).

Extending the mentioned example of vegetated areas and the possibilities of extracting terrain points, LiDAR can be considered an advantageous 3D geodata gathering method because it is nonselective in what is captured. Not only can vegetation be removed from the point cloud, but also different object types (e.g., buildings, roads) and derivatives (e.g., different digital elevation models) can be extracted from the same LiDAR dataset depending on the geoarchaeological research question. Subsequently, a proper definition of which objects belong to the terrain and which objects are off terrain is indispensable for a successful study (Fig. 11.4). If, for example, the terrain without any vegetation or anthropogenic structures is of interest, all objects being off terrain can be removed. On the other hand, if other objects such as building remnants, plantations, etc. are in focus of a study, they can be extracted on demand from the same point cloud. Thus, it is crucial to have access to and to back up the original point cloud as a basis for potential follow-up analysis.

Another core benefit of LiDAR data is that scenes are covered in real 3D as opposed to 2.5D. A common example for 2.5D data is a raster dataset describing the surface of a landscape with XY coordinates representing the single raster cells and one single Z value representing the elevation of the respective raster cell. In case that a scene contains overhanging geometries as, for example, caves, rooms, rock cliffs with niches, bridges, and arcs, modeling the scene with the 2.5D raster approach would not result in a satisfactory representation. LiDAR point clouds on the other hand comprise XYZ coordinates of each scanned point without being restricted to one Z value at a certain position. The point cloud example shown in Fig. 11.5a comprises real 3D data: the wall of the scanned building consists of vertically "stacked" measurements. Also, the bell in the small tower is captured hovering over the main roof, but at the same time it is covered by the small bell tower roof so that three objects are captured at the same XY position.

Finally, survey-grade data quality of high-resolution and scene coverage can be gathered. Datasets captured with high-end LiDAR sensors allow, for example, for direct measurements of even delicate details directly within the point cloud (Fig. 11.5b, c). Additionally, high-resolution point clouds seamlessly covering an area offer the option to derive products like high-

Fig. 11.4 Necessity of appropriate terrain definition using the example of an archaeological site near Koumasa, Crete, Greece. The *red lines* delineate the terrain according to three exemplary definitions: (**a**) all objects in the scene including walls, vegetation, etc. belong to the terrain, (**b**) vegetation does not belong to the terrain, (**c**) vegetation and walls are defined as off terrain

resolution digital terrain models (DTMs) which, again, are highly valuable input data in geoarchaeological research (Höfle and Wagener 2012; Zielhofer et al. 2012; Schneider et al. 2014; Chap. 13).

Regarding the drawbacks of LiDAR, the acquisition of high-quality datasets requires high-end equipment, leading to high costs both in case a scanning device is bought and in case data acquisition is performed by a service provider. Time and subsequently costs are also consumed for processing the data, especially if new methods and algorithms have to be tailored by specialized staff for tackling the respective research question. It has to be kept in mind that off-the-shelf algorithms and workflows for processing 3D geodata are still hardly available. However, open source tools such as CloudCompare or the Point Cloud Library are intensively developed and available for a wide range of applications not least in geoarchaeology (e.g., Bevan et al. 2014). Finally, computing equipment with high performance in terms of processing and storage has to be provided as LiDAR datasets easily comprise billions of points, mostly with additional features beyond the XYZ coordinates attached to each point.

11.4 A Typical Workflow for LiDAR Data Capturing and Processing

A typical workflow from preparing a scanning campaign to visualizing the final results is outlined in the following part. Figure 11.6 sums up the main steps of an idealized LiDAR data capturing and processing chain. Depending on the specific application, some of the listed steps may be replaced by other tasks or skipped at all.

In case LiDAR data are not readily available, the initial step of a LiDAR workflow is to acquire proper datasets. Thorough planning of a laser scanning campaign is crucial as the gathered data are the basis for all subsequent analyses (Fernandez-Diaz et al. 2014). The requirements which the dataset has to fulfill, for example, regarding spatial resolution or full-waveform recording (Doneus et al. 2010), as well as the size of the examined objects, extent and accessibility of the research area, etc., lead to the selection of a proper LiDAR sensor and the suitable platform. Additionally, planning of paths of dynamic systems, the prospection and potential distribution of scan positions in case of static LiDAR, and the method for georeferencing the data are imperative for a successful campaign.

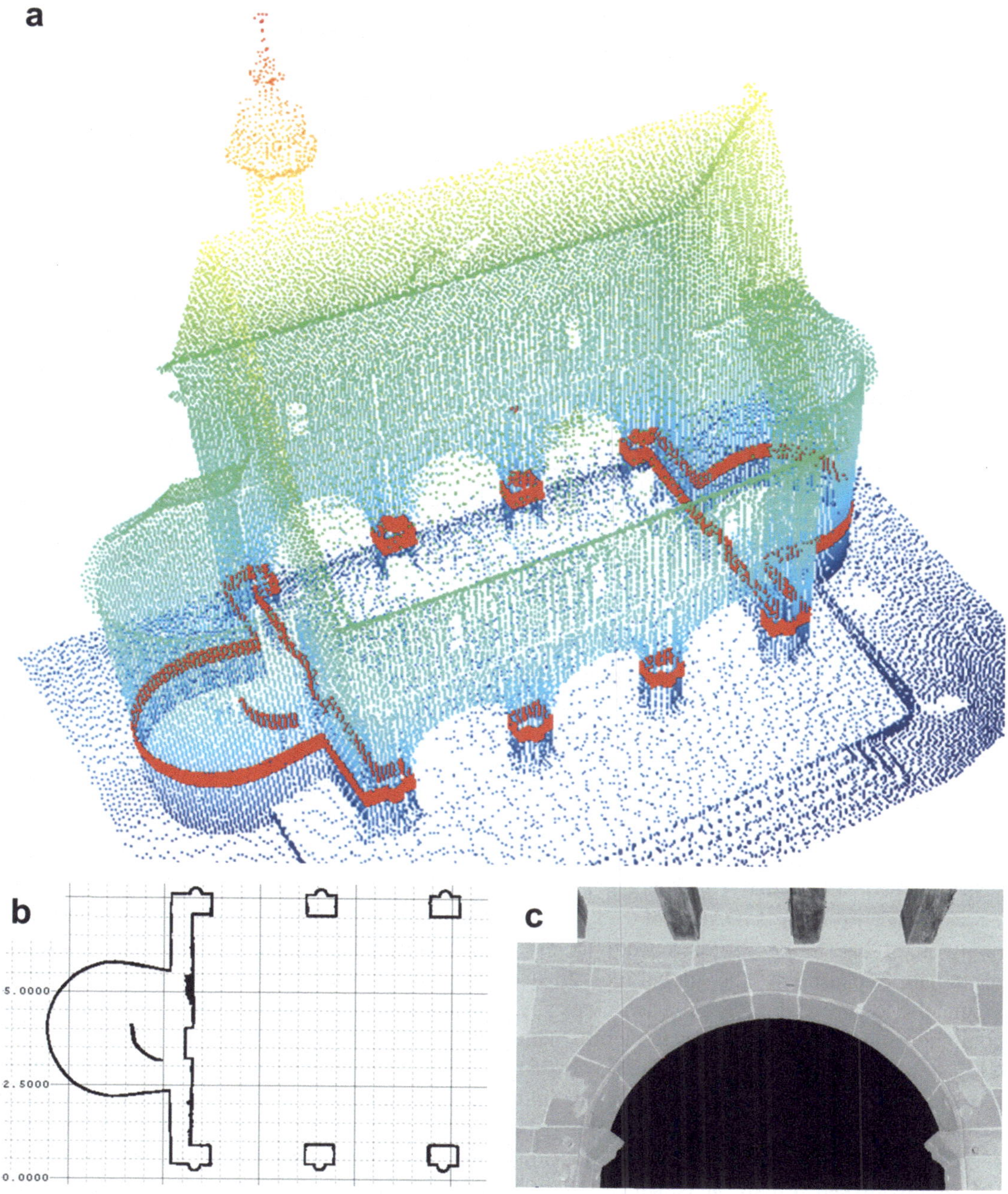

Fig. 11.5 Terrestrial laser scanning point cloud of the Lorsch Abbey gatehouse. (**a**) Coarse point cloud colored according to height (*blue* = 0 m, *orange* = ca. 20 m) with high-resolution point cloud cross section inserted in *red*, (**b**) plan view of the cross section with scale grid (distance of *solid lines*: 2.5 m), (**c**) detailed view on roof beams and masonry of the western façade (colored according to backscatter signal amplitude)

Chapter 12 addresses the topic of campaign planning as well as LiDAR and reference data acquisition in more detail.

The second bundle of tasks in a LiDAR workflow comprises the steps which are necessary to prepare the raw data for the analyses. As

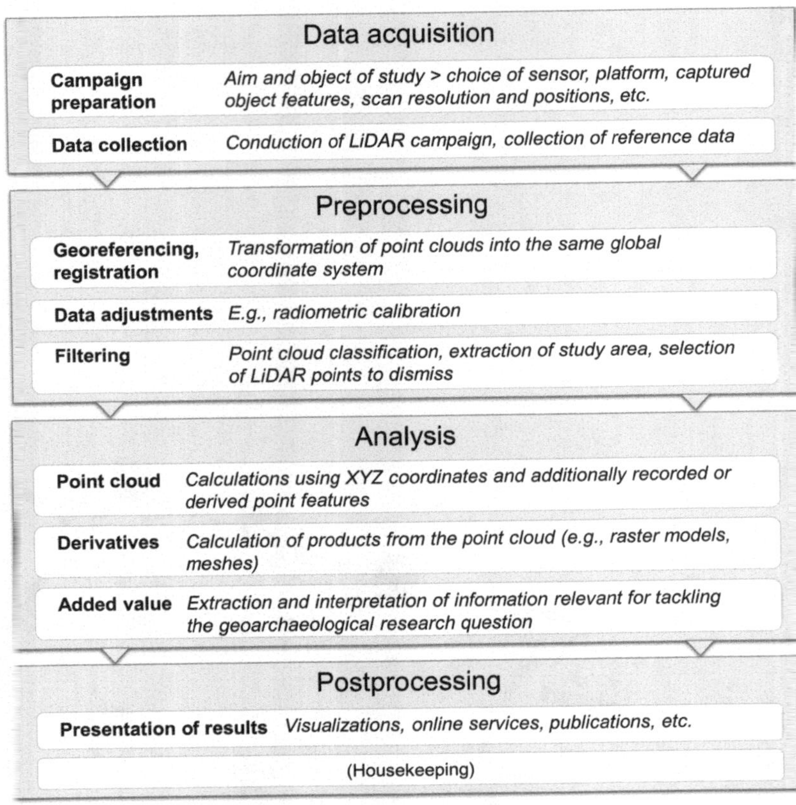

Fig. 11.6 Typical workflow of a LiDAR data processing chain

the geographic location is the core feature and value of geodata, global coordinates have to be assigned to each LiDAR point, i.e., the point cloud has to be georeferenced. In case of dynamic platforms, the global coordinate is mostly assigned directly to each point by combining the scanner measurements with Global Navigation Satellite System (GNSS) measurements for capturing the position, and inertial measurement unit (IMU) measurements for capturing the movements of the platform (El-Sheimy 2009; Lichti and Skaloud 2010; Fernandez-Diaz et al. 2014). Georeferencing of static measurements can also be done either directly or indirectly via tie points (Fig. 11.7). A tie point is a distinct object such as a building corner or a highly reflecting target placed in the scene, which can be seen clearly in the point cloud. Georeferencing the point cloud is then done by surveying the tie points in the target global coordinate system and by transforming the entire dataset via the tie points' local and global coordinate pairs. Similarly, corresponding tie points can be used to register multiple scan positions, i.e., to transform them into a common coordinate system.

Another preprocessing step is the radiometric calibration of LiDAR point clouds. Radiometric calibration is necessary if the subsequent analyses rely on radiometric features like backscatter strength. The recorded backscatter strength of surfaces depends on factors like scanning range and incidence angle (Höfle and Pfeifer 2007). If the same target is scanned from different distances between scanner and object surface, for example, because of different flying heights or scan positions, the signal backscatter will be different although the same

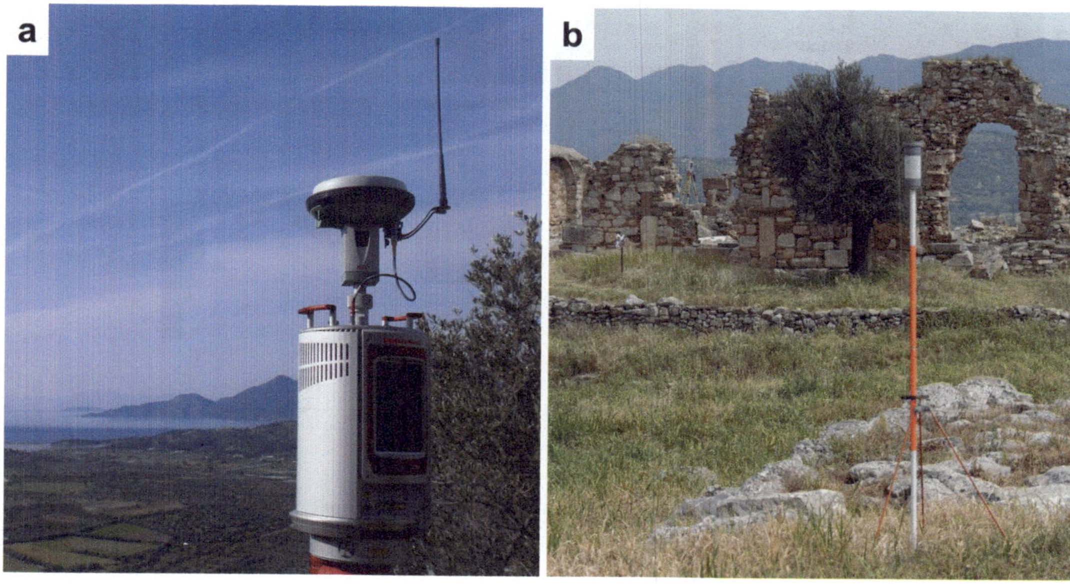

Fig. 11.7 (**a**) Long-range terrestrial laser scanner Riegl VZ-6000 with integrated compass and a Leica GS15 RTK GNSS mounted on top for direct georeferencing, (**b**) artificial tie point target of known geometry placed on top of a signal pole for indirect georeferencing (Hoffmeister et al. 2010, image courtesy Christian Seitz)

surface, having corresponding reflectance properties, is captured. The points covering the same surface or surface type have, thus, to be adjusted to a reference value. To achieve a calibration of data within a project, the reference value can be extracted from the data itself (e.g., by picking homogeneously backscattering areas, Höfle and Pfeifer 2007). For inter-campaign analyses, a reference target with known reflectivity has to be captured in the single campaigns, allowing for an absolute calibration to the target reference reflectivity. Alternatively, a calibration function can be derived for the scanning device and then applied on the point clouds (Koenig et al. 2013; Höfle 2014).

In a further preprocessing task, the point cloud can be filtered, for example, to identify points which can be dismissed (reflections on water surfaces, pollen or insects in the air, pedestrians, etc.) or points which are treated separately in the analyses. A common filtering task is, for example, to assign points to the classes "vegetation" or "terrain." Classification approaches range from very simple methods examining only the height of points above ground to more sophisticated algorithms including different surface interpolation methods (Sithole and Vosselman 2004), different 3D neighborhood definitions (e.g., cylinders and spheres, Höfle et al. 2009), neighborhood dimensionality in different neighborhood extents (Brodu and Lague 2012), or the combination of radiometric and geometric features (Höfle 2014).

After the preprocessing is completed, the point clouds can be visually inspected or analyzed to extract the information which the data were captured for. The analyses can be conducted directly in the point clouds (e.g., further classifications similar to the above-mentioned vegetation filtering) or based on products derived from the point clouds (e.g., digital terrain models, shadings, viewshed analyses). Chapter 12 gives a detailed introduction to the analyses of LiDAR data; case studies can be found in Chaps. 13 and 16.

Finally, the results of a project are normally presented via different media. The last step in the outlined LiDAR workflow is, thus, the visualization of data and analysis results. Similar to

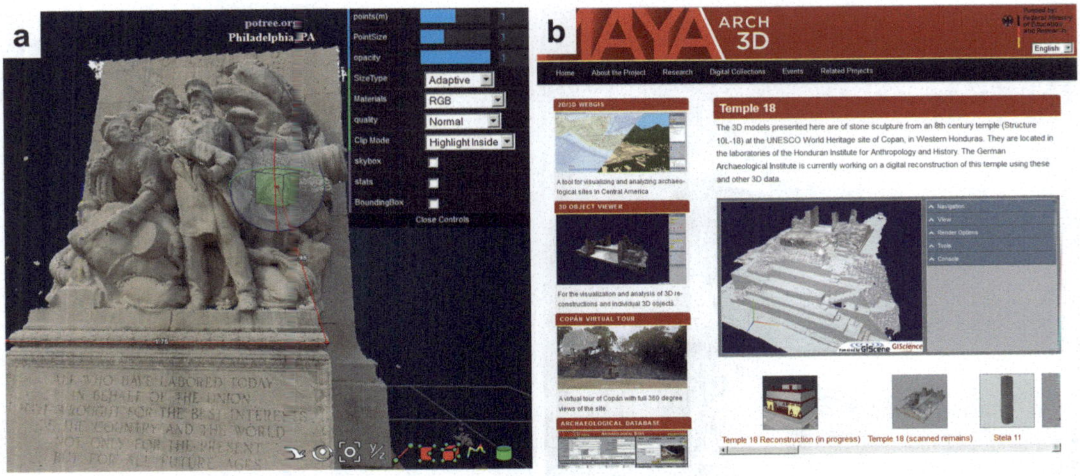

Fig. 11.8 (**a**) Online point cloud viewer with geometric measurement tools (courtesy Markus Schütz, potree.org), (**b**) virtual research environment for the documentation and analysis of complex archaeological sites, use case: ancient Maya city of Copán, Honduras (mayaarch3d.org)

the analyses, the visualization of data can be based on the point cloud itself or on its derivatives. Figure 11.8 exhibits examples for an advanced point cloud viewer including tools for extracting geometrical features,and a comprehensive online geoarchaeological information system comprising the visualization of archaeological sites including their landscape context and clickable information provided by a geodatabase containing archaeological data (Auer et al. 2014).

It should be mentioned that the workflow steps are not always clearly separated. Classifying a point cloud into terrain and vegetation points, for example, can be a part of data preprocessing if the vegetation class is excluded from the further analyses. However, if the vegetation points play a role in the further processing steps, the classification can also be considered as a part of the analysis. Similarly, the visualization of data and results does not necessarily constitute the final step of working with LiDAR data, as there is a wide range of analyses which can be done within a visualization (Fig. 11.8). However, the outline should give an impression of the necessary steps which are to be considered when capturing and processing LiDAR datasets.

11.5 Working on Different Scales: Selected Case Studies of LiDAR Applied in Geoarchaeology

In the following, a small selection of case studies using LiDAR in an archaeological context is presented. The examples cover a wide range of scales in terms of the spatial extent of examined objects or sites. Correspondingly, the scanning setup differs from explicitly capturing single objects, to scanning a complete archaeological area with sensors and operators being physically on-site, to collecting data off-site, for example, with an airborne LiDAR campaign.

Europe's oldest Jewish cemetery, the *Heiliger Sand* in Worms, Germany, was captured with two sensors to cover different scales: firstly, a terrestrial laser scan of millimeter accuracy and precision of the entire cemetery, leading to a detailed terrain model as well as accurate positions and aspects of the headstones. Secondly, micrometer accuracy structured light scans of selected headstones were performed to extract inscriptions not any more visible in images or to the bare eye (Mara et al. 2010; Krömker 2013; Fig. 11.9). A similar approach of combining two sensor systems covering

Fig. 11.9 Complementary use of two 3D sensors for capturing the Jewish cemetery *Heiliger Sand* near Worms, Germany. Foreground, structured light scanner Breuckmann smartSCAN 3D-HE color for scanning the headstones in micrometer accuracy; background, terrestrial laser scanner Riegl VZ-400 for scanning the entire site (image courtesy Hubert Mara)

different scales (ALS for site prospection, TLS for capturing details on-site) is presented in Chap. 13 in this book.

Terrestrial LiDAR (TLS) was also used to visualize an abris and its surroundings in the Peruvian Andes (Forbriger et al. 2011). From the 3D model, information about the abris' settlement history can be derived by examining its geometry and the geometry of stone rings erected around the abris. Due to its location close to perennial water sources, which are generally rare in this arid region, the role of the abris regarding transhumance can be better understood.

In another application of TLS, the gunpowder magazine of the Heidelberg Castle in Germany, consisting of a tower broken to pieces due to an explosion, was captured in 3D. The TLS point cloud was transferred to a mesh to examine geometric features like curvature, radius, etc. of the largest part of the broken construction. Additionally, a virtual "3D puzzle" was solved by relocating the large fragment to its original position. In the study it was proved that the broken part of the tower is actually an original part of the building and no historicizing reconstruction (Forbriger et al. 2013).

Shifting the focus on data captured with scanners located off-site, a range of geoarchaeological studies uses airborne laser scanning (ALS) datasets. For example, to assess the strategic importance of three locations around Thurant Castle, Germany, a viewshed analysis is conducted in a digital terrain model based on ALS data by Höfle and Wagener (2012). Based on the mutual visibility between castle and the three points as well as the visibility from the three points to the surrounding landscape, the strategic value of the locations for placing a tower was reassessed.

The characteristic topography of objects is also a feature often used in geoarchaeological studies: based on ALS data, Hesse (2014) shows the potential of high-resolution DTMs to detect and map traces of conflicts such as bomb craters, trenches, defensive structures, etc. Similarly, archaeological sites are prospected on the basis of ALS data as shown, for example, in Hesse (2013) and Johnson and Oumiet (2014).

Introducing a new airborne laser bathymetry (ALB) system, Doneus et al. (2013) map a partly submerged Roman settlement on the island of Sveti Petar, Croatia. The scene was captured seamlessly in one single ALB campaign, covering the archaeological remains on land and in water (Fig. 11.10).

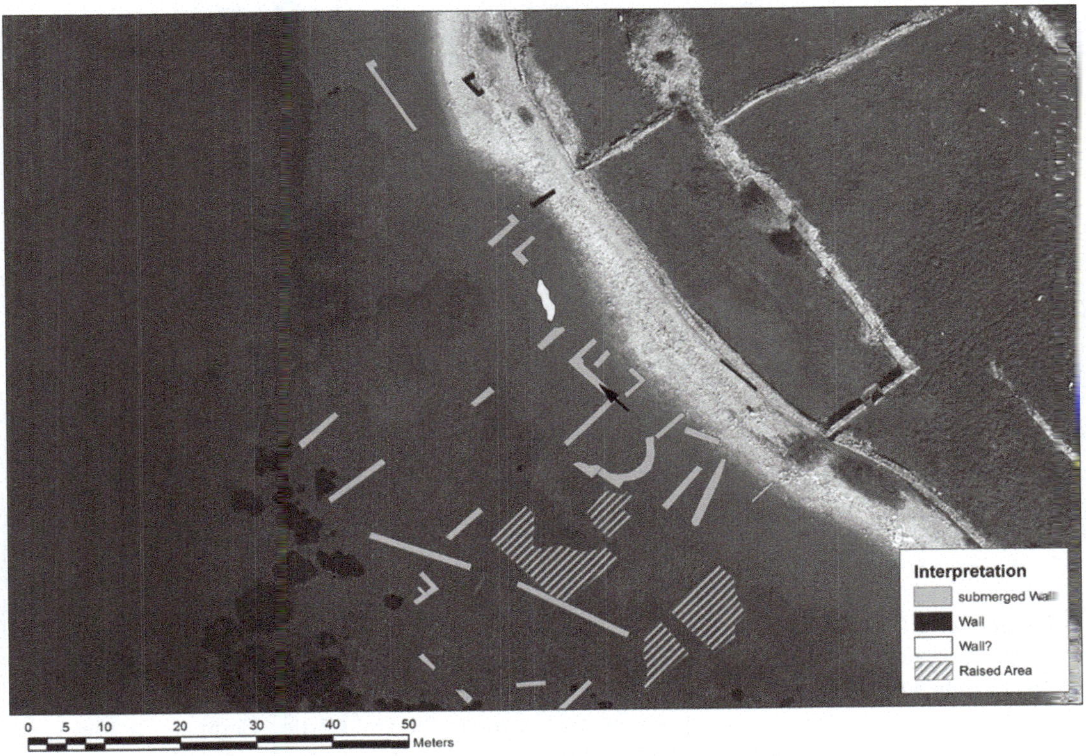

Fig. 11.10 Locations of subaerial and submerged archaeological remains of a Roman villa at Sveti Petar Island, Croatia, as derived from ALB data. Reprinted from Doneus et al. (2013), Copyright (2013), with kind permission from first author and Elsevier

11.6 Conclusions: Bridging Methods and Scales

The mentioned examples emphasize that LiDAR data are highly valuable for archaeological studies. By including the third dimension, new features can be derived with new analysis methods. In the last part of this introduction, some spotlights reaching beyond the exclusive use of LiDAR in geoarchaeology are presented, followed by remarks on emerging research fields and current developments in LiDAR research.

The toolbox for generating 3D geodata steadily grows. Interesting low-cost devices and approaches are available, allowing for the complementary use with LiDAR. The structured light sensor Kinect for Xbox 360, for example, originally a component of a gaming console system, delivers point clouds of scenes up to the size of rooms (Meister et al. 2012; Bondarev et al. 2013). Promising are also tests with the second Kinect generation (Kinect for Xbox One), which captures scenes in 3D according to the time-of-flight principle (Sarbolandi et al. 2015). A further source for 3D geodata is the photogrammetric approach combining structure from motion and dense image matching, which allows to derive a point cloud from photographs taken with uncalibrated consumer-grade cameras, for example, from terrestrial positions or unmanned aerial vehicles (UAVs) (Remondino 2013; Chap. 10).

Multi-sensor and multi-platform approaches are promising for collecting comprehensive 3D geodata with different methods complementing each other (e.g., Krömker 2013, Chap. 13). Also, going beyond the exclusive use of 3D geodata opens new possibilities. Bennett et al. (2013) show, for example, that combining complementary information derived from spectral data, aerial images, and ALS-based DTMs enhances the detection of archaeological features and contextual information. Hesse (2015) uses high-resolution multi-temporal satellite imagery to prospect damages on Peruvian geoglyphs and examines damaged sites in detail on the basis of 3D geodata produced from pole-based imagery. The fusion of 3D geodata also plays a crucial role in the study of Siart et al. (2013): the authors combine terrestrial LiDAR data of the surface of a doline complex in eastern Crete, Greece, with 3D subsurface models derived from electrical resistivity tomography and refraction seismics into a digital terrain and subsurface model (DTSM). The DTSM allows to derive valuable information about the doline geometries, which in turn can be linked to archaeological topics like settlement history or agricultural use of the catchment area (Siart et al. 2009; Siart 2018).

Latest developments in the application of LiDAR in geoarchaeology comprise, for example, multi-wavelength laser scans as used in Briese et al. (2013) for archaeological prospection and mapping. Depending on the wavelength, remains of Roman Carnuntum, Austria, are shown to be detectable in orthophotos derived from multi-wavelength ALS data.

Although available now for some years, the potential of LiDAR full-waveform analysis for advancing geoarchaeological research has yet to be fully tapped. As shown in Doneus et al. (2010), the extraction of terrain points can be enhanced by analyzing the full waveform in an ALS dataset. Products derived from the terrain points like DTMs reach a higher accuracy, which in turn is beneficial for subsequent archaeological analyses and interpretations.

Geoarchaeological studies can strongly benefit from the combination of the available methods and their respective advantages, for example, by relying on the survey-grade data quality produced with high-end laser scanners and capturing occluded or non-accessible parts of a scene with low-cost approaches. Each of the described methods, sensors, and platforms for capturing 3D geodata as a supporting tool for answering archaeological questions has its advantages and shortcomings and, thus, proper field of application (Fig. 11.1). If a digital terrain model void of vegetation is the aim of a LiDAR campaign, a bird's-eye perspective and, thus, airborne laser scanning would be the preferred capturing approach. If radiometric features are valuable for an analysis, a sensor being able to capture radiometric information has to be chosen.

LiDAR can, thus, be considered a method for capturing 3D geodata of high value for archaeological research. By including the third dimension, new geometric and radiometric features of objects can be derived and analyzed. Archaeologists can select data capturing and analysis methods out of a comprehensive toolbox which is steadily growing and which offers tailored solutions best fitting to the object in focus of a geoarchaeological study.

References

Auer M, Agugiaro G, Billen N, Loos L, Zipf A (2014) Web-based visualization and query of semantically segmented multiresolution 3D models in the field of cultural heritage. ISPRS Ann Photogramm Remote Sens Spat Inf Sci II-5:33–39. https://doi.org/10.5194/isprsannals-II-5-33-2014

Bennett R, Welham K, Hill RA, Ford A (2013) Using LiDAR as part of a multi-sensor approach to archaeological survey and interpretation. In: Opitz RS, Cowley DC (eds) Interpreting archaeological topography. 3D data, visualisation and observation. Oxbow Books, Oxford

Beraldin JA, Blais F, Lohr U (2010) Laser scanning technology. In: Vosselman G, Maas HG (eds) Airborne and terrestrial laser scanning. CRC Press, Boca Raton

Bevan A, Li X, Martinón-Torres M, Green S, Xia Y, Zhao K, Zhao Z, Ma S, Cao W, Rehren T (2014) Computer vision, archaeological classification and China's terracotta warriors. J Archaeol Sci 49:249–254. https://doi.org/10.1016/j.jas.2014.05.014

Boeder V, Kersten T, Thies Th, Sauer A (2011) Mobile laser scanning on board hydrographic survey vessels – applications and accuracy investigations. In: Proceedings of FIG working week 2011, Marrakech, Marocco, 18–22 May 2011

Böhler W, Heinz G (1999) Documentation, surveying, photogrammetry. In: Proceedings of XVII CIPA symposium, Olinda, Brazil, 3–6 Oct 1999

Bondarev E, Heredia F, Favier L, Lingni M, de With PHN (2013) On photo-realistic 3D reconstruction of large-scale and arbitrary-shaped environments. IEEE consumer communications and networking conference, Las Vegas, 11–14 Jan 2013. https://doi.org/10.1109/CCNC.2013.6488511

Briese C (2010) Extraction of digital terrain models. In: Vosselman G, Maas HG (eds) Airborne and terrestrial laser scanning. CRC Press, Boca Raton

Briese C, Pfennigbauer M, Ullrich A, Doneus M (2013) Multi-wavelength airborne laser scanning for archaeological prospection. Int Arch Photogramm Remote Sens Spat Inf Sci XL-5/W2:119–124

Brodu N, Lague D (2012) Terrestrial LiDAR data classification of complex natural scenes using a multi-scale dimensionality criteria: applications in geomorphology. J Photogramm Remote Sens 68:121–134. https://doi.org/10.1016/j.isprsjprs.2012.01.006

Challis K, Howard AJ (2013) The role of LiDAR intensity data in interpreting environmental and cultural archaeological landscapes. In: Opitz RS, Cowley DC (eds) Interpreting archaeological topography. 3D data, visualisation and observation. Oxbow Books, Oxford

Diskin S, Heyvaert VMA, Pavlopoulos K, Schütt B (2013) Geoarchaeology: a toolbox of approaches applied in a multidisciplinary research discipline. Quat Int 308–309:1–3 https://doi.org/10.1016/j.quaint.2013.09.004

Doneus M, Briese C, Studnicka N (2010) Analysis of full-waveform ALS data by simultaneously acquired TLS data: towards an advanced DTM generation in wooded areas. Int Arch Photogramm Remote Sens Spat Inf Sci XXXVIII(7B):193–198

Doneus M, Doneus N, Briese C, Pregesbauer M, Mandlburger G, Verhoeven G (2013) Airborne laser bathymetry – detecting and recording submerged archaeological sites from the air. J Archaeol Sci 40 (4):2136–2151. https://doi.org/10.1016/j.jas.2012.12.021

Elbering SO, Khoshelham K (2015) Automatic extraction of railroad centerlines from mobile laser scanning data. Remote Sens 7(5):5565–5583. https://doi.org/10.3390/rs70505565

El-Sheimy N (2009) Georeferencing component of LiDAR systems. In: Shan J, Toth CK (eds) Topographic laser ranging and scanning. Principles and processing. CRC Press, Boca Raton

Fernandez-Diaz JC, Carter WE, Shrestha RL, Glennie CL (2014) Now you see it... now you don't: understanding airborne mapping LiDAR collection and data product generation for archaeological research in Mesoamerica. Remote Sens 6:9951–10001. https://doi.org/10.3390/rs6109951

Forbriger M, Müller L, Siart C, Schittek K, Höfle B, Bubenzer O, Reindel M, Eitel B (2011) Terrestrial laser scanning in geoarchaeology – capturing one of the oldest settlement places in the high Andes of southern Peru. Geophys Res Abstr 13(EGU2011-12236)

Forbriger M, Mara H, Rieck B, Siart C, Wagener O (2013) Der "Gesprengte Turm" am Heidelberger Schloss. Untersuchung eines Kulturdenkmals mithilfe hoch auflösender terrestrischer Laserscans. Denkmalpflege in Baden-Württemberg - Nachrichtenblatt Landesdenkmalpflege 3(2013):165–168

Ghilardi M, Desruelles S (2009) Geoarchaeology: where human, social and earth sciences meet with technology. SAPIENS 2(2)

Heritage G, Hetherington D (2007) Towards a protocol for laser scanning in fluvial geomorphology. Earth Surf Process Landf 32:66–74. https://doi.org/10.1002/esp.1375

Heritage G, Large A (2009) Principles of 3D laser lcanning. In: Heritage G, Large A (eds) Laser scanning for the environmental sciences. Wiley-Blackwell, Chichester

Hesse R (2013) The changing picture of archaeological landscapes: lidar prospection over very large areas as part of a cultural heritage strategy. In: Opitz RS Cowley DC (eds) Interpreting archaeological topography. 3D data, visualisation and observation. Oxbow Books, Oxford

Hesse R (2014) Geomorphological traces of conflict in high-resolution elevation models. App Geogr 46:11–20. https://doi.org/10.1016/j.apgeog.2013.10.004

Hesse R (2015) Combining structure-from-motion with high and intermediate resolution satellite images to document threats to archaeological heritage in arid environments. J Cult Herit 16(2):192–201. https://doi.org/10.1016/j.culher.2014.04.003

Hoffmeister D, Bolten A, Curdt C, Waldhoff G, Bareth G (2010) High resolution crop surface models (CSM) and crop volume models (CVM) on field level by terrestrial laserscanning. Proc SPIE 7840:78400E. https://doi.org/10.1117/12.872315

Höfle B (2014) Radiometric correction of terrestrial LiDAR point cloud data for individual maize plant detection. IEEE Geosci Remote Sens Lett 11 (1):94–98. https://doi.org/10.1109/LGRS.2013.2247022

Höfle B, Pfeifer N (2007) Correction of laser scanning intensity data: data and model-driven approaches. ISPRS J Photogramm Remote Sens 62(6):415–433. https://doi.org/10.1016/j.isprsjprs.2007.05.008

Höfle B, Wagener O (2012) Burgen in der Landschaft - Inszenierung und Entzifferung anhand neuer Methoden. In: Wagener O (ed) Symbole der Macht? Aspekte mittelalterlicher und frühneuzeitlicher Architektur. Peter Lang, Frankfurt am Main

Höfle B, Mücke W, Dutter M, Rutzinger M, Dorninger P (2009) Detection of building regions using airborne LiDAR – a new combination of raster and point cloud based GIS methods. In: Proceedings of 21st AGIT symposium Angewandte Geoinformatik, Salzburg, Austria, 8–10 July 2009

Johnson KM, Ouimet WB (2014) Rediscovering the lost archaeological landscape of southern New England using airborne light detection and ranging (LiDAR). J Archaeol Sci 43:9–20. https://doi.org/10.1016/j.jas.2013.12.004

Koenig K, Höfle B, Müller L, Hämmerle M, Jarmer T, Siegmann B, Lilienthal H (2013) Radiometric correction of terrestrial LiDAR data for mapping of harvest residues density. ISPRS Ann Photogramm Remote Sens Spatial Inf Sci II-5(W2):133–138. https://doi.org/10.5194/isprsannals-II-5-W2-133-2013

Krömker S (2013) Neue Methoden zur besseren Lesbarkeit mittelalterlicher Grabsteine am Beispiel des Heiligen Sands in Worms. In: Heberer P (ed) Die SchUM-Gemeinden Speyer - Worms - Mainz. Auf dem Weg zum Welterbe. Schnell & Steiner, Regensburg

Liang X, Hyyppä J, Kukko A, Kaartinen H, Jaakkola A, Yu X (2014a) The use of a mobile laser scanning system for mapping large forest plots. IEEE Geosci Remote Sens Lett 11(9):1504–1508. https://doi.org/10.1109/LGRS.2013.2297418

Liang X, Kukko A, Kaartinen H, Hyyppä J, Yu X, Jaakkola A, Wang Y (2014b) Possibilities of a personal laser scanning system for forest mapping and ecosystem services. Sensors 14(1):1228–1248. https://doi.org/10.3390/s140101228

Lichti D, Skaloud J (2010) Registration and calibration. In: Vosselman G, Maas HG (eds) Airborne and terrestrial laser scanning. CRC Press, Boca Raton

Mara H, Krömker S, Jakob S, Breuckmann B (2010) GigaMesh and Gilgamesh – 3D multiscale integral invariant cuneiform character extraction. In: Artusi A, Joly M, Lucet G, Pitzalis D, Ribes A (eds) Proceedings of the 11th international VAST symposium on virtual reality, archaeology and cultural heritage, Paris, France

Meister S, Kohli P, Izadi S, Hämmerle M, Rother C, Kondermann D (2012) When can we use KinectFusion for ground truth acquisition? In: Proceedings of IEEE/RSJ international conference on intelligent robots and systems, Vilamoura, Portugal, 7–12 Oct 2012

Petrie G, Toth CK (2009a) Introduction to laser ranging, profiling, and scanning. In: Shan J, Toth CK (eds) Topographic laser ranging and scanning. Principles and processing. CRC Press, Boca Raton

Petrie G, Toth CK (2009b) Airborne and spaceborne laser profilers and scanners. In: Shan J, Toth CK (eds) Topographic laser ranging and scanning. Principles and processing. CRC Press, Boca Raton

Petrie G, Toth CK (2009c) Terrestrial laser scanners. In: Shan J, Toth CK (eds) Topographic laser ranging and scanning. Principles and processing. CRC Press, Boca Raton

Pfeifer N, Mandlburger G (2009) LiDAR data filtering and DTM generation. In: Shan J, Toth CK (eds) Topographic laser ranging and scanning. Principles and processing. CRC Press, Boca Raton

Pfenningbauer M, Riegl U, Rieger P, Amon P (2014) UAS based laser scanning for forest inventory and precision farming. In: Proceedings of the international workshop on remote sensing and GIS for monitoring of habitat quality, Vienna, Austria, 24–25 Sept 2014

Rapp G Jr, Hill CL (2009) Geoarchaeology. Yale University Press, New Haven

Remondino F (2013) Worth a thousand words – photogrammetry for archaeological 3D surveying. In: Opitz RS, Cowley DC (eds) Interpreting archaeological topography. 3D data, visualisation and observation. Oxbow Books, Oxford

Renfrew C (1976) Archaeology and the earth sciences. In: Davidson DA, Shackley ML (eds) Geoarchaeology. Earth Science and the Past. Duckworth, London

Risbøl O (2013) Cultivating the 'wilderness' – how lidar can improve archaeological landscape understanding. In: Opitz RS, Cowley DC (eds) Interpreting archaeological topography. 3D data, visualisation and observation. Oxbow Books, Oxford

Sarbolandi H, Lefloch D, Kolb A (2015) Kinect range sensing: structured-light versus time-of-flight Kinect. Comput Vision Image Underst. https://doi.org/10.1016/j.cviu.2015.05.006

Schneider A, Takla M, Nicolay A, Raab A, Raab T (2014) A template-matching approach combining morphometric variables for automated mapping of charcoal kiln sites. Archaeol Prospect 22(1):45–62. https://doi.org/10.1002/arp.1497

Siart C (2018) Merging the views: highlights on the fusion of surface and subsurface geodata and their potentials for digital geoarchaeology. In: Siart C, Forbriger M, Bubenzer O (eds) Digital geoarchaeology – new techniques for interdisciplinary human-environmental research. Springer, Heidelberg, pp 253–266

Siart C, Ghilardi M, Holzhauer I (2009) Geoarchaeological study of karst depressions integrating geophysical and sedimentological methods: case studies from Zominthos and Lato (Central and East Crete, Greece). Géomorphol Relief Process Environ 4:17–32. https://doi.org/10.4000/geomorphologie.7709

Siart C, Forbriger M, Nowaczinski E, Hecht S, Höfle B (2013) Fusion of multi-resolution surface (terrestrial laser scanning) and subsurface geodata (ERT, SRT) for karst landform investigation and geomorphometric quantification. Earth Surf Process Landf 38(10):1135–1147. https://doi.org/10.1002/esp.3394

Sithole G, Vosselman G (2004) Experimental comparison of filter algorithms for bare-Earth extraction from airborne laser scanning point clouds. ISPRS J Photogramm Remote Sens 59(1–2):85–101. https://doi.org/10.1016/j.isprsjprs.2004.05.004

Stilla U, Jutzi B (2009) Waveform analysis for small-footprint pulsed laser systems. In: Shan J, Toth CK (eds) Topographic laser ranging and scanning. Principles and processing. CRC Press, Boca Raton

Wallace L, Lucieer A, Watson C, Turner D (2012) Development of a UAV-LiDAR system with application to forest inventory. Remote Sens 4:1519–1543. https://doi.org/10.3390/rs4061519

Zielhofer C, Clare L, Rollefson G, Wächter S, Hoffmeister D, Bareth G, Roettig C, Bullmann H, Schneider B, Berke H, Weninger B (2012) The decline of the early Neolithic population center of 'Ain Ghazal and corresponding earth-surface processes, Jordan Rift Valley. Quat Res 78(3):427–441. https://doi.org/10.1016/j.yqres.2012.08.006

Zlot R, Bosse M, Greenop K, Jarzab Z, Juckes E, Roberts J (2014) Efficiently capturing large, complex cultural heritage sites with a handheld mobile 3D laser mapping system. J Cult Herit 15(6):670–678. https://doi.org/10.1016/j.culher.2013.11.009

3D Laser Scanning for Geoarchaelogical Documentation and Analysis 12

Dirk Hoffmeister

Abstract

Geoarchaeology is an interdisciplinary research area that applies geoscientific concepts, methods and knowledge for analysis of archaeological sites, as well as the reconstruction of past environments. Geoarchaeologic sites can be documented, and specific research questions can be solved by the implementation of available remote sensing methods, in particular by laser scanning. Therefore, the general workflow of applying laser scanning and particularly steps for data integration are shown, as well as specific analysis and visualization steps. However, there are still problems to be solved, which are, for example, the storage, exchange, quality control and metadata description of 3D models, as well as specific problems with each method.

The exemplary workflow in this chapter describes the steps of field campaign planning, data acquisition steps, preprocessing steps and different analysis. The main issues in field campaign preparation are training on the instrument and establishing common field procedures, as well as getting to know and discuss the aims of the project. Data acquisition in the field can be divided in the crucial steps of scan position estimation, setting the resolution and knowing the errors as well as conducting different registration tasks. These steps directly affect the necessary preprocessing steps, such as registration, filtering and further adjustments to the point clouds. The final data set can be used in the following analysis steps that are divided in an iconic reconstruction and symbolic modelling. For all of these steps, different software packages are listed. Finally, different analysis results are depicted.

Keywords

Laser scanning • Point clouds • Archaeology • Geoarchaeology • Workflow • 3D

D. Hoffmeister (✉)
GIS & Remote Sensing, Institute of Geography,
University of Cologne, Cologne, Germany
e-mail: dirk.hoffmeister@uni-koeln.de

© Springer International Publishing AG 2018

C. Siart et al. (eds.), *Digital Geoarchaeology*, Natural Science in Archaeology,
DOI 10.1007/978-3-319-25316-9_12

12.1 Introduction

Geoarchaeology is an interdisciplinary research area that applies geoscientific concepts, methods and knowledge for analysis of archaeological sites and aims to reconstruct past environments (Brückner and Gerlach 2007). Closely related to that is cultural heritage (CH) documentation, which is important in order to preserve the current state of archaeology, nowadays most probably in 3D (Remondino and Rizzi 2010). These partly reconstructed 3D models are also analysed in order to understand past living conditions (Guidi et al. 2014). All these sites can be documented, and specific research questions can be solved by the implementation of available remote sensing methods. Recent developments of remote sensing methods in different scales are described by Remondino (2011), and the historic development of remote sensing for geoarchaeology is shown by Nunez Andres and Buill Pozuelo (2009).

However, there are still problems with regard to the storage, exchange, quality control and metadata describing the data and the involved processes. Additionally, specific problems with each remote sensing method exist (Remondino 2011). The lack of software and knowledge for 3D data handling generally diminishes the potential of 3D data, and the data is still often reduced to 2D drawings (Rüther et al. 2012).

In particular laser scanning (LS) has shown a great potential for the accurate, fast and nearly full covered 3D site documentation. As shown in the introduction of Chap. 11 the different TLS, ALS and MLS systems provide highly accurate and detailed point clouds on different scales. These point clouds allow in numerous ways to establish a documentation of a site and corresponding analysis. However, in order to achieve a comprehensive data set, several specifics, i.e. data integration, need to be regarded.

Thus, a detailed overview of the necessary and additional steps for geoarchaeologic research, as well as possible analysis, will be presented in this chapter. In the following, the general workflow of applying terrestrial laser scanning and steps for data integration will be shown by examples. Airborne and mobile laser scanning are also shortly considered. Finally, different analysis based on the established 3D data sets will be shown in Sect. 12.3 followed by a short discussion and conclusion.

12.2 The General Workflow for Geoarchaeologic Research

The general workflow of applying terrestrial laser scanning and steps for data integration consists of three main parts: (1) the preparation and planning, (2) the data acquisition and (3) the analysis step. Each of the previously described LS systems requires a specific survey planning in terms of point density, point accuracy, coverage and registration and georeferencing processes. Additional data, like images, GNSS and total station surveying points and other 3D models, as well as airborne and satellite remote sensing needs to be regarded in order to achieve the surveying goals and integrate all data in a common reference system. This allows the analysis of the data set within its context.

Survey planning is particularly important for more expensive MLS and ALS campaigns, as these need, for instance, specific allowances, specific weather conditions and more thorough survey pattern planning (Beraldin et al. 2010). A comprehensive description on ALS for geoarchaeologic research is provided by Fernandez-Diaz et al. (2014).

12.2.1 Field Campaign Preparation

Before a successful field campaign can be achieved, it is really important to get familiar with the full acquisition workflow. TLS devices are still quite expensive and complex in contrast to single-point measurement methods, such as GNSS (global navigation satellite system) or a total station. Thus, a thorough evaluation of specific demands and available devices, components and software should be conducted in advance, probably with the support of manufacturers' demonstrations. Charlton et al. (2009) show, besides the elaborated device purchase, in detail the necessity of training with a TLS device and the corresponding software prior to a field campaign. The authors describe in detail "what can go wrong" (Charlton et al. 2009: 44) during field work, starting from missing cables to software problems. In particular, much attention needs to be spent on the registration and georeferencing procedures, which are explained later on in this chapter and are depending on the available equipment.

It has to be mentioned that laser scanning has different error sources, which can be divided into systematic, external and registration errors. Systematic errors exist as a consequence of imperfections in manufacturing and assembly of the complex instruments. Atmospheric effects and backscattering effects are summarized as external errors, and registration errors are described below. Systematic error sources are based on the range measurement itself, errors in the horizontal and vertical direction and polygonal mirror errors (Lichti 2010). All errors are detectable by comparing laser scanner readings to tachymetric surveying observations of calibration fields. The comparison of these measurements allows to calibrate the instruments by the establishment of (statistical) models for correction. As an example, the relative offset of the laser axis to the trunnion axis for a phase-shift scanner (e.g. Faro 880) shows an 8 mm inaccuracy depending on the vertical angle (Lichti 2007).

For the planning of a field campaign, as depicted in Fig. 12.1, general goals and problems should be discussed. For instance, the size and complexity of the areas of interest can be clarified by analysing pictures and digital orthophotos, as well as other data, which is online available. Google Earth is a very common preplanning tool, which also allows an easy exchange between the involved persons. The needed deliverables need to be identified. For example, it is important to know which resolution and coverage is needed. In general, access, permissions and supplies should be recognized. Depending on these factors, the correct method or device could be chosen. For large-scale

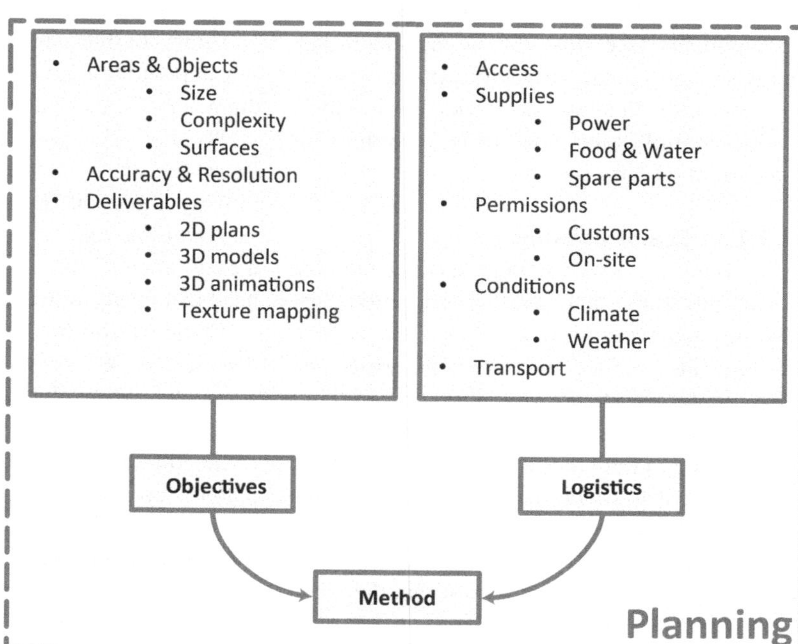

Fig. 12.1 Important planning issues for a field campaign (Hoffmeister 2014)

projects, some devices might be more suitable than others.

In addition to that, all common field trip issues (Fig. 12.1), such as general climate and current weather and local transport and admissions, as well as supplies need to be clarified. Further, the transport of the devices to the study area needs to be carefully undertaken, as these are partly sensitive devices concerning temperature, shocks, dust and humidity. Usually appropriate cases are delivered with these instruments. On an international level, one has to regard security and custom regulations of own and foreign countries. A Carnet ATA (temporary admission) might be necessary for the equipment.

12.2.2 Data Acquisition

On site, there are at least two major tasks for the data acquisition. The first one is to record the data by the device and to achieve a nearly full coverage in a certain resolution. The second is to keep in mind the registration tasks. A further task, which is depending on the goals and the device, is to calibrate and control the attached camera as well as to calibrate the intensity measurements (see this volume in Chap. 11). Likewise, additional measurements might be necessary for data integration tasks. For instance, specific control points need to be measured for transformation purposes.

12.2.2.1 Scan Positions

As applying an active remote sensing method, one might think that weather conditions are regardless, but rain and dust, as well as high sunlight can produce measurement errors and noise. In terms of security, the implied laser safety class of the laser scanner needs to be regarded (e.g. class 3R might injure the retina). In addition, it must be avoided that the device might injure people directly by falling or trapping by cables. Likewise, careful attention at moving the device from one to another scan position must be taken.

The surveying positions for a TLS device are depending on its field of view. Generally, positions with a good overlap should be considered. Lifted positions are preferred as these show mostly an extended field of view and avoid flat angles of incidence (Buckley et al. 2008; Rüther et al. 2012). The registration method always needs to be regarded. Therefore, targets placing for registration purposes (see below) should be planned and controlled. As a very practical method, two or more persons can be used for representing different scan positions. In a further step, a discussion is needed about visible areas and targets (Rüther et al. 2012). In general a checklist with all necessary steps is advisable and combinable with a log of the different scan positions, chosen resolution and used targets (Charlton et al. 2009).

12.2.2.2 Resolution

Each single laser beam is diverging by range. The resulting footprint size or spot diameter is one part of the theoretical resolution of a point cloud (Baltsavias 1999). The theoretical resolution in a distance perpendicular to the sensor further depends on the sample step width between each emitted laser pulse. In addition, the precision of each single laser measurement has to be taken into consideration.

Under- or oversampling occurs, when the chosen step width is smaller or larger than the beam diameter at a particular distance (Buckley et al. 2008). For ALS and MLS systems, this is an issue for the direction perpendicular to the moving direction. For 3D coverage, sampling rate and moving speed are crucial. For TLS measurements, the step width of the moving head can result in this over- or under-sampling.

However, this theoretical resolution at a given distance is of course not suitable for real surveying, where objects are in various distances, have a different surface structure and are tilted in some direction. As a result, irregular spot patterns arise in an unstructured point cloud.

12.2.2.3 Registration

For the registration of point clouds, several methods exist, which can be distinguished in direct and indirect measurements of each scan position. The term registration is used for the necessary transformations between different coordinate systems. Each scan position has an own coordinate system that needs to be transformed into at least a project-specific coordinate system that consists of all scan positions. Likewise, the whole project can be transformed into a global coordinate system, which is for geoarchaeologic research involving the whole context of a site an important step.

Additional sensors like a mounted RGB camera have also a different coordinate system. Every transformation is usually conducted by a matrix multiplication with a matrix consisting of nine parameters. Six of these are used for the rotation and three for the translation (shift) (Lichti and Skaloud 2010). These transformations can be established and enhanced by approaches based on the iterative closest point (ICP) method. Most of the LS software is handling the data in the single-precision floating-point format in order to reduce memory occupation and for precise visualization. Thus, an easy translation (basically a subtraction of big coordinates) between global coordinates and a project coordinate system is mostly conducted.

Targets of different texture and shape (spheres, cylinders, checkerboard patterns on sheets), called tie points, can be used for the indirect determination of the scan position (Fig.12.2). Usually the latter targets are automatically recognized by modern TLS, but specific high-resolution scans might be necessary. At least three targets in two overlapping scan positions are needed, but more are advantageous (Fan et al. 2015). This makes this approach time-consuming or impossible in complex areas, but it is generally more accurate as the distance and angles to each target are measured in a high accuracy. However, in order to enable a transformation to another geodetic or geographic system, several of these targets need to be additionally measured, normally by six or more equally distributed targets.

These large amounts of targets are avoided by the direct determination of the scan position (Fig. 12.3), which means to measure or calculate the actual position of the laser beam origin. Sometimes this approach is called backsighting. This can be conducted by measuring a fixed target on the TLS device (e.g. a surveying prism), by measuring the position of the TLS by an attached GNSS antenna or by measuring the base point of the levelled scanner by any of the previous devices (Mårtensson et al. 2012). For all these possibilities, the position of at least one further target (Hoffmeister et al. 2016) or the orientation measured by a compass (Rüther et al. 2012) needs to be known. In order to derive the actual laser beam origin, additional calculations on the measured point might be necessary regarding the dimensions between the fixed target, antenna centre or base point and the laser beam output. For that purpose dimensional drawings are usually delivered with the manual of the scanner. The advantage of the latter approach is the direct georeferencing of the data and less needed targets. The GNSS-based approach shows a minor accuracy, as RTK GPS systems usually have an accuracy of about 1 cm and DGPS devices of about 20–30 cm. An approach with a

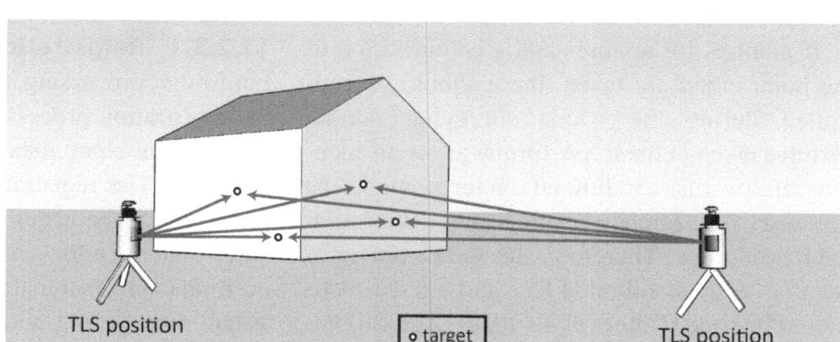

Fig. 12.2 Indirect registration method (Hoffmeister 2014)

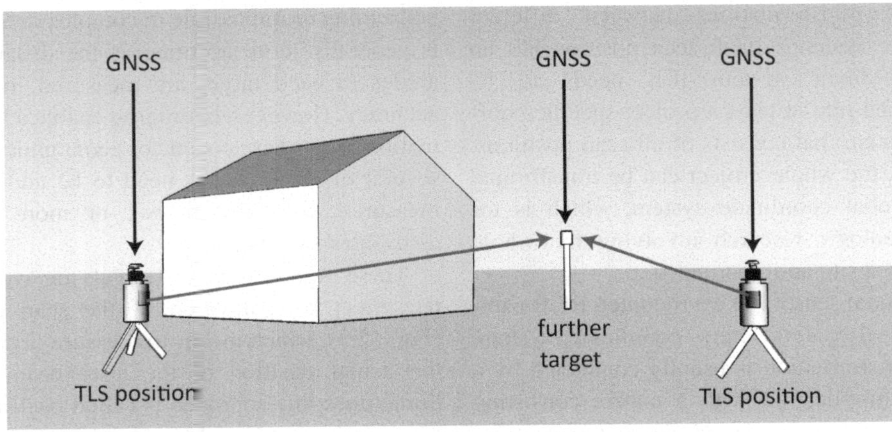

Fig. 12.3 Direct georeferencing method (GNSS or total station measurements suitable) (Hoffmeister 2014)

total station device and a prism attached to the scanner is more accurate but needs additional measurements for transformation of the local coordinate system.

In contrast, working inside a cave or in deep woods avoids using GNSS surveying methods. Thus, a combination of several methods is mandatory. For instance, several control points or scan positions can be measured directly with the GNSS and correspondingly with a total station. Further targets can then be acquired with the total station, and all these points can be transformed into global coordinates afterwards.

In this process it should be regarded how the site was previously measured or other data was recorded in order to achieve data integration. For instance, excavations of archaeologists tend to have an established local surveying network, which can be reused, or similar points for a transformation need to be measured in both systems.

If pictures for textures and a colourisation of the point cloud are taken, these should be controlled during the whole surveying period. Pictures taken in the RAW format allow an adaption afterwards. Additional pictures might be necessary for complete coverage and with similar light conditions. Therefore, the same camera or cameras with a calibrated lens and a fixed focus should be used (Rüther et al. 2012). The calibration or correction of intensity values should also be taken into account, if these values are of further interest, for example, for segmentation purposes.

Overall, results of the data acquisition should be regularly checked, for instance, after the work at each scan position is finished and after a day or specific part of an object is recorded in order to check for completeness or errors. Field notes with brief instructions help to avoid missing all necessary steps.

12.2.3 Preprocessing

After a backup of all data and a digitization of the notes taken during the field campaign, the preprocessing step has to be conducted. This means that the acquired data has to be processed to a final, consistent point cloud, which reflects the area or object of interest (Fig. 12.4). This involves the registration and filtering of the data.

12.2.3.1 Registration

For further processing, the most important step is the registration process, which results in a consistent point cloud measured from different scan positions. The registration comprises the transformation of the acquired data from one coordinate system into another, e.g. several scan positions can be transformed into a single, consistent system and additionally into a geodetic system. Point clouds from a single scan are usually in a right-handed Cartesian coordinate

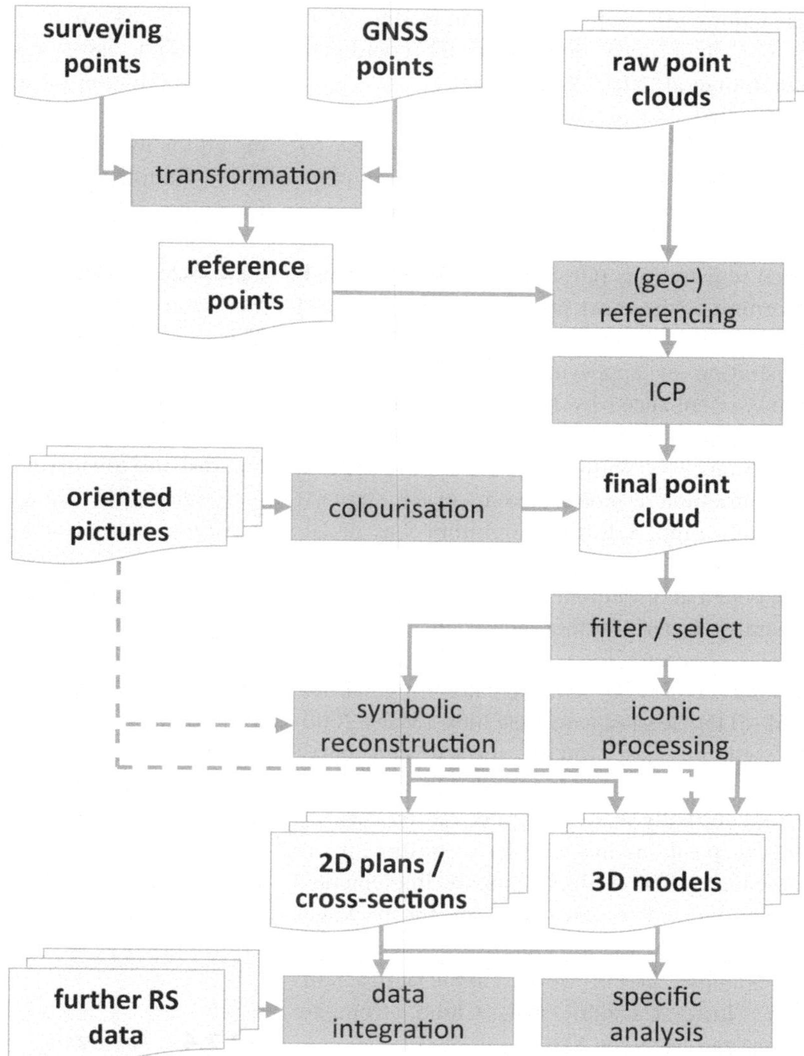

Fig. 12.4 General workflow of point cloud processing (Hoffmeister et al. 2014)

system, where the laser beam output point is the origin (Lichti and Skaloud 2010). One translation and rotation are necessary for each scan position, and a second transformation is needed, when the data has to be integrated into a geodetic system. This step is handled differently in the available software, and each software has additional settings. As stated before, generally coordinates need to be assigned to each scan position, to different targets, or the registration is manually conducted by selecting similar points in two partly overlapping point clouds. Partly these steps are automated and similar targets are automatically found, or external coordinates are assigned. Notes and drawings from the survey might help to sort the targets and scan positions.

For ALS and MLS approaches, INS data and GNSS positions are incorporated to obtain the orientation of right-handed frames (2D lines). All relations between the sensor, the INS, the GNSS antenna and the carrier system need to be described accurately to solve the different translations. All of these items can lead to errors for the final derivation of a 3D coordinate (Beraldin et al. 2010; Wehr 2009).

The first and partially rough estimations of point cloud registration can be enhanced by algorithms that are capable of merging point clouds. These approaches can be divided into iterative closest-point methods and feature-based registrations (Grant et al. 2012). The ICP algorithm is the most popular algorithm, based on the work of Besl and McKay (1992). The registration of approximately registered point clouds is enhanced by iteratively searching the closest points of similar regions until the Euclidean distance is within a threshold, whereas a transformation is conducted in every step. At the same time, a further algorithm was developed, where a surface is used as the master fitting area (Chen and Medioni 1992). This approach is generally faster. Further approaches based on these methods are, for example, the incorporation of objects (Bae 2009), triangulated irregular network (TIN) least-square matching (Maas 2000) as well as least square surface and curve matching (Gruen and Akca 2005). The alignment results normally in a standard deviation between all scan positions in the order of millimetres to centimetres. These algorithms are implemented in different alterations in software packages provided by the manufacturer of the scanner or, for example, in Polyworks (InnovMetric Software Inc., Canada) and Cloud Compare (Girardeau-Montaut 2015; Fig. 12.4).

12.2.3.2 Filtering

For all purposes, cleaning of the point clouds is one important step. This involves removing obstacles, vegetation and points out of the areas of interest. For the generation of digital terrain models, several filters are available (Pfeifer and Mandlburger 2009). A classification of the point clouds in different classes is also possible (Brodu and Lague 2012). In particular, for TLS point clouds with archaeological remains, development of filters is needed (Rüther et al. 2012), as most of this work is done manually. This step is conducted for each single scan position, as doing this step on the whole project point cloud can cause crashes on the computer. As the point cloud might have a very varying density of points, a thinning for a more equally distributed point cloud is affordable. Presented algorithms regard the local neighbouring points for this purpose (Puttonen et al. 2013; Nothegger and Dorninger 2009).

12.2.3.3 Further Adjustments

Data adjustments should be carefully undertaken. This accounts for intensity values that need a correction or calibration (Höfle 2014; Höfle and Pfeifer 2007) as well as for images taken, which can be corrected by colour balancing and general contrast and tonality adjustments.

Further 3D data, such as tachymetric surveying data of archaeological findings, RTK GPS points of drillings or georadar profiles, as well as topographic maps, airborne and satellite imagery should be integrated (Fig. 12.4). Again this is conducted by establishing transformations between all coordinate systems. In addition, areas of special interest that are maybe recorded in a higher resolution, e.g. by a structured-light scanner or photogrammetry, can be integrated. In this case, tie points are helpful but sometimes forbidden on walls in archaeological areas. A coarse registration by manually selected points and a subsequent alignment enable an integration (Hoffmeister et al. 2016).

12.2.4 Analysis

In a final step, registered and filtered point clouds can be already presented as a result. RGB values gained by an attached camera or manually joined pictures give a nearly realistic look, as shown in Chap. 11 (Fig. 12.2a). Cross-sections, images of details and fly-through and walk-through videos can be based on this final point cloud. However, in order to establish 3D models and 2D plans, several additional steps are necessary, which are depicted in Fig. 12.4. In general two different options are available, which can be called the iconic processing and the symbolic reconstruction of an object. Iconic processing, as a data-driven approach, is the direct use of 3D points. In contrast, the symbolic interpretation is a model approach, using the data for reconstruction of primitives that fit (partially) into the point cloud. Typical file formats for exchange are ASCII, Text, LAS and E57 type for point clouds as well as STL, PLY and OBJ for 3D objects. DXF format is used for cross-sections and outlines for plans. Raster-based data is normally stored as a tiff file.

The symbolic reconstruction can be conducted mostly manually, normally in computer-aided design (CAD) software, but (semi-)automatic algorithms are also available (Lindenbergh 2010). For instance, specific elements, such as detailed floor plans or cross-sections of important areas, can be reconstructed. In the presented case of Table 12.1, the software PointSense Heritage (Kubit GmbH, Germany) in combination with AutoCAD Civil 3D 2013 (Autodesk Inc., USA) was used. Further software in this area is listen in Table 12.1.

Iconic processing can be subdivided in 2.5D and true 3D approaches. A typical data-driven processing result is a raster data set representing a digital elevation model (DEM), where the additional height information is assigned to each 2D coordinate cell of a raster (Briese 2010; Vosselman and Klein 2010). This 2.5D representation is of importance due to the available amount of image processing methods, but is associated with data loss, in particular regarding multiple surfaces of complex objects derived from MLS or TLS. The 2.5D representation is derived by binning, interpolation or the setting up of TIN surfaces. Binning simply uses the mean, median, lowest or any other simplification rule of point values in a certain previously set cell size. Interpolation methods are able to statistically derive surfaces and are able to fill gaps, whereas the interpolation varies from more simple (e.g. Inverse Distance Weighting, IDW) to geo-statistical analysis, like kriging (Brunsdon 2009).

Triangulation produces TINs from the unstructured point clouds for a meshed, closed surface. The most popular possibility is the Delaunay triangulation, which relies on the assumption that no further point lies within the circumference of any derived triangle. However, a 2D data structure is assumed for this method. Thus, every triangulation process of this type considers a plane, usually the XY plane. A further triangulation method, which establishes watertight surfaces of 3D objects, is the Poisson surface reconstruction (Kazhdan et al. 2006). Those operations all result in polygonal meshes, which are a popular representation of 3D objects (Campbell and Flynn 2001). The vertices or edges of all triangles are stored.

Objects that are smoothed surface approximations can be called free-form objects, which are compact descriptions of complex objects. These objects can be stored as non-uniform rational B-splines (NURBS), which are an industrial norm for surfaces, as they are mathematically defined and compressed. Usually, NURBS are applied in modelling approaches of automotive parts, but any other free-form surface can also be modelled.

Irregular objects and objects with uncommon characteristics need manual processing. This is supported by numerous functions in specific programs. In preparation mainly for automatic or semi-automatic segmentation, and modelling purposes, an unstructured point cloud has to be structured (Vosselman and Klein 2010). For this purpose and enhanced, faster representations of point clouds, octree or k-d tree algorithms are used (Elseberg et al. 2013). Databases are a new option for storing and working with larger amounts of data (van Oosterom et al. 2015).

Hence, an automatic segmentation of point clouds, which is the sorting of points according to a specific criterion, such as spatial coherence, is possible. The segmentation enables to extract information and can be used to automate or support, modelling and filtering (Vosselman and Klein 2010).

Texture mapping, as the process to join images to the corresponding 3D models, allows to establish photorealistic visualizations of an object and adds more detail from the high-resolution pictures. This method is particularly important for cultural heritage documentation. However, the image acquisition for more suitable

Table 12.1 Overview of available software sorted by the main purpose, with operation modus and recent website

Software	Operation modus	Website
Raw data (nearly full workflow)		
*RiSCan Pro	Stand-alone, Windows	riegl.com
*Faro Scene	Stand-alone, Windows	faro.com
Cloud Compare	Stand-alone, Windows, Unix	danielgm.net/cc
Polyworx	Stand-alone, Windows	innovmetric.com
*Leica Cyclone	Stand-alone, Windows	leica-geosystems.de
*Trimble RealWorks	Stand-alone, Windows	trimble.com/3d-laser-scanning/realworks
VisionLidar	Stand-alone, Windows	geo-plus.com/visionlidar_lidar_software
JRC Reconstructor	Stand-alone, Windows	gexcel.it
VR Mesh	Stand-alone, Windows	vrmesh.com
CAD tools		
LupoScan	Stand-alone, Windows	lupos3d.de
Point Sense	Add-on for AutoCAD, Windows	faro-3d-software.com
ReCap 360	AutoCAD add-on, Windows	autodesk.com/products/recap-360/
Phidias	MicroStation add-on, Windows	phocad.de
Pointools	MicroStation add-on, Windows	bentley.com
Triangulation (iconic)		
Geomagic	Stand-alone, Windows	geomagic.com
MeshLab	Stand-alone, Windows, Mac	meshlab.sourceforge.net
Airborne laser scanning		
OPALS	Stand-alone, Windows, Unix	geo.tuwien.ac.at/opals
Trimble Inpho (SCOP++)	Stand-alone, Windows	trimble.com
Global Mapper (+ LiDAR module)	Stand-alone, Windows	bluemarblegeo.com
Quick Terrain Modeler	Stand-alone, Windows	appliedimagery.com
LAStools	C++ programming API	cs.unc.edu/~isenburg/lastools/
Extensions in QGIS, GrassGIS, SagaGIS and ESRI products		
Visualization		
SketchUp Pro	Stand-alone, Windows, Mac	sketchup.com
Blender	Stand-alone, Windows, Mac, Unix	blender.org
Unity	Stand-alone, Windows, Mac	unity3d.com
Maya	Stand-alone, Windows, Mac, Unix	autodesk.com/products/maya
Voxler and Surfer	Stand-alone, Windows	goldensoftware.com
Coltop 3D	Stand-alone, Windows	terranum.ch/products/coltop3d

Software of LS device manufactures is marked by an asterisk. Please note: this list might be not complete, and categories are depending on the author's view

texture mapping should be taken separately (Al-Khedera et al. 2009; Rüther et al. 2012), as images taken by internal or mounted cameras on the laser scanner devices might be not sufficient. Zalama et al. (2011) present an approach for the automatic texture mapping of additionally captured images, and the approach of Al-Khedera et al. (2009) based on the work of Alshawabkeh and Haala (2005) diminishes different occlusion effects.

12.3 Possible Analysis and Results

As mentioned in the introduction, geoarchaeologic remote sensing shows two aims, the documentation and analysis of a site. The section on possible results now shows these two aims. The accurate documentation of any object in geoarchaeology and cultural heritage documentation allows in general to determine sizes, extents and volumes of objects. Therefore, most of the software (Table 12.1) enables to conduct measurements or to calculate the volume of a selected, closed object. The 2D cross-sections and maps are the most common results. Perspective views of the derived 3D models or the final point cloud are further visualizations of this method. All this can be presented already in the Internet. One interesting project is the Zamani project (zamaniproject.org), which has the aim to document sites from the African continent (Rüther et al. 2009). As a further example, a detailed plan was established for the archaeological site of 'Ain Jamman, Jordan (Fig. 12.5a), by applying a symbolic reconstruction. The detected walls cover an area of ~30 by 40 m, and several rooms can be distinguished. Additionally, cross-sectional profiles of details, like windows, doors and stairs, can be derived, as well as the distribution of single stones (Fig. 12.5b).

For a geoarchaeologic documentation and analysis, the surrounding area is of interest. An integration of a site in the corresponding context is usually achieved by a high-resolution digital terrain model (DTM) (see this volume in Chap. 11). For instance, a high-resolution DTM allows to estimate past processes. In Zielhofer et al. (2012), it was shown that an archaeological site in Jordan was not destroyed by landslides, as the corresponding hill showed no indication of landslide movement. As a further example for the advantages of high resolution, the 3D model of the Sodmein Cave, Egypt, and the surrounding cliff wall is presented (Fig. 12.5c). The fault lines and the vertical displacements (>2 m) were estimated in the model and match to previous investigations (Moeyersons et al. 2002). The cave volume (~12,200 m^3) and the nearly similar debris volume in front of the cave (~13,200 m^3) were computed by incorporation of virtual planes for closure. Profiles along and across the cave were established from the 3D model and show the dimensions of the cave (Fig. 12.5d). The archaeological measurements of each artefact part were measured by a total station, and by applying a corresponding transformation, all these results can be integrated.

ALS data enables to extend or reveal new archaeological sites. Bewley et al. (2005) early reported the advantages of digital surface models derived by ALS surveys, which allow to artificially be illuminated in order to find new features of the Stonehenge area. Hill-shading from multiple illumination angles and a subsequent principal component analysis generally is able to reveal new insights into known archaeological areas (Devereux et al. 2008). Besides the aforementioned approach, several other analysis methods of digital terrain models are available, such as the openness factor (Doneus 2013), and should be selected carefully depending on terrain type and purpose to detect and document features and sites from ALS data (Štular et al. 2012). Further examples of ALS surveys show similar results (Fernandez-Diaz et al. 2014; Hare et al. 2014; Chase et al. 2014). The detection of sites in forests is enhanced by the analysis of the recorded full waveform of the repeated laser pulse signal for bare-earth extraction, as it is more reliable and accurate (Doneus et al. 2008). Furthermore, bathymetric ALS incorporating a laser in the visible green wavelength (~305 nm) is able to detect and document sites in shallow water (Doneus et al. 2013).

The full 3D information is of interest for geoarchaeologic investigations of caves. The information can be used to shape protection areas (Elez et al. 2013). Likewise, a flooding simulation for the Cussac Cave, France, was conducted in order to prove detected inundation levels of the cave (Jaubert et al. 2016). A geomorphometric approach was used by Gallay et al. (2016) in order to examine the palaeohydrography of the Domica Cave, Slovakia. Rüther et al. (2009) concluded, based on a 3D model of the Wonderwerk Cave, South Africa, and the surrounding area, that the existence of a further entrance was highly unlikely. Hoffmeister et al. (2016) applied a similar approach to the Ardales Cave in southern Spain in order to reveal possible areas of further entrances.

As a further analysis tool and for representation purposes, physically accurate, authentic illumination simulations (Chalmers et al. 2006) can be applied to 3D models in order to receive information about past living conditions (Happa et al. 2010). Masuda et al. (2010) showed that specific areas of the Fugoppe Cave, Japan, were

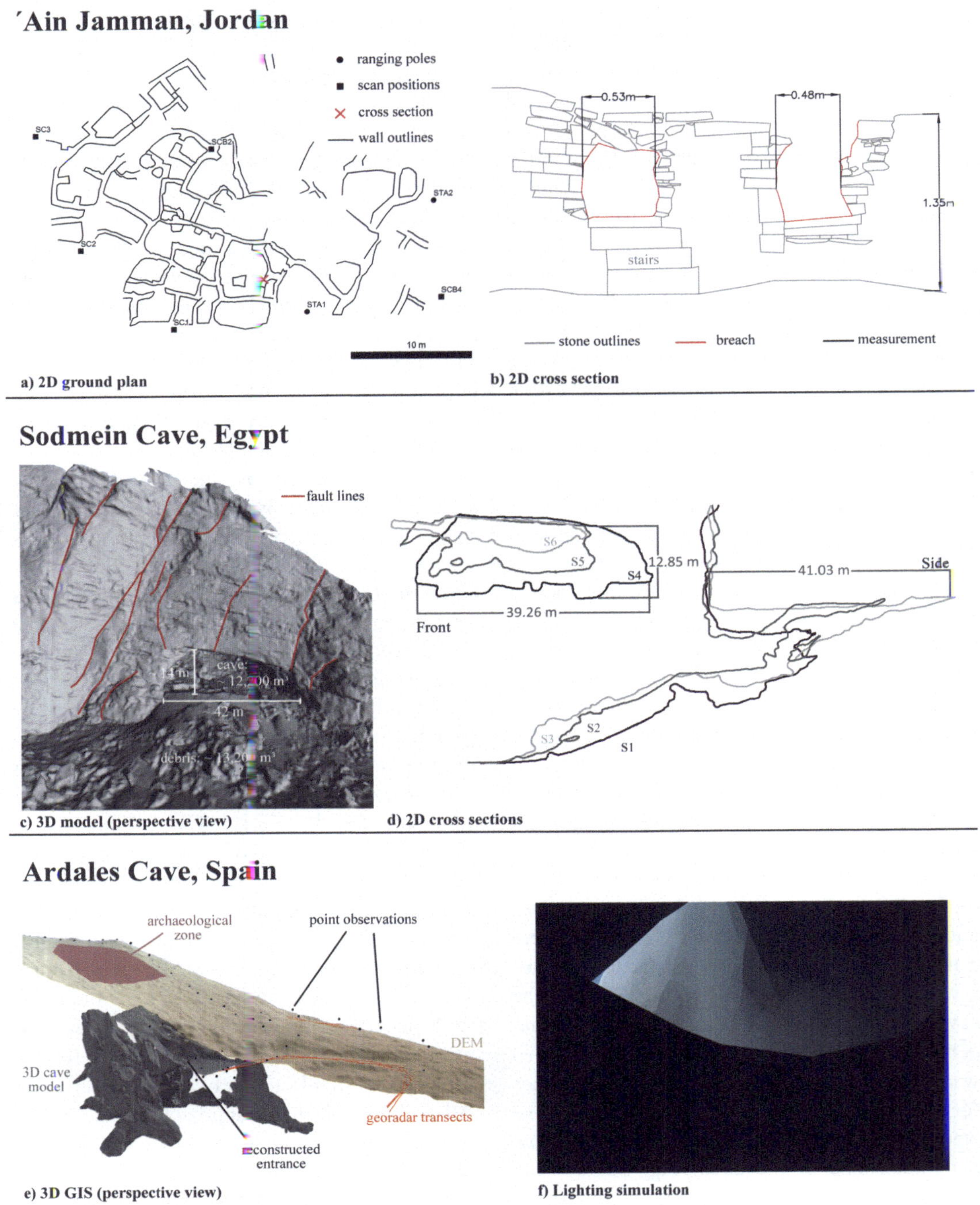

Fig. 12.5 Examples for different results and visualizations (Hoffmeister et al. 2014)

potentially reached by sunlight, which enabled painting. In this and other cases, the software suite Radiance was used, which uses the finite element algorithm radiosity (Ward 1994). In contrast, ray-tracing algorithms more easily allow to stochastically render global illumination (Happa et al. 2010) and are able to simulate also other types of surfaces such as glossy or specular materials (Happa et al. 2009). In Hoffmeister et al. (2016), path tracing was used for an illumination simulation, which demonstrated where sunlight might have reached the cave by virtually removing the modern entrance building and by reconstructing the ancient entrance (Fig.12.5f). Likewise, illumination of tallow lamp lights can be accurately simulated by this approach (Hoffmeister 2017).

In addition to that, 3D models can be integrated into geographic information systems (GIS). This allows to integrate these models with other geodata and enables further analysis (Katsianis et al. 2008). Likewise, this data serves as a basis for visualization systems (Rodriguez-Gonzalvez et al. 2012). This is also shown in Fig. 12.5e, where a perspective view of the previously described Ardales Cave is depicted with the integrated archaeological and geomorphological data.

Generally, for visualization and animation purposes, computer gaming and rendering software can be used (Table 12.1). As an example, Rua and Alvito (2011) imported a simplified version of reconstructed models into a gaming engine in order to create the ambience. Likewise, fauna and flora as well as avatars were implied. Similar visualizations are provided in Guidi et al. (2014). These methods can be used for virtual exhibitions and to enhance existing ones (Bruno et al. 2010).

For further analysis and exchange, reconstructed buildings are exported as an interoperable exchange information, CityGML objects set by the Open Geospatial Consortium (OGC) (Dore and Murphy 2012; Costamagna and Spanò 2013). In general the use of interoperability standards will allow further exchange and analysis.

12.4 Conclusion and Outlook

Laser scanning is an accurate and reliable tool to support geoarchaeological surveys and cultural heritage documentation with high-resolution 3D data (Remondino and Rizzi 2010). In particular, it is possible to easily survey important cave sites, which are hardly conductible by other remote sensing methods (Buchroithner and Gaisecker 2009; Rüther et al. 2009).

For geoarchaeologic surveys, data integration plays an important role. Thus, accurate georeferencing is necessary in order to relate the data to spaceborne remote sensing and all other relevant data. Applying an RTK GPS is very common for that purpose. In addition, local and previously measured data should be integrated by establishing transformations between all surveying systems. Thus, it is possible to visualize and analyse the full data set in its context.

It should be kept in mind that the acquisition of 3D data can become complex, when different types of registration at a tackling site are needed. For instance, for a full coverage of larger objects, tens of scan positions might be necessary. It was additionally shown that more efforts need to be used in order to derive information and to analyse the data. A factor of 1:3 should be taken into account for the time needed to process the data. Thus, a survey of 3 days takes 9 full workdays for an experienced user. Besides these main points to regard, training and tests of the equipment are further very important task. Still the filtering of point clouds is a problem, as mostly available filters are designed to derive digital terrain models but not walls or facades.

A further problem for all 3D models is a lack of standards. There are, for instance, several file types for the exchange of point clouds, as well as the derived models, such as OBJ, STL and PLY. Likewise, the documentation of accuracy and model resolution is hardly conducted. Thus, keeping and distributing the final point cloud is important.

During the past years, photogrammetry shows by the structure from motion approach a new fast and easy way to derive accurate 3D data (De Reu et al. 2013; Plets et al. 2012). In combination with UAVs, this method is capable to survey areas very fast and reliable. A combination with TLS measurements was early conducted in 2007 (Lambers et al. 2007), and recent examples show the advantages of mixing these approaches (Xu et al. 2014; Ortiz et al. 2013).

References

Al-Khedera S, Al-Shawabkeh Y, Haala N (2009) Developing a documentation system for desert palaces in Jordan using 3D laser scanning and digital photogrammetry. J Archaeol Sci 36(2):537–546. https://doi.org/10.1016/j.jas.2008.10.009

Alshawabkeh Y, Haala N, (2005) Automatic multi-image photo-texturing of complex 3D scenes. In: CIPA 2005 ICOMOS, Torino, Italy, pp 68–73

Bae KH (2009) Evaluation of the convergence region of an automated registration method for 3D laser scanner point clouds. Sensors 9(1):355–375. https://doi.org/10.3390/s90100355

Baltsavias EP (1999) Airborne laser scanning: basic relations and formulas. ISPRS J Photogramm Remote Sens 54(2–3):199–214

Beraldin J-A, Blais F, Lohr U (2010) Laser scanning technology. In: Vosselman G, Maas HG (eds) Airborne and terrestrial laser scanning, 1st edn. Whittles Publishing, Dunbeath, pp 1–42

Besl PJ, McKay ND (1992) A method for registration of 3-D shapes. IEEE Trans Pattern Anal Mach Intell 14(2):239–256

Bewley RH, Crutchley SP, Shell CA (2005) New light on an ancient landscape: lidar survey in the Stonehenge world heritage site. Antiquity 79(305):636–647

Briese C (2010) Extraction of digital terrain models. In: Vosselman G, Maas HG (eds) Airborne and terrestrial laser scanning, 1st edn. Whittles Publishing, Dunbeath, pp 135–168

Brodu N, Lague D (2012) 3D terrestrial lidar data classification of complex natural scenes using a multi-scale dimensionality criterion: applications in geomorphology. ISPRS J Photogramm 68:121–134. https://doi.org/10.1016/j.isprsjprs.2012.01.006

Brückner H, Gerlach R (2007) Geoarchäologie. In: Gebhardt H, Glaser R, Radtke U, Reuber P (eds) Geographie—Physische Geographie und Humangeographie. Elsevier, München, pp 513–516

Bruno F, Bruno S, De Sensi G, Luchi ML, Mancuso S, Muzzupappa M (2010). From 3D reconstruction to virtual reality: A complete methodology for digital archaeological exhibition. J Cult Herit 11:42–49. doi:10.1016/j.culher.2009.02.006

Brunsdon C (2009) Geostatistical analysis of lidar data. In: Heritage GL, Large ARG (eds) Laser scanning for the environmental sciences, 1st edn. Wiley-Blackwell, Chichester, pp 66–81

Buchroithner MF, Gaisecker T (2009) Terrestrial laser scanning for the visualization of complex dome in an extreme alpine cave system. Photogramm Fernerkun 4(4):329–339. https://doi.org/10.1127/1432-8364/2009/0025

Buckley SJ, Howell JA, Enge HD, Kurz TH (2008) Terrestrial laser scanning in geology: data acquisition, processing and accuracy considerations. J Geol Soc Lond 165:625–638

Campbell RJ, Flynn PJ (2001) A survey of free-form object representation and recognition techniques. Comput Vis Image Underst 81(2):166–210. https://doi.org/10.1006/cviu.2000.0889

Chalmers A, Roussos I, Ledda P (2006) Authentic illumination of archaeological site reconstructions. Conference on colour in graphics, imaging, and vision, 2006 (1): pp 431–434

Charlton ME, Coveney SJ, McCarthy T (2009) Issues in laser scanning. In: Heritage GL, Large ARG (eds) Laser scanning for the environmental sciences, 1st edn. Wiley-Blackwell, Chichester, pp 35–48

Chase A, Chase D, Awe J, Weishampel J, Iannone G, Moyes H, Yaeger J, Brown K, Shrestha R, Carter W, Fernandez Diaz J (2014) Ancient Maya regional settlement and inter-site analysis: the 2013 west-central Belize LiDAR survey. Remote Sens 6(9):8671–8695. https://doi.org/10.3390/rs6098671

Chen Y, Medioni G (1992) Object modeling by registration of multiple range images. Image Vis Comput 10(3):145–155

Costamagna E, Spanò A (2013) CityGML for architectural heritage. In: Abdul Rahman A, Boguslawski P, Gold C, Said M (eds) Developments in multidimensional spatial data models. Lecture notes in geoinformation and cartography. Springer, Berlin. https://doi.org/10.1007/978-3-642-36379-5_14

De Reu J, Plets G, Verhoeven G, De Smedt P, Bats M, Cherretté B, De Maeyer W, Deconynck J, Herremans D, Laloo P, Van Meirvenne M, De Clercq W (2013) Towards a three-dimensional cost-effective registration of the archaeological heritage. J Archaeol Sci 40(2):1108–1121. https://doi.org/10.1016/j.jas.2012.08.040

Devereux BJ, Amable GS, Crow P (2008) Visualisation of LiDAR terrain models for archaeological feature detection. Antiquity 82(316):470–479

Doneus M (2013) Openness as visualization technique for interpretative mapping of airborne lidar derived digital terrain models. Remote Sens 5(12):6427–6442

Doneus M, Briese C, Fera M, Janner M (2008) Archaeological prospection of forested areas using full-waveform airborne laser scanning. J Archaeol Sci 35(4):882–893. https://doi.org/10.1016/j.jas.2007.06.013

Doneus M, Doneus N, Briese C, Pregesbauer M, Mandlburger G, Verhoeven G (2013) Airborne laser bathymetry—detecting and recording submerged archaeological sites from the air. J Archaeol Sci 40(4):2136–2151. https://doi.org/10.1016/j.jas.2012.12.021

Dore C, Murphy M (2012) Integration of historic building information modeling (HBIM) and 3D GIS for recording and managing cultural heritage sites. In: 18th international conference on virtual systems and multimedia (VSMM), 2012, 2–5 Sept 2012, pp 369–376. https://doi.org/10.1109/VSMM.2012.6365947

Elseberg J, Borrmann D, Nüchter A (2013) One billion points in the cloud—an octree for efficient processing of 3D laser scans. ISPRS J Photogramm Remote Sens 76:76–88. https://doi.org/10.1016/j.isprsjprs.2012.10.004

Elez J, Cuezva S, Fernandez-Cortes A, Garcia-Anton E, Benavente D, Cañaveras JC, Sanchez-Moral S (2013) A GIS-based methodology to quantitatively define an Adjacent Protected Area in a shallow karst cavity: The case of Altamira cave. J Environ Manage 118:122–134. doi:10.1016/j.jenvman.2013.01.020

Fan L, Smethurst JA, Atkinson PM, Powrie W (2015) Error in target-based georeferencing and registration in terrestrial laser scanning. Comput Geosci 83:54–64. https://doi.org/10.1016/j.cageo.2015.06.021

Fernandez-Diaz J, Carter W, Shrestha R, Glennie C (2014) Now you see it. . . Now you don't: understanding airborne mapping LiDAR collection and data product generation for archaeological research in Mesoamerica. Remote Sens 6(10):9951–10001. https://doi.org/10.3390/rs6109951

Gallay M, Hochmuth Z, Kaňuk J, Hofierka J (2016) Geomorphometric analysis of cave ceiling channels mapped with 3-D terrestrial laser scanning. Hydrol Earth Syst Sci 20(5):1827–1849. https://doi.org/10.5194/hess-20-1827-2016

Girardeau-Montaut D (2016) Cloudcompare (version 2.7) (gpl software)—EDF R&D, Telecom Paris-Tech. http://www.danielgm.net/cc/. Accessed 15 July 2016

Grant D, Bethel J, Crawford M (2012) Point-to-plane registration of terrestrial laser scans. ISPRS J Photogramm Remote Sens 72:16–26. https://doi.org/10.1016/j.isprsjprs.2012.05.007

Gruen A, Akca D (2005) Least squares 3D surface and curve matching. ISPRS J Photogramm Remote Sens 59(3):151–174. https://doi.org/10.1016/j.isprsjprs.2005.02.006

Guidi G, Russo M, Angheleddu D (2014) 3D survey and virtual reconstruction of archaeological sites. Digit Appl Archaeol Cult Herit 1(2):55–69. https://doi.org/10.1016/j.daach.2014.01.001

Happa J, Artusi A, Dubla P, Bashford-Rogers T, Debattista K, Hulusi V, Chalmers A (2009) The virtual reconstruction and daylight illumination of the Panagia Angeloktisti. In: Paper presented at the proceedings of the 10th international conference on virtual reality, archaeology and cultural heritage, St. Julians, Malta

Happa J, Mudge M, Debattista K, Artusi A, Gonçalves A, Chalmers A (2010) Illuminating the past: state of the art. Virtual Reality 14(3):155–182. https://doi.org/10.1007/s10055-010-0154-x

Hare T, Masson M, Russell B (2014) High-density LiDAR mapping of the ancient city of Mayapán. Remote Sens 6(9):9064–9085. https://doi.org/10.3390/rs6099064

Hoffmeister D (2014) Feasibility studies of terrestrial laser scanning in coastal geomorphology, agronomy, and geoarchaeology. Dissertation, University of Cologne. http://kups.ub.uni-koeln.de/5497/. Licence at https://creativecommons.org/licenses/by-nc/2.5/

Hoffmeister D (2017) Simulation of tallow lamp light within the 3D model of the Ardales cave, Spain. Quat Int. 430A:22–29. https://doi.org/10.1016/j.quaint.2016.05.010

Hoffmeister D, Zellmann S, Kindermann K, Pastoors A, Lang U, Bubenzer O, Weniger GC, Bareth G (2014) Geoarchaeological site documentation and analysis of 3D data derived by terrestrial laser scanning. ISPRS Ann Photogramm Remote Sens Spat Inf Sci II-5:173–179. https://doi.org/10.5194/isprsannals-II-5-173-2014

Hoffmeister D, Zellmann S, Pastoors A, Kehl M, Cantalejo P, Ramos J, Weniger G-C, Bareth G (2016) The investigation of the Ardales Cave, Spain – 3D documentation, topographic analyses, and lighting simulations based on terrestrial laser scanning. Archaeol Prospect 23(2):75–86. https://doi.org/10.1002/arp.1519

Höfle B (2014) Radiometric correction of terrestrial LiDAR point cloud data for individual maize plant detection. IEEE Geosci Remote Sens 11(1):94–98. https://doi.org/10.1109/LGRS.2013.2247022

Höfle B, Pfeifer N (2007) Correction of laser scanning intensity data: data and model-driven approaches. ISPRS J Photogramm Remote Sens 62(6):415–433. https://doi.org/10.1016/j.isprsjprs.2007.05.008

Jaubert J, Genty D, Valladas H, Camus H, Courtaud P, Ferrier C, Feruglio V, Fourment N, Konik S, Villotte S, Bourdier C, Costamagno S, Delluc M, Goutas N, Katnecker É, Klaric L, Langlais M, Ledoux L, Maksud F, O'Farrell M, Mallye JB, Pierre M, Pons-Branchu E, Régnier É, Théry-Parisot I (2016) The chronology of human and animal presence in the decorated and sepulchral cave of Cussac (France). Quat. Int. 1–20. https://doi:10.1016/j.quaint.2016.01.052

Katsianis M, Tsipidis S, Kotsakis K, Kousoulakou A (2008) A 3D digital workflow for archaeological intra-site research using GIS. J Archaeol Sci 35(3):655–667. https://doi.org/10.1016/j.jas.2007.06.002

Kazhdan M, Bolitho M, Hoppe H (2006) Poisson surface reconstruction. In Polthier K, Sheffer A (eds) Proceedings of the 4th eurographics symposium on geometry processing

Lambers K, Eisenbeiss H, Sauerbier M, Kupferschmidt D, Gaisecker T, Sotoodeh S, Hanusch T (2007) Combining photogrammetry and laser scanning for the recording and modelling of the late intermediate period site of Pinchango Alto, Palpa, Peru. J Archaeol Sci 34(10):1702–1712. https://doi.org/10.1016/j.jas.2006.12.008

Lichti DD (2007) Error modelling, calibration and analysis of an AM-CW terrestrial laser scanner system.

ISPRS J Photogramm Remote Sens 61(5):307–324. https://doi.org/10.1016/j.isprsjprs.2006.10.004

Lichti DD (2010) Terrestrial laser scanner self-calibration: correlation sources and their mitigation. ISPRS J Photogramm Remote Sens 65(1):93–102. https://doi.org/10.1016/j.isprsjprs.2009.09.002

Lichti DD, Skaloud J (2010) Registration and calibration. In: Vosselman G, Maas HG (eds) Airborne and terrestrial laser scanning, 1st edn. Whittles Publishing, Dunbeath, pp 83–134

Lindenbergh R (2010) Engineering applications. In: Vosselman G, Maas HG (eds) Airborne and terrestrial laser scanning, 1st edn. Whittles Publishing, Dunbeath, UK, pp 237–269

Maas HG 2000 Least-squares matching with airborne laserscanning data in a TIN structure. In: Schenk T, Vosselman G (eds) ISPRS archives—volume XXXIII part B3, proceedings XIX ISPRS conference, Amsterdam, the Netherlands, 16–26 July 2000. pp 548–555

Mårtensson S-G, Reshetyuk Y, Jivall L (2012) Measurement uncertainty in network RTK GNSS-based positioning of a terrestrial laser scanner. J Appl Geod 6 (1):25–32. https://doi.org/10.1515/jag-2011-0013

Masuda T, Yamada Y, Kuchitsu N, Ikeuchi K (2010) Illumination simulation for archaeological investigation. In: Digitally archiving cultural objects. Springer, New York, pp 419–439. https://doi.org/10.1007/978-0-387-75807_20

Moeyersons J, Vermeersch PM, Van Peer P (2002) Dry cave deposits and their palaeoenvironmental significance during the last 115 ka, Sodmein Cave, Red Sea Mountains, Egypt. Quat Sci Rev 21(7):837–851. https://doi.org/10.1016/S0277-3791(01)00132-9

Nothegger C, Dorninger P (2009) 3D filtering of high-resolution terrestrial laser scanner point clouds for cultural heritage documentation. Photogramm Fernerkun 1:53–63. https://doi.org/10.1127/0935-1221/2009/0006

Nunez Andres MA, Buill Pozuelo F (2009) Evolution of the architectural and heritage representation. Landsc Urban Plan 91(2):105–112. https://doi.org/10.1016/j.landurbplan.2008.12.006

Ortiz J, Gil ML, Martínez S, Rego T, Meijide G (2013) Three-dimensional modelling of archaeological sites using close-range automatic correlation photogrammetry and low-altitude imagery. Archaeol Prospect 20(3):205–217. https://doi.org/10.1002/arp.1457

Pfeifer N, Mandlburger G (2009) Lidar data filtering and DTM generation. In: Shan J, Toth CK (eds) Topographic laser ranging and scanning. Taylor & Francis Group, Boca Raton, FL, pp 308–333

Plets G, Gheyle W, Verhoeven G, De Reu J, Bourgeois J, Verhegge J, Stichelbaut B (2012) Three-dimensional recording of archaeological remains in the Altai mountains. Antiquity 86(333):884–897

Puttonen E, Lehtomäki M, Kaartinen H, Zhu L, Kukko A, Jaakkola A (2013) Improved sampling for terrestrial and mobile laser scanner point cloud data. Remote Sens 5(4):1754–1773. https://doi.org/10.3390/rs5041754

Remondino F (2011) Heritage recording and 3D modeling with photogrammetry and 3D scanning. Remote Sens 3(6):1104–1138. https://doi.org/10.3390/rs3061104

Remondino F, Rizzi A (2010) Reality-based 3D documentation of natural and cultural heritage sites—techniques, problems, and examples. Appl Geomat 2(3):85–100. https://doi.org/10.1007/s12518-010-0025-x

Rodriguez-Gonzalvez P, Mancera-Taboada J, Gonzalez-Aguilera D, Munoz-Nieto A, Armesto J (2012) A hybrid approach to create an archaeological visualization system for a Palaeolithic Cave. Archaeometry 54(3):565–580. https://doi.org/10.1111/j.1475-4754.2011.00638.x

Rua H, Alvito P (2011) Living the past: 3D models, virtual reality and game engines as tools for supporting archaeology and the reconstruction of cultural heritage—the case-study of the Roman villa of Casal de Freiria. J Archaeol Sci 38(12):3296–3308. https://doi.org/10.1016/j.jas.2011.07.015

Rüther H, Chazan M, Schroeder R, Neeser R, Held C, Walker SJ, Matmon A, Horwitz LK (2009) Laser scanning for conservation and research of African cultural heritage sites: the case study of Wonderwerk Cave, South Africa. J Archaeol Sci 36(9):1847–1856. https://doi.org/10.1016/j.jas.2009.04.012

Rüther H, Bhurtha R, Held C, Schroder R, Wessels S (2012) Laser scanning in heritage documentation: the scanning pipeline and its challenges. Photogramm Eng Remote Sens 78(4):309–316

Štular B, Kokalj Ž, Oštir K, Nuninger L (2012) Visualization of lidar-derived relief models for detection of archaeological features. J Archaeol Sci 39 (11):3354–3360. https://doi.org/10.1016/j.jas.2012.05.029

van Oosterom P, Martinez-Rubi O, Ivanova M, Horhammer M, Geringer D, Ravada S, Tijssen T, Kodde M, Gonçalves R (2015) Massive point cloud data management: design, implementation and execution of a point cloud benchmark. Comput Graph 49:92–125. https://doi.org/10.1016/j.cag.2015.01.007

Vosselman G, Klein R (2010) Visualisation and structuring of point clouds. In: Vosselman G, Maas HG (eds) Airborne and terrestrial laser scanning, 1st edn. Whittles Publishing, Dunbeath, UK, pp 45–82

Ward GJ (1994) The RADIANCE lighting simulation and rendering system. Paper presented at the proceedings of the 21st annual conference on computer graphics and interactive techniques

Wehr A (2009) Lidar systems and calibration. In: Shan J, Toth CK (eds) Topographic laser ranging and scanning. Taylor & Francis Group, Boca Raton, FL, pp 129–172

Xu Z, Wu L, Shen Y, Li F, Wang Q, Wang R (2014) Tridimensional reconstruction applied to cultural heritage with the use of camera-equipped UAV and terrestrial laser scanner. Remote Sens 6(11):10413–10434. https://doi.org/10.3390/rs61110413

Zalama E, Gomez-Garcia-Bermejo J, Llamas J, Medina R (2011) An effective texture mapping approach for 3D models obtained from laser scanner data to building documentation. J Comput Aided Civ Infrastruct Eng 26(5):381–392. https://doi.org/10.1111/j.1467-8667.2010.00699.x

Zielhofer C, Clare L, Rollefson G, Wachter S, Hoffmeister D, Bareth G, Roettig C, Bullmann H, Schneider B, Berke H, Weninger B (2012) The decline of the early Neolithic population center of 'Ain Ghazal and corresponding earth-surface processes, Jordan Rift Valley. Quat Res 78(3):427–441. https://doi.org/10.1016/j.yqres.2012.08.006

Visual Detection and Interpretation of Cultural Remnants on the Königstuhl Hillside in Heidelberg Using Airborne and Terrestrial LiDAR Data

13

Karl Hjalte Maack Raun, Michelle Pfeiffer, and Bernhard Höfle

Abstract

LiDAR-based data acquisition, also referred to as laser scanning (LS), is an exceedingly used procedure for investigating site-specific details and spatial context. The main product of LiDAR scanning is digital elevation models (DEMs) derived from recorded 3D point clouds. Two major outcomes of DEMs are digital terrain models (DTM) of the bare earth and digital surface models (DSM) with canopy details. For detection and management of information from the past, especially, the DTM reveals important information for understanding, investigating, and managing sites and landscapes of cultural heritage interest.

In this case study, the advantages and disadvantages between airborne and terrestrial LiDAR DTM data are assessed. The investigations resulted in a differentiated perspective on scale of view and concluded that highest resolution is not always the best practice for visual detection of cultural heritage monuments in areas with complex canopy details, such as in dense vegetation. Because dense vegetation can disturb and distort terrain and surface segmentation to such a degree, the information retrieved might be more difficult to understand compared to visualization with less canopy details. Thus, the best practice is presently established in a combination of the two approaches. Minor details are lost in the ALS data, but large-scale context elude us by only using TLS data.

Our study revealed both major and minor details of infrastructure in the landscape from the eighteenth and nineteenth century. Many of the structures and details in the landscape have not been described or

K.H.M. Raun (✉) • M. Pfeiffer
Junior Research Group Digital Humanities and Digital Cultural Heritage at the Interdisciplinary Center for Scientific Computing and Cluster of Excellence Asia & Europe in Global Context, Heidelberg University, Heidelberg, Germany
e-mail: hjalte.raun@asia-europe.uni-heidelberg.de; michelle.pfeiffer@asia-europe.uni-heidelberg.de

B. Höfle
GIScience & 3D Spatial Data Processing Research Group, Institute of Geography, Heidelberg University, Im Neuenheimer Feld 368, 69120 Heidelberg, Germany

Heidelberg Center for the Environment, Im Neuenheimer Feld 229, Heidelberg, Germany
e-mail: hoefle@uni-heidelberg.de

© Springer International Publishing AG 2018

C. Siart et al. (eds.), *Digital Geoarchaeology*, Natural Science in Archaeology,
DOI 10.1007/978-3-319-25316-9_13

documented before and provide new understanding on cultural activity on the Königstuhl hillside in Heidelberg. Hereby especially two cellar structures were of particular interest to the investigations carried out.

Keywords
LiDAR • TLS and ALS comparison • Archaeological prospection • Scale of view • Visual object detection

13.1 Introduction

Understanding LiDAR Data, LiDAR Data in Archaeology, the LiDAR Truth, and the Necessity of Strategic Scanning
In order to perform comprehensive investigations of spatial context and cultural and temporal impact on landscape, it is necessary to understand and analyze the procedures and methods to retrieve correct ground truth of comparable data and site information. Consequently, techniques and methods need as much attention as results. Results are already manipulated data and as such often strongly related to specific research questions. Hence, data retrieval and manipulation need proper assessment and analysis before any conclusions can be finale.

Utilization of LiDAR data could easily become the standard from which cultural heritage monument detection and management could be initiated for a cost-effective approach for large-scale handling and processing. But it is equally important to understand external spatial relations and context in order to understand intra-site details. Such a spatial understanding of landscape can be retrieved and executed by combining large-scale airborne laser scanning (ALS) and small-scale terrestrial laser scanning (TLS). This will be exemplified in a case study at the Königstuhl hillside, Heidelberg (Fig. 13.1), with further elaborations as how to understand the monuments detected and their spatial and cultural context.

Monuments of cultural heritage are within this study defined as structures of cultural heritage which visually and physically change and shape the landscape.

Large-scale archaeological surveying and prospection is a difficult, highly disputed and problematic field within archaeology (Opitz and Cowley 2013). However, with the incorporation of ALS, there is a viable solution for quantitative macro-scaled airborne assessment and furthermore with qualitative micro-scaled documentation and assessment through TLS (Doneus and Briese 2006). Such a dual process can help assist in managing and identifying cultural heritage within a vast variety of different landscapes (cf. Doneus et al. 2010).

By logic of acquisition techniques and resolution outcome, ALS data are presently more purposeful for macro-scaled analysis, whereas TLS data are more local by virtue. Naturally, these boundaries are only stipulated by present notions of scanning logistics and technological capabilities to different interpolated levels by scale of view in relation to resolution and *points per square meter* (*ppsm*). This study will use ALS for macro-scaled analysis and TLS for local comparison on site-specific details regarding objects, structures, and landscape.

The main objectives are comparison of acquisition, handling and processing of TLS data. Comparison will be made between different interpolated TLS manipulations of cell size in order to assess the decrease and increase in spatial information and conclude upon optimal procedures for manual visual object detection through LiDAR data. Results will be compared with known spatial and cultural information from written records.

The goals will be an improved understanding of possibilities for cultural heritage detection within densely vegetated slopes. Further comprehending the amount of potential

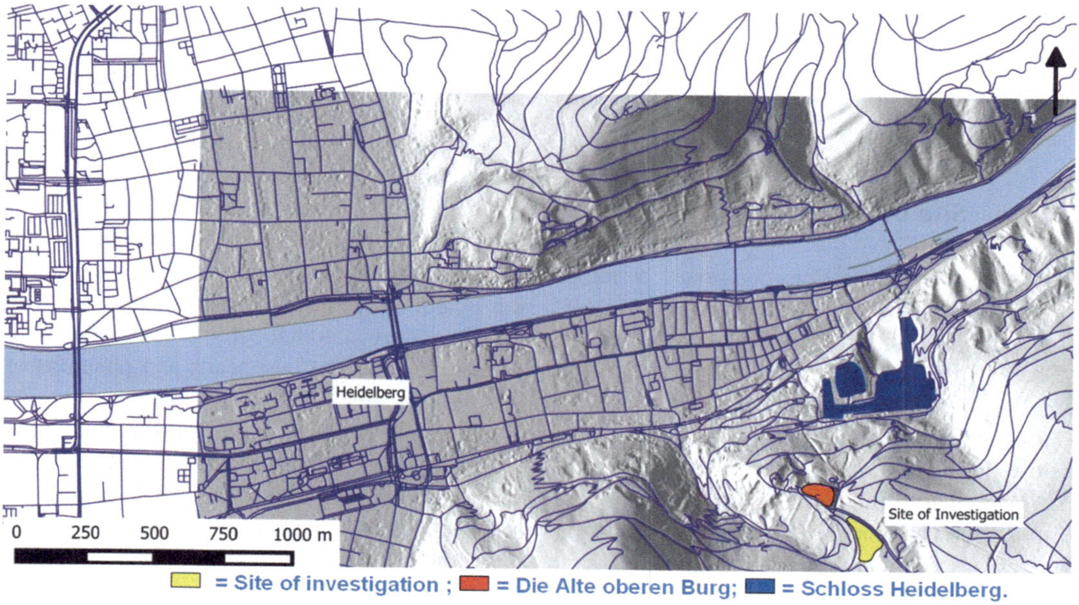

Fig. 13.1 The site of investigation lies above the historic center of Heidelberg on the Königstuhl hillside. Data source: © OpenStreetMap contributors and Landesamt für Geoinformation und Landentwicklung (LGL)

misleading information between digital truths and ground truths are essential in trying to understand generated digital landscapes.

13.1.1 State of the Art

Prospection by LiDAR has become a widely used tool within archaeology, but utilization of point cloud data for digital elevation models (DEM) do not derive objective truth (cf. Zakšek et al. 2011; Hesse 2013; 2014). Hence, data acquisition and manipulation need to be fully assessed in order to produce standardized output and normalized comparable data. Otherwise, filtering and interpolation might generate misleading data artifacts or remove and hide archaeological structures due to diverging definitions of landscape (Crutchley 2010; Devereux et al. 2008). Especially in areas not understood based on ground truth, such as remote and forested areas not easily accessible, it is important to understand the capabilities and potential misleading information in the data (Doneus 2013b). Thus, benchmark studies, such as this project at the Königstuhl of Heidelberg, help the general comprehension on areas of slope and dense forest vegetation. Other projects focus more specifically on certain algorithms and procedures in order to achieve results through data manipulation, such as visualization for interpretive mapping through the DEM derivatives "Openness" and "Sky-View factor" (Doneus 2013a; Zakšek et al. 2011).

Within management and detection of cultural heritage through LiDAR data, there are also currently undergoing some promising prospects for the future. From the state office for cultural heritage management, Baden-Württemberg, Jörg Bofinger, and Ralf Hesse (2011) continue an encouraging ALS-based detection and management project, with many new insights into archaeological use of LiDAR. From the Norwegian Computing Center and the Norwegian Institute for Cultural Heritage, several large-scale projects have been carried out with many recommendations and insights into LiDAR and remotely sensed archaeology (Grøn et al. 2003; Trier and Zortea 2012; Trier et al. 2013). Further, the project

"ArchLand" develops a European scaled perspective on remote sensing data on new methods and applications from which many standards can be derived (Opitz and Cowley 2013).

13.1.2 Site of Investigation

The site of investigation comprises a minor unnoticeable area on the Königstuhl hillside, south of the Molkenkur Hotel, and the earlier "Obere Burg" of Heidelberg (Fig. 13.1). The area is in between the two modern roads of Gaiberger Weg and Klingenteichstraße, just below the present-day Kleinkaliber-Sportverein Alt Heidelberg, and the old stone quarry, the Kammerforststeinbruch.

Previous fieldwork revealed unknown cultural heritage in the shape of terrace systems, pathways, a house foundation, border stones, wall structures, and two cellar structures. In order to get an overview of details within the area of investigation, it was decided to scan the area with a terrestrial LiDAR scanner, as well as compare the area by airborne LiDAR data in means of assessing best practice for visual object detection.

13.2 Methods

Scanning, Registration, and Interpolation

The main steps for any field scanning campaign involve three steps: pre-processing, processing, and post-processing.

Pre-processing includes research questions to investigate, strategic planning, and data acquisition. Processing involves data handling and registration for further visual manipulation. Finally, post-processing procedures embrace answering research questions, comparison, and evaluation based on automatic, semiautomatic, and manual extraction of information.

The pre-processing procedures included TLS sample data recording during a measurement campaign in June 2014 with dense vegetation cover by leaf-on conditions. Data were captured with a Riegl VZ-400 terrestrial laser scanner with a high-resolution calibrated fish-eye camera for RGB colors (Fig. 13.2).

Reflector tiepoints were strategically placed to co-register the different scanning positions with as much overlap as possible. A minimum of 4 tiepoints were visible from every scanning positions. The average amount of tiepoints from the scanning positions was 5.58 tiepoints. The maximum amount of tiepoints used at one of the scanning positions was 8 tiepoints. The tiepoints were automatically selected but with manual selection of best scans and positions of tiepoints on-site.

In order to compensate for dense vegetation in the underwood, it was decided to pursue more scanning positions rather than high amount of scanned details. However, two scanning positions in front of one of the cellar structures were prioritized to scan with a high amount of points of 3 mm at 10 m distance. Overview of the scanning positions can be seen in Fig. 13.3.

The result was 12 different scanning positions of 14 scans in total (Fig. 13.3) with the two additional scans in front of the cellar structure by point density changes between 8 and 3 mm at 10 m. The DTM was created by selecting minimum z-value per raster cell, resulting in some areas having vegetation as minimum z-value and consequently being included as terrain within the DTM. The 12 normal scanning positions were set at a resolution of 8 mm per point at 10 m distance. The additional two high-resolution scans were of 3 mm per point at 10 m distance. In total 230,555,115 points were recorded for the 12 scanning positions with a resolution of 8 mm at 10 m, and the two additional scan positions included 24,469,696 points of 3 mm at 10 m. In total, the area scanned consist of c. 1.5 ha sloped hillside with dense vegetation, containing 255,024,811 points.

The processing procedures included data handling and manipulation to be better understood. The retrieved point clouds were processed in RISCAN PRO, operating and processing software for Riegl 3D laser scanners. The single scan positions were co-registered in RISCAN PRO by applying the so-called multi-station adjustment with an average error of 1.17 cm.

Fig. 13.2 The Riegl VZ-400 on-site with dense vegetation in the underwood

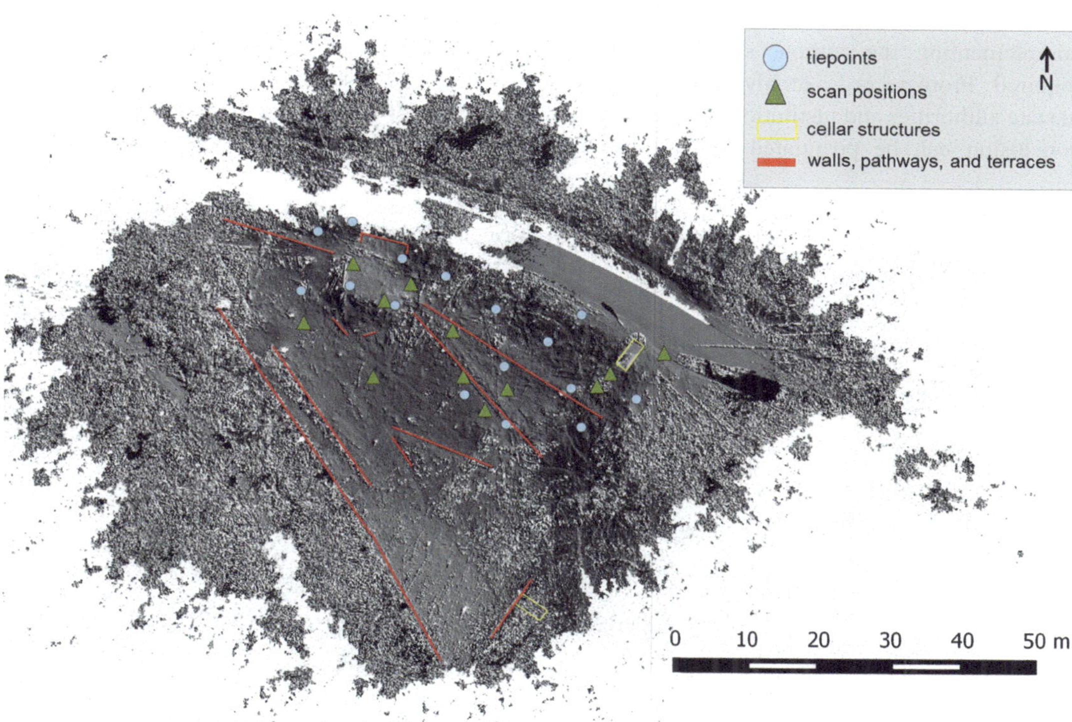

Fig. 13.3 Combined TLS scans with scanner and tiepoint positions marked

Individual ASCII text files were exported for each scan to be further processed in OPALS, *Orientation and Processing of Airborne Laser Scanning data* (cf. Mandlburger et al. 2009). From OPALS, data was interpolated to DEMs of different cell size of 1 m, 0.5 m, and 0.1 m. Different structured cells were attempted for interpolation, but it was decided that a grayscale hillshade relief was the best means for manual visual object detection of small and large structures at the same time. Especially the minor pathways were best seen by shading for indication of minor height differences.

13.3 Scanning Results

Interpretation and Understanding Context
The results of field, scanning, and remote investigations revealed many ground truths regarding monuments of cultural heritage interest. However, to comprehend the specific details of cultural and temporal context of the detected structures, further sources and investigation complementing the remote sensing data are required. From records, archives, and inquiries to state authorities and institutions, no remarks of conclusion can be postulated for the specific cultural and temporal context. Nonetheless, in the later stages of post-processing data, certain evidence and hints of evidence have arrived which might conclude some of the temporal and cultural context of the structures located at the site of investigation.

Two cellar structures, a house structure, some border stones, wall parts, and several road and terrace systems were located on-site (Figs. 13.3 and 13.5). A closer view of the house structure and one of the cellar structures can be seen in Fig. 13.4.

13.3.1 Contextualizing the Scanned Structures

The structures detected show contextual dependency (Figs. 13.5 and 13.7). The minor road systems run on both sides of the house structure and continue south toward the main concentration of details with the two cellar structures (Fig. 13.7). It is likely that most of the minor roads and terraces on-site are in relation with the cellar structures. All the structures on-site are made of the same local Bunter sandstone bedrock slabs of various sizes.

The major wall facing the modern road at the bottom of the slope (Fig. 13.5) is defined by a

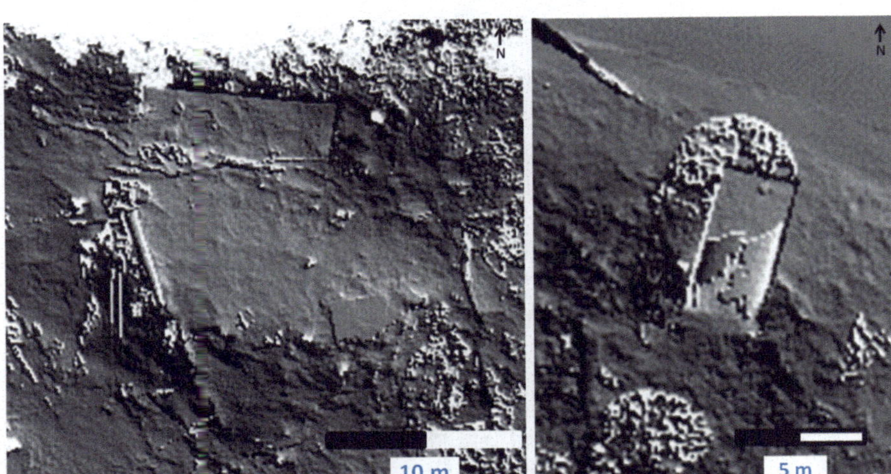

Fig. 13.4 *Left*: *top view* of house structure details from TLS. *Right*: *top view* of cellar structure from TLS. Hillshade raster, 0.1 m cell size

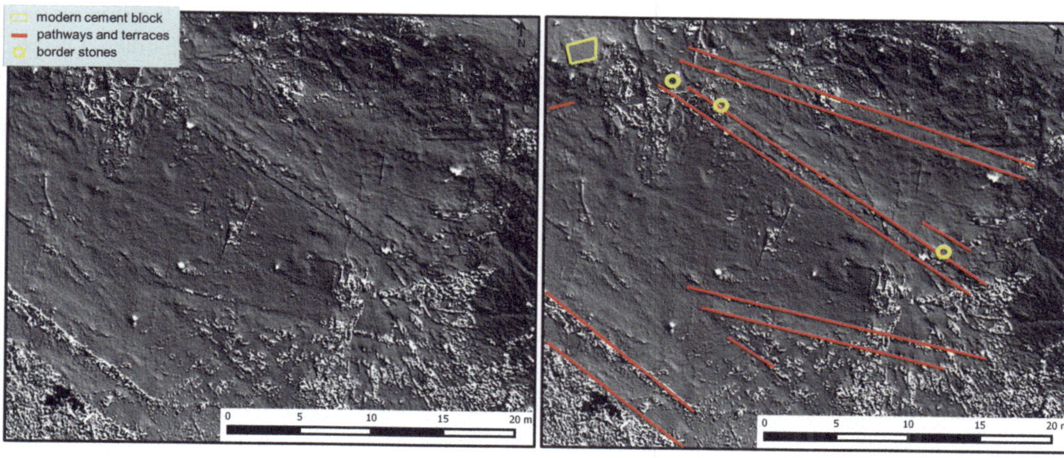

Fig. 13.5 The TLS scan with border stones, terraces, and pathways highlighted. Hillshade raster, 0.1 m cell size

small wall and terrace ca. 3 m away from the major wall, leaving a perfect distance between these two components for heavy transportation of horse and oxen-driven transportation, thus also the likely logistical path for transportation between the Molkenkur and the Klingentor down the valley of the Klingengraben.

From the site of investigation, the roads running southwest are cut by the vast amount of soil from the previous active stone quarry in use between 1812 and 1900 (Derwein 1940: 279). Beneath the quarry soil, the road and terrace systems seem to continue. The present new Gaiberger Weg, built in 1812–1848 (Derwein 1940: 279), replaces the old Kohlhöfersteig from 1755 (Derwein 1940: 183), and the former path of the Kohlhöfersteig is likely to be one of the plateaus documented at the site of investigation. The temporal and cultural definitions of the structures on-site are therefore not easily defined. However, many phases of activity can be defined to the area and structures documented in the TLS data. Pathways and roads have a long-continued existence near the Molkenkur and the "Obere Burg," but context changes, and consequently things and ideas change according to context. Subsequently the Kohlhöfersteig likely replaced a path that existed before 1755, in the same manner as the Gaiberger Weg replaced the Kohlhöfersteig.

One wall face runs perpendicular with the soil from the stone quarry, respecting the quarry activity and thus likely to be of a similar time horizon. The aforementioned wall structure contains one of the cellar structures and consequently indicates that the cellar structures are built before or meanwhile the stone quarry is active. Within the site of investigation, three border stones were also documented (Fig. 13.5). Most of the border stones erected on the Odenwald mountains near Heidelberg are from the sixteenth to the nineteenth century (Mertens et al. 2013: 658–663). The main forest border at the end of the eighteenth and beginning of the nineteenth century runs at the southern edge of the area of investigation toward east and the Plättelsweg (Derwein 1940, Plan V). Meaning the area of investigation could be part of this forest border definition, but not by border stones running north and south as documented in Fig. 13.5. Rather the border would run along the perpendicular wall going east to west. The perpendicular wall likely respects the forest border defined in 1791 (Mertens et al. 2013: 658–659), but evidence of such is covered by the excess amount of soil from the previous stone quarry.

The two cellar structures seem to be similar. The lower cellar structure within the perpendicular wall is collapsed by the added amount of soil from the stone quarry. It is therefore impossible to get a complete understanding of this structure. The upper cellar near Gaiberger Weg is, however, still intact. The upper cellar is 4.9 m × 1.8 m × 2.1 m, constructed of heterogeneous local Bunter sandstone slabs. At the end wall, two niches are built. Just above the niches, a "chimney" is situated, possibly for ventilation. Within the cellar there is a constant cool temperature perfect for storing purposes. No material evidence was located on the surface within the cellar to determine exact purpose, but the cellar is perfect for many different storing purposes, such as beverages or food.

The two cellar structures did for a long period constitute a large puzzle for understanding the area of investigation. A clear conclusion on how to define the cellar structures was not easy to attain, because the cellar structures do not seem to be described or documented in any of the written records or more modern observations. Further, none of the historical maps indicated any activity in the vicinity besides minor pathways. From Lorentzen's map of 1907 (Fig. 13.6), many of the historic pathways are visible. From the beginning of the nineteenth century, city activity extended to the area around the lower part of the Klingenteichstraße. Of special interest is the establishment of several beer cellars in the area by the Brauerei Kleinlein and Brauerei Jäger. The beer cellars are today present at Klingenteichstraße 5, 20, and 26 (Mertens et al. 2013: 330). Especially, the wall seen at Klingenteichstraße 26 is a perfect duplication of the large wall within the area of investigation. From the beginning and the mid-nineteenth century, a comprehensive settlement of the lower Klingenteichstraße takes place, combined with a renovation of the Klingenteichstraße. Consequently, much of the stonework in the area is similar all the way from the Klingentor and along Klingenteichstraße toward the Molkenkur. The cellar structures within the area of investigation are likewise very similar with direct comparison and similarity from top to bottom of the Klingenteichstraße. Thus it is likely that the present visible structures at the site of investigation are part of the comprehensive construction and renovation from the beginning and mid-nineteenth century settlement and infrastructure activity (Mertens et al. 2013: 324) with significant focus on logistics to and from

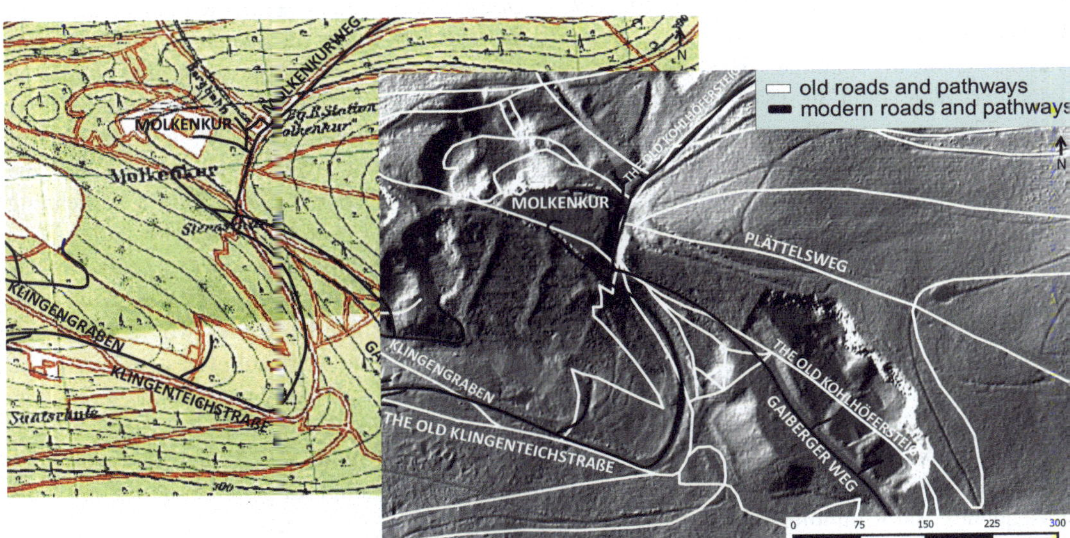

Fig. 13.6 *Left*: Lorentzen's map of 1907 with several indications of pathways near the Molkenkur and beyond. Pathways are extracted and compared on DTM. Data source: Lorentzen 1907. *Right*: airborne DTM with indications of old and modern pathways from Lorentzen 1907. Hillshade raster, Azi. 45°, 270 angle, 1 m cell size. Data source: LGL

the old Kammerforst stone quarry. But likely the comprehensive construction activity should also be seen in relation to the breweries of Brauerei Jäger and Braurei Kleinlein from Klingentor toward the east around the old Biersiedersteig (Derwein 1940: 112) under the Schloß Hang and further toward the southern Bierhelderhof, the Kohlhof, and beyond toward the south. Especially the activity on the Schloß Hang consisted historically of several cellar structures for the storage of beer (Derwein 1940: 112). Likely this activity in the end also extended to the other side of the ridge and thus included the area from the Molkenkur and below. Whether or not more beer cellars are still hidden within the landscape of the Königstuhl hillside is difficult to say, but should be explored further.

13.4 Technical Conclusion

Best Practice Reflection, Comparison, and Scale of View

Regarding the structures detected in the area of investigation, the cellar structures are naturally not visible in the ALS data due to the structures being mainly subsurface structures. Hence any ALS reconnaissance would not reveal subsurface structures but only structures with elevation details above terrain. ALS data will therefore naturally always be incomplete in the detection of potential monuments below terrain. The house structure is revealed in the ALS data from the site of investigation (Fig. 13.7), but not all the road and terrace structures are. The lack of information in the ALS data is due to a lack of points per square meter (*ppsm*), and thus the apparent geometric resolution of the DTM of 1 m cell. Consequently the scale of view in the acquired ALS data is not sufficient for cultural heritage monument detection within the area of investigation and sloped dense vegetation. The amount of details revealed in the ALS data is consequently not sufficient for complete information extraction and manual visual detection of cultural heritage monuments.

In order to understand amount of details necessary to detect cultural heritage in the area of investigation, the following study has investigated many different interpolated DTMs at different levels of detail. However, the most remarkable changes occur in the difference of cell size in the interpolation process. The increase and decrease in amount of information is not linear with the amount of ppsm and potential amount of information and details in the landscape. Meaning, too much or too little information can be equally disturbing. An example of such can be seen in Fig. 13.8.

A close-up of the road and terrace structures, as seen in Fig. 13.8, indicates some of the changes in different cell size when interpolating. The changes in level of detail reveal that some details can be seen in the 1 m cell, but the amount of information is too low to distinguish them as being cultural traces left in the landscape. In the 0.5 m cell, the road and terrace structures can be distinguished as not being part of the natural landscape and stand out as clear lines. In the 0.1 m cell, road and terrace structures are present and distinguishable as cultural traces left in the natural landscape. However, the amount of other details in the landscape also increases in the 0.1 m gridded interpolation. The visualization therefore becomes more blurred because more detail is revealed and information given. Thus, the high amount of detail in the interpolation with the highest amount of *ppsm* and information demonstrates not to be the most relevant or efficient for manual visual detection of objects and structures. The 0.5 m DTM reveals the same information in a simpler and faster procedure.

The results show that the highest amount of data is not necessarily the best approach. In conclusion, it is more relevant to focus on increased scanning positions instead of amount of detail recorded at each scanning position when documenting in dense vegetation. Focused and structured procedures of scanning will in the long run produce the highest amount of information and thus give the most complete picture of the area of investigation. Future ALS resolution consequently needs to include resolution capable of

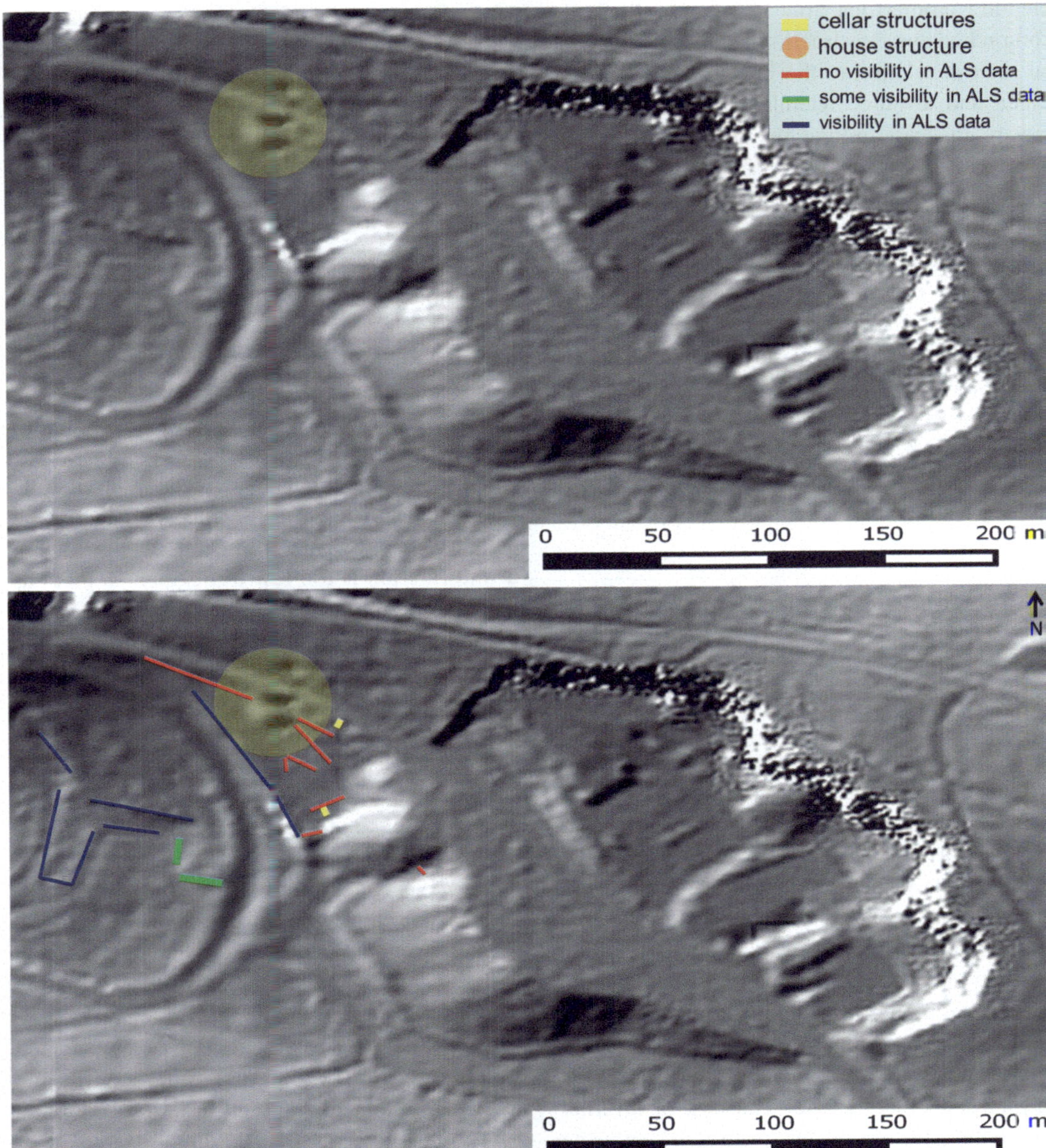

Fig. 13.7 Area of investigation visually represented by ALS data. Hillshade raster, 1 m cell size. Data source: LGL

producing comprehensive 0.5 m gridded interpolations in order to become the effective means of large-scale cultural heritage detection.

But one approach cannot replace the other. Within the area of investigation, it is almost impossible to get a complete overview of the details and structures on-site. One of the major pathways within the area of investigation was not detected before a closer investigation of the TLS data was initiated. Since then, this specific pathway has been confirmed as a ground truth, but the dense vegetation and collapsed trunks made it almost impossible to detect by the initial fieldwork. It was only by knowing exact details from the TLS data that it was possible to confirm this digitally detected plateau as part of the remaining

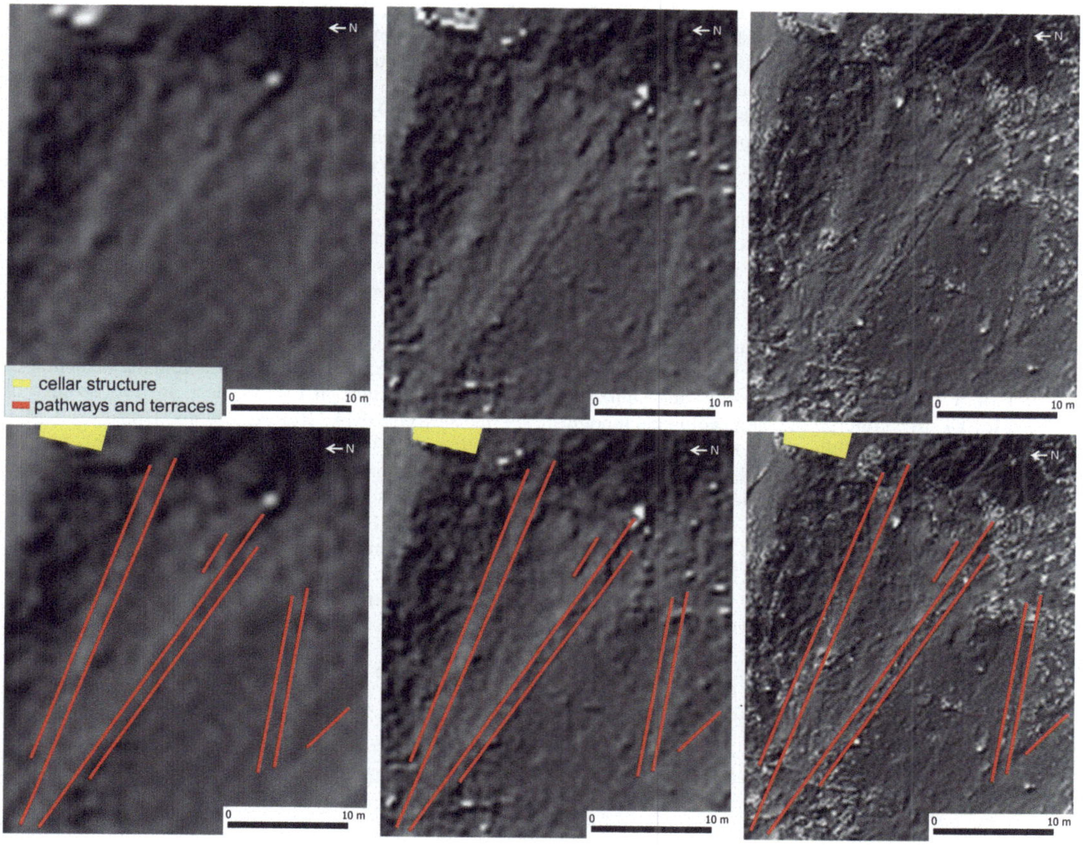

Fig. 13.8 Combined TLS scans with different cell size. From *left* to *right*: 1 m, 0.5 m, 0.1 m. Pathways and cellar structure

cultural complex. However, many other details were equally difficult to determine within the TLS data and necessitated prior knowledge or later ground confirmation of its existence. Thus, all three data sources were necessary in order to construct a comprehensive overview of the cultural activities within the area of investigation, and none of them were completely capable of replacing the other.

References

Bofinger J, Hesse R (2011) As far as the laser can reach... Laminar analysis of LiDAR detected structures as a powerful instrument for archaeological heritage management in Baden-Württemberg, Germany. In: Cowley DC (ed) Remote sensing for archaeological heritage management. EAC Occasional Paper 5:161–171

Crutchley S (2010) The light fantastic: using airborne lidar in archaeological survey. In: ISPRS TC VII symposium – 100 years ISPRS IAPRS, Vienna, Austria, vol XXXVIII, part 7B, pp 193–198, 5–7 July 2010

Derwein H (1940) Die Flurnamen von Heidelberg – Eine Stadtgeschichte. Carl Winter's Universitätsbuchhandlung, Heidelberg

Devereux BJ, Amable GS, Crow P (2008) Visualisation of lidar terrain models for archaeological feature detection. Antiquity 82:470–479

Doneus M (2013a) Openness as visualization technique for interpretative mapping of airborne lidar derived digital terrain models. Remote Sens 5:6427–6442

Doneus M (2013b) Die hinterlassene Landschaft – Prospektion und Interpretation in der Landschaftsarchäologie. Mitteilungen der Prähistorischen Kommission 78

Doneus M, Briese C (2006) Full waveform airborne laser scanning as a tool for archaeological reconnaissance. In: From space to place. Proceedings of the 2nd

international conference on remote sensing in archaeology, BAR international series, 1568, pp 99–105
Doneus M, Briese C, Studnicka N (2010) Analysis of full-waveform ALS data by simultaneously acquired TLS data: towards an advanced DTM generation in wooded areas. In: ISPRS TC VII symposium – 100 years ISPRS, IAPRS, Vienna, Austria, vol. XXXVIII, part 7B, pp 193–198, 5–7 July 2010
Grøn O, Aurdal L, Christensen F, Tømmervik H, Loska A (2003) Locating invisible cultural heritage sites in agricultural fields – evaluation of methods for satellite monitoring of cultural heritage sites, results 2003. Digital Copy Studio AS, Oslo
Hesse R (2013) The changing picture of archaeological landscapes: lidar prospection over very large areas as part of a cultural heritage strategy. In: Opitz RS, Cowley DC (eds) Interpreting archaeological topography, Occasional publication of the AARG vol 5. Oxbow Books, Oxford, pp 171–183
Hesse R (2014) 2 and 3 dimensions. Blog [Online]. https://blog23d.wordpress.com/. Accessed 17 May 2015
Lorentzen T (1907) Heidelberg und Umgebung. Map, von Knoblauch & Dürr, Stuttgart. http://www.ub.uni-heidelberg.de/helios/fachinfo/www/math/heidelberg/PlanHD/lore-1907.jpg. Accessed 17 May 2015
Mandlburger G, Otepka J, Karel W, Wagner W, Pfeifer N (2009) Orientation and processing of airborne laser scanning data (OPALS) – concept and first results of comprehensive ALS software. ISPRS workshop, laser scanning '09, vol 38. Paris, France, pp 55–60
Mertens M, Naumann D, Plate U (2013) Kulturdenkmale in Baden-Württemberg. Stadtkreis Heidelberg. Vol 1. Jan Thorbecke Verlag der Schwabenverlag AG Ostfildern
OpenStreetMap Data is available under the Open Database License. http://www.openstreetmap.org/copyright
Opitz R, Cowley D (2013) Interpreting archaeological topography: 3D data, visualisation and observation. Oxbow Books, Oxford
Trier ØD, Zortea M (2012) Semi-automatic detection of cultural heritage in LIDAR data. In: Proceedings of the 4th GEOBIA, Rio de Janeiro, Brazil, 7–9 May 2012, pp 123–128
Trier ØD, Zortea M, Tonning C, Loska A (2013) Grave mounds discovered by automatic heap detection method. In: Procedings of the 4th EARSeL workshop on cultural and natural heritage matera, Italy, 6–7 June 2013, pp 1–16
Zakšek K, Oštir K, Kokalj Z (2011) Sky-view factor as a relief visualisation technique. Remote Sens 3:398–415

Part IV

Geophysical Methods and Data Fusion

An Introduction to Geophysical and Geochemical Methods in Digital Geoarchaeology

14

Apostolos Sarris, Tuna Kalayci, Ian Moffat, and Meropi Manataki

Abstract

Archaeological geophysics is a range of techniques for the minimally invasive, remote investigation of the physical parameters of the near-surface environment. This suite of methods is complementary to archaeological survey or excavation as it can provide information about the stratigraphy of the survey area, locate anthropogenic traces of the past, document their spatial dimensions and—under ideal conditions—explore the physical properties of subsurface materials. Both material culture items such as a building foundations and indirect indications of anthropogenic activity such as subsurface disturbance or evidence of burning are excellent direct targets for geophysical investigations since they can be differentiated on the basis of their material properties from the wider soil context. In addition to directly locating archaeological material, geophysical techniques can make an important contribution to geoarchaeological investigations by elucidating the site stratigraphy and mapping its lateral geometry. In some cases, such as when locating prehistoric material buried offshore or within open Palaeolithic sites, the reconstruction of past landscapes may make a more important contribution to archaeological investigations than the direct geophysical detection of archaeological materials.

Different material culture items have characteristic physical properties (such as electrical resistance or conductivity, magnetic susceptibility) and so require different instrumentation for effective detection. The main techniques for archaeological prospection include magnetometer, resistance meter, magnetic susceptibility meter, ground-penetrating radar and electromagnetic induction meter. Apart from that, seismic methods (reflection and refraction seismics), gamma spectroscopy and gravity techniques are also used in certain circumstances.

A. Sarris (✉) • T. Kalayci • I. Moffat • M. Manataki
GeoSat ReSeArch Lab, Foundation for Research and Technology, Hellas (F.O.R.T.H.), Rethymno, Crete, Greece
e-mail: asaris@ret.forthnet.gr

© The Author(s) 2018

C. Siart et al. (eds.), *Digital Geoarchaeology*, Natural Science in Archaeology,
DOI 10.1007/978-3-319-25316-9_14

Unfortunately, there is no standard approach for the application of one specific geophysical method for all archaeological materials in all geological environments. The success of geophysical prospection techniques depends on a combination of soil and sediment characteristics as well as depth below surface and preservation of archaeological findings. In order to achieve the most reliable results and enhance the chance of detecting archaeological material, an integrated, multi-method approach is suggested.

In addition to field surveying, the effective processing of measured geophysical data is a crucial part of the interpretation process. Data processing aims to enhance signals of interest in order to better delineate archaeological and geological features. It helps to produce more interpretable results and therefore facilitates and fosters collaboration between geophysicists and archaeologists.

Keywords
Geophysics • Landscape archaeology • Palaeolandscape reconstruction • Soil stratigraphy

14.1 Introduction

Geophysical methods are an important component of geoarchaeological investigations due to their ability to non-invasively image the subsurface of an archaeological landscape. New developments in multi-sensor and positioning technology have facilitated the use of these methods over large areas, allowing archaeological questions to be addressed on a landscape scale. They are particularly useful in geoarchaeological investigations for defining site stratigraphy, mapping site disturbance and reconstructing palaeolandscapes.

Geophysical methods also make a significant contribution to archaeological investigations beyond geoarchaeology. Excellent detailed introductions to the application of specific geophysical techniques to archaeology in general are available for ground-penetrating radar (Manataki et al. 2015; Conyers 2013, 2015b), magnetometry (Armstrong and Kalayci 2015; Aspinall et al. 2008), resistivity (Schmidt 2013) and electromagnetic techniques (Simon and Moffat 2015).

14.2 An Overview of Geophysical Methods

14.2.1 Ground-Penetrating Radar

Ground-penetrating radar (GPR) is a widely used technique in geoarchaeology, which allows the detection of features in the subsurface that cause a change in the propagation of electromagnetic energy. A typical GPR includes a transmitting antenna (Tx) and a receiver antenna (Rx). The Tx transmits electromagnetic waves with a fixed frequency into the subsurface. The Rx receives the portion of the energy, which is reflected by variations in material properties of the subsurface and registers the amplitude of this response for mapping purposes (Fig. 14.1).

Electromagnetic wave propagation is complex as it includes reflections, refractions and diffractions caused by changes in relative dielectric permittivity (ε) and/or conductivity (σ). Variations in these values reflect geological and/or geomorphologic features (e.g. changes in

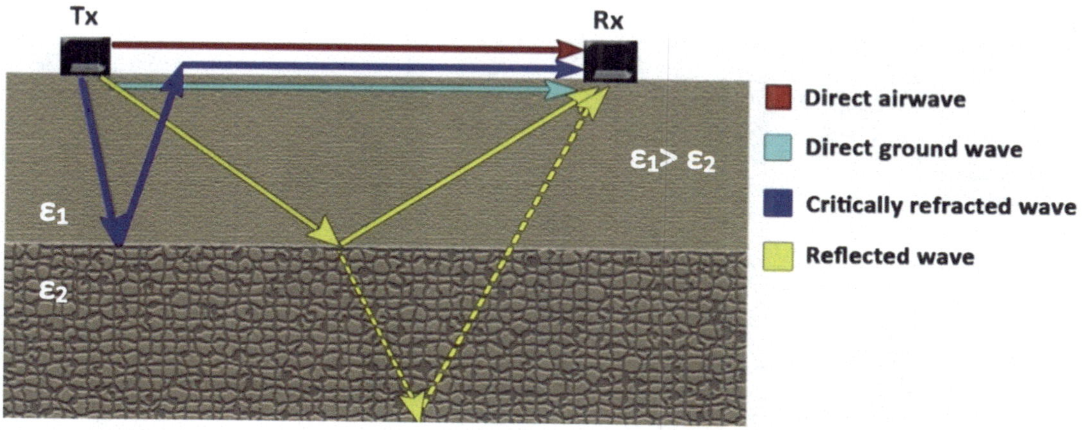

Fig. 14.1 Signal paths between the transmitter (Tx) and the receiver (Rx) for two layers with different electrical properties (ε stands for the dielectric permittivity). The *dashed yellow line* indicates that part of the signal's energy was refracted and then reflected (Annan 2009)

lithology, faults, strike and dip of beds, cavities, bedding features, loose sediments vs. bedrock etc.), changes in hydrological conditions (degree of fluid saturation and water chemistry) or the presence of archaeological materials. GPR can be deployed with a variety of antenna frequencies allowing features on a wide range of scales to be imaged. In general, antennas with a frequency range of approximately 200–500 MHz are appropriate for most archaeological studies as they provide a depth of investigation up to about 3 m and an acceptable resolution (Fig. 14.2a) (Jol 1995; Conyers 2015a, b, pp. 14–15). Lower-frequency antennas can be used for deeper investigations (such as the location of geomorphic features), and higher frequency antennas are suitable for shallower, higher resolution surveys such as investigation of mortar thickness. Recent developments in GPR technology include multi-channel systems (Fig. 14.2b) coupled with GPS navigation systems for fast surveys, ultra-wide-band antennas (Trinks et al. 2010) and stepped-frequency antennas (Linford et al. 2010). Drone-mounted GPR systems are also expected to become available within the coming years (Merz et al. 2015).

Usually, GPR surveys are undertaken on a grid of regularly spaced parallel lines, which are commonly post-processed together into a 3D volume. Radargrams represent the amplitude of the reflected signals as a function of the time taken for the electromagnetic wave to travel from the Tx to the Rx via the subsurface. When the velocity of propagation of the electromagnetic radiation within the ground can be estimated, it is possible to convert the time of arrival to a depth estimate. GPR data can be viewed as 1D soundings (traces), 2D profiles (radargrams), amplitude 'slices' through the 3D volume, 3D cubes or 3D representation of the signal amplitude. In general, amplitude slicing is more effective for mapping archaeological features with obvious linear geometry and a high amplitude response than for stratigraphic features which are often more geometrically complex. For the investigation of geoarchaeological research questions, the combined analysis of individual traces, 2D profiles and slice maps is most likely to be more effective than the exclusive use of slice maps (Conyers 2015a, b).

GPR applications in geoarchaeology range from determination of the depth to bedrock, mapping of palaeolandscape features (such as channels and shorelines, Fig. 14.3), locating of anthropogenic earthwork features like canals or moats and identification of soil disturbance due to habitation (Fig. 14.4). In addition, GPR allows mapping unmarked graves, which often can be identified through soil anomalies caused by excavation of tombs. Unfortunately, the high

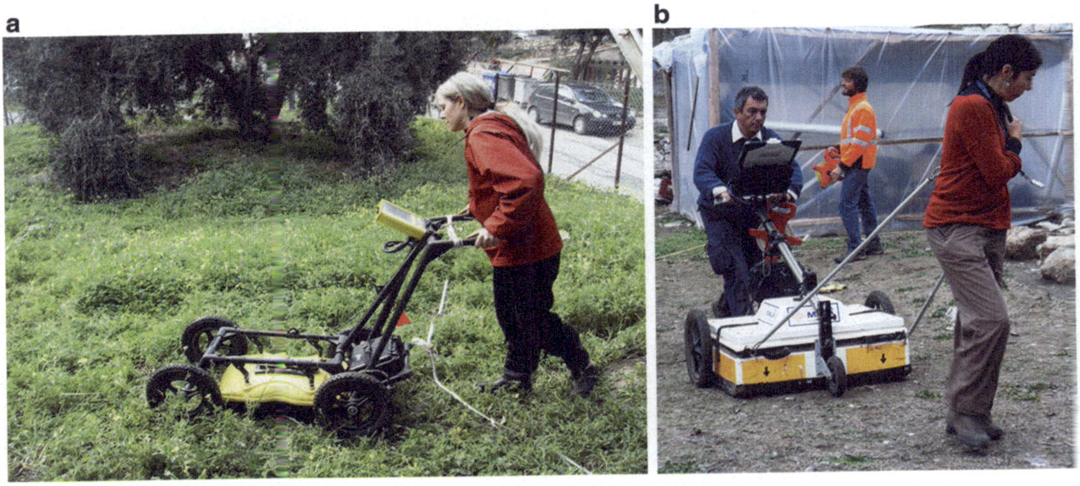

Fig. 14.2 (**a**) Noggin Plus (Sensors & Software Ltd.) 250 MHz GPR system in a survey at the ancient theatre of Chersonessos in Crete. (**b**) MALA MIRA multi-antenna GPR configuration with eight 400 MHz antenna applied in a survey at Feres (Velestino), Central Greece

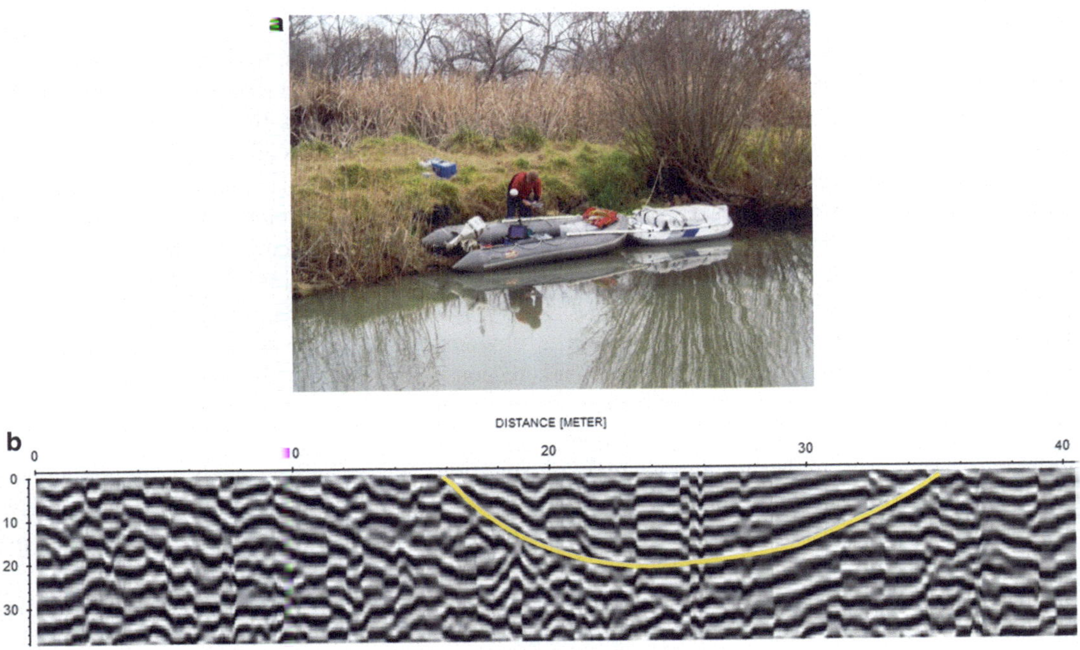

Fig. 14.3 (**a**) Mala X3M 100 Mhz GPR system being used for mapping sedimentary environments within a lake. Lake Alexandrina, South Australia. (**b**) A palaeochannel (see the *yellow line*) imaged in a 2D profile using a Mala X3M 250 MHz ground-penetrating radar from the Umbum Creek, Central Australia

attenuation of radar energy in saltwater precludes the application of GPR in marine surveys, but it is suitable for use in limnic environments (Fig. 14.3).

A major advantage of GPR for geo-archaeological studies is that it can image sedimentary structures, such as cross bedding. Thus, it is an effective technique to be applied in areas

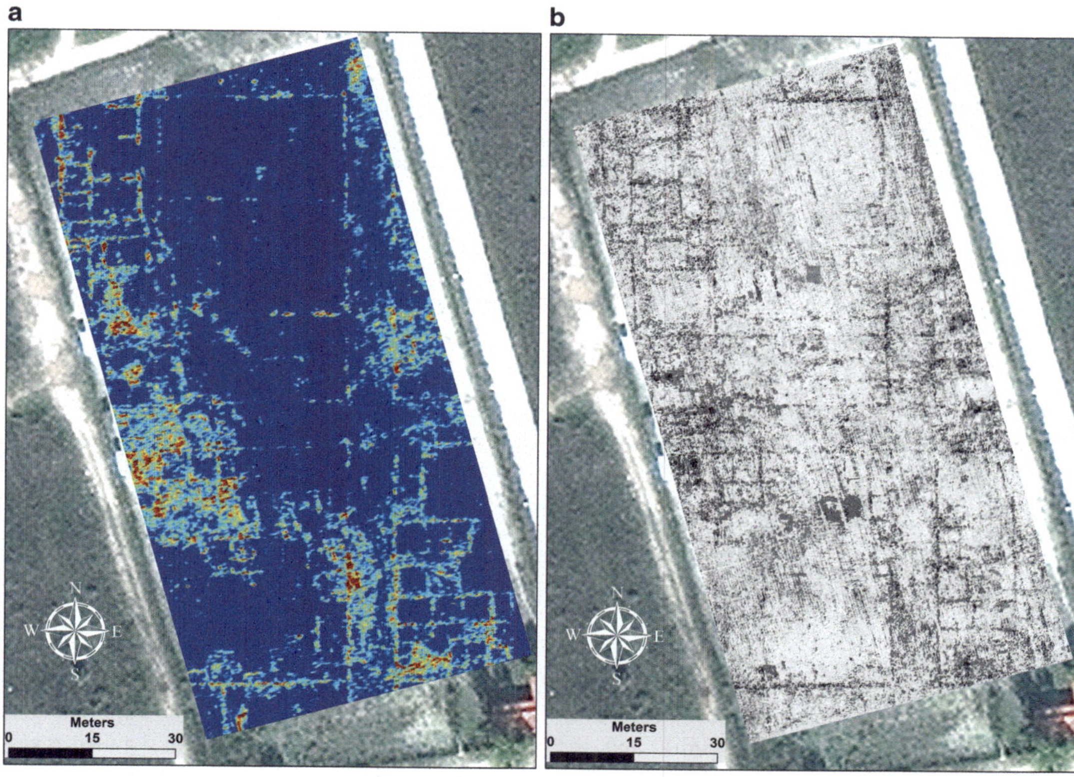

Fig. 14.4 (a) Results of GPR slices at about 60–70 cm below the surface resulting from Noggin Plus and MALA. (b) GPR systems from Demetrias archaeological site, Central Greece. Measurements were taken along transects 0.5 m apart

characterized by changes in depositional facies that are not accompanied by lithological changes. Moreover, as the GPR antenna can be moved continuously during data acquisition, the method allows faster data acquisition and higher survey speed compared to methods such as resistivity tomography (see below) that require stakes to be inserted in the ground. GPR also benefits from the shielding commonly placed around antennas, which reduces the interference from features on the surface (particularly metal) as opposed to other methods, such as electromagnetic induction or magnetometry, making it suitable for use in urban settings (Sarris and Papadopoulos 2012; Papadopoulos et al. 2009).

14.2.2 Electromagnetic Induction Methods

Electromagnetic induction (EMI) can be used to calculate the apparent conductivity and magnetic susceptibility of the subsurface. In the most common type of EM instruments, a transmitter coil generates a primary electromagnetic field for a given frequency (~8–90 kHz) (Fig. 14.5). The interaction of the field with subsurface elements generates an electrical current. This electrical current, in return, generates a secondary electromagnetic field which is sensed by a receiver coil in the instrument alongside the primary field. The magnitude of secondary field is divided into two

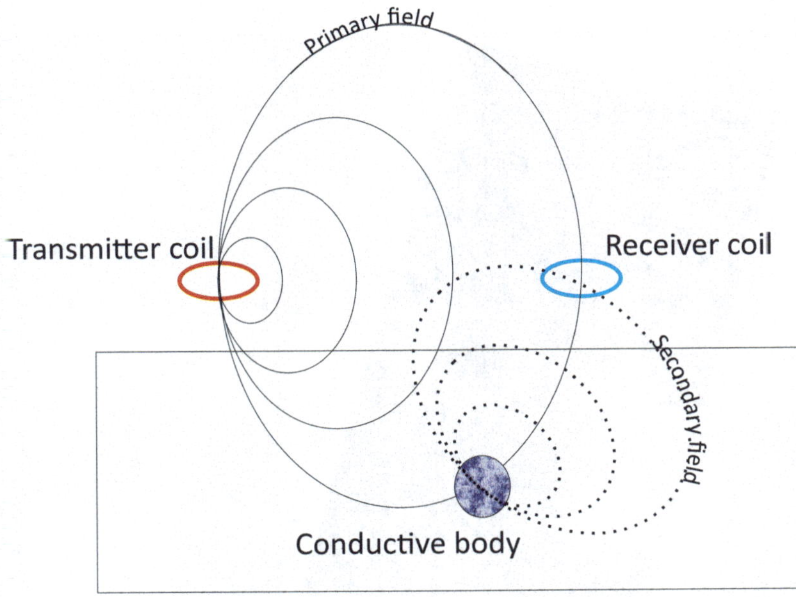

Fig. 14.5 Diagrammatic representation of the principle of operation of the electromagnetic induction methods

orthogonal components: the quadrature and the in-phase. The quadrature provides information on apparent conductivity (in mS/m), and the in-phase is related to the magnetic susceptibility (emu) of materials. The depth of investigation is mainly controlled by the separation between the transmitter and the receiver coils (Fig. 14.6; Simon et al. 2015a, b).

EM requires no direct contact with the ground surface allowing large areas to be surveyed relatively quickly (Fig. 14.6 and 14.7). Conductivity can be a good measure of changes in rock and soil units, because it is influenced by composition, the porosity, degree of water saturation and fluid chemistry. Magnetic susceptibility may provide information about erosion, pedogenesis and anthropogenic burning as all of these processes may form or concentrate magnetic minerals (French 2003).

Another subfield of the use of electromagnetic induction in geoarchaeology is the specific investigation of magnetic susceptibility by dedicated instruments. Corresponding instruments are more sensitive to changes in magnetic susceptibility than general-purpose electromagnetic induction instruments and can be used to survey on the surface, on archaeological sites and excavations and even down in drill holes (Dalan and Banerjee 1998). However, such applications are often complemented by more detailed magnetic measurements (Dalan 2007), as explained in the section on magnetic methods below.

14.2.3 Electrical Resistance Techniques

Electrical resistivity investigations work by measuring either the self-potential (passive) or the direct current resistivity or induced polarization (active) of the subsurface. The resistivity of geological and archaeological materials is controlled by their physical properties as well as their water chemistry and degree of saturation (Hecht and Fassbinder 2006).

Electrical resistivity surveys are undertaken between probes of a known separation. The larger the separation between the electrodes, the greater the penetration depth and the smaller the resolution. Various electrode configurations can be used, including the Twin, Wenner, Square, pole-dipole, dipole-dipole and Schlumberger arrays (Loke 2000; Fig. 14.8). Each configuration provides particular advantages for specific

	GEM2		CMD Mini Explorer			
	Configuration	Depth	Configuration	Depth (T_1)	Depth (T_2)	Depth (T_3)
Electrical Conductivity	HCP	2.5m	HCP	0.5m	1m	2m
			VCP	0.3m	0.7m	1.3m
Magnetic susceptibility	HCP	1.7m	HCP	0.2m	0.5m	1m
			VCP	0.3m	0.7m	1.3m

Fig. 14.6 GEM2 Geophex (**a**) and CMD Mini Explorer–CF Instruments (**b**) configurations and coil arrays for the investigation of the soil conductivity and soil magnetic susceptibility with respect to the coil separation and orientation of coils (horizontal HCP and vertical VCP)

situations, depending on the size, geometry, orientation, depth and relative resistivity contrast between the target and background (Schmidt 2013). Multi-probe, wheeled and tractor-driven systems have also been developed for smooth and conductive landscapes.

Resistivity measurements can be carried out in various ways. Vertical electric soundings, in which the electrode separation is gradually increased at a fixed point, are used to obtain a model of the depth to bedrock at a fixed location. Resistivity mapping, the most common configuration for archaeological prospection, makes use of a fixed electrode separation which is moved as an array along parallel profiles within a grid. This approach produces a plan-view image of the distribution of the resistivity at a specific depth. Finally, multi-probe arrays [commonly referred to as electrical resistivity tomography (ERT)] use multiple equidistant electrodes that measure both the lateral and vertical variations of the subsurface resistivity as a 2D profile (Fig. 14.9). Having a number of parallel ERT transects, it is possible to generate 3D volumetric images of the resistivity. Resistivity, especially when used in the ERT configuration, together with seismic techniques, is excellent for imaging changes in lithology and geology (Laigre et al. 2012; Scapozza and Laigre 2014). ERT is also valuable when either deep depositional targets (e.g. ancient ports covered by alluvial deposits, ditches and palaeochannels) or offshore archaeological features are the target of investigations (Sarris et al. 2014; Tonkov 2014; Simyrdanis et al. 2015).

14.2.4 Magnetic Methods

Magnetic prospection (or magnetometry) works by measuring disturbances to the earth's magnetic field caused by the presence of iron minerals. Magnetization of bodies consists of two components, namely, the induced and the remanent magnetization. While the former is

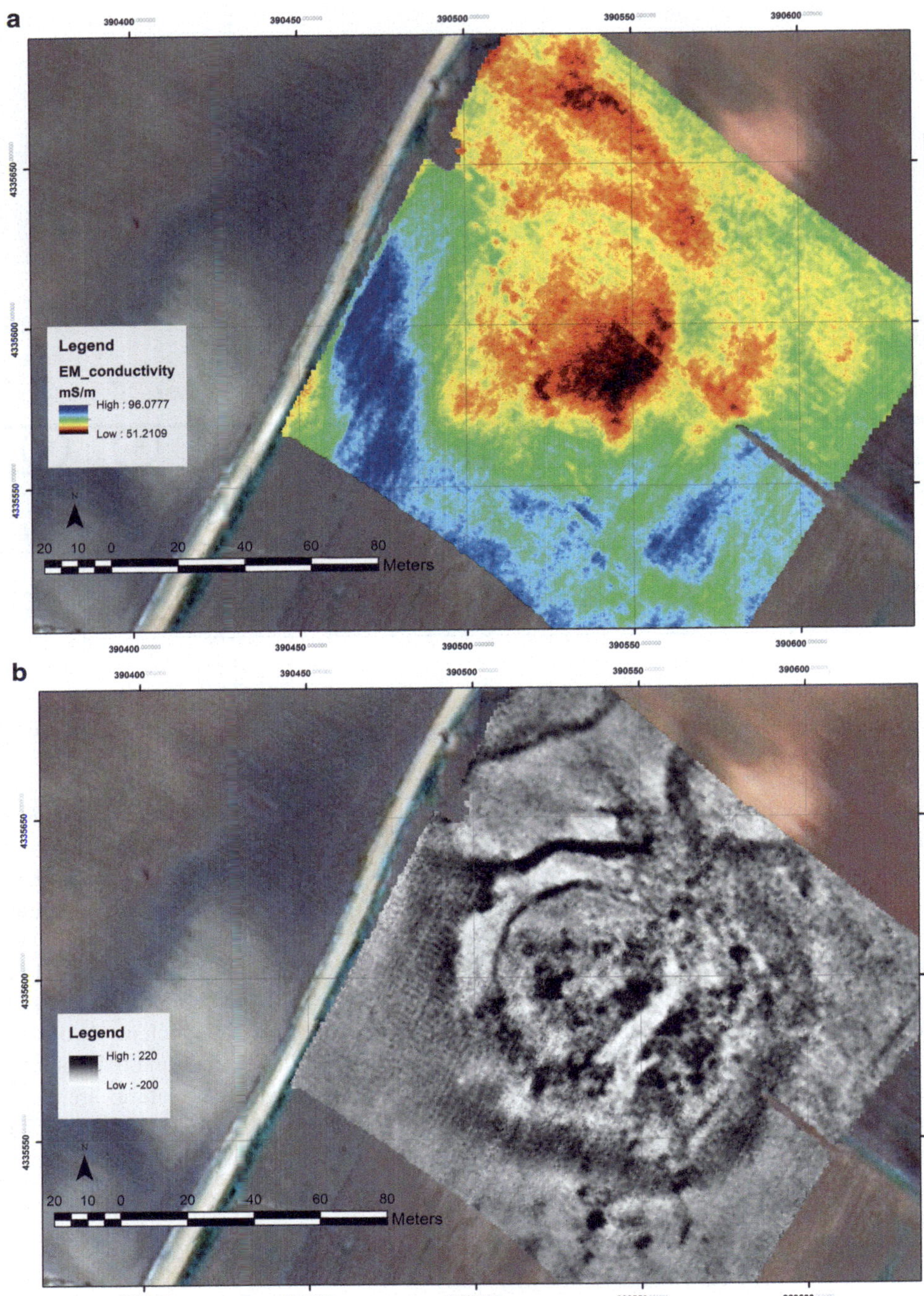

Fig. 14.7 Indicative results from the EMI survey (GEM2) at the Neolithic site of Almyros 2 at Thessaly, Greece representing the soil conductivity (**a**), which suggests a higher conductivity to the southern side of the tell (indicative of the flooding zone) and the magnetic susceptibility (**b**) that outline the surrounding ditches and some of the details of the structural remains of the core of the settlement (Sarris et al. 2015a)

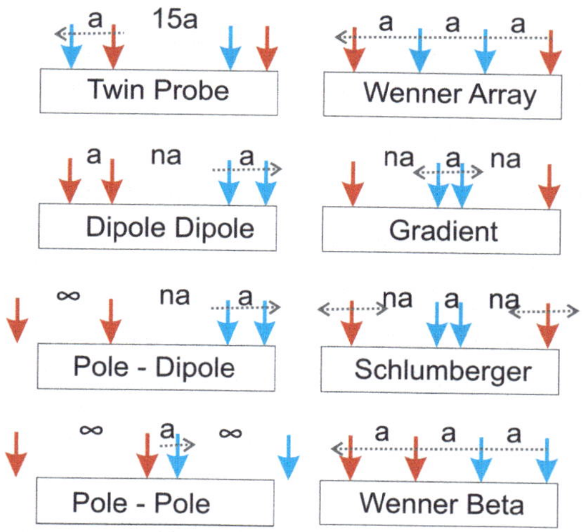

Fig. 14.8 Indicative electrode configurations used in a soil resistance survey. *Blue arrows* represent potential electrodes, and *red arrows* represent current electrodes. The distance between them varies according to the electrode array. In the mapping mode, either all or some of the electrodes move within the area of interest

created by the modern magnetic field, the latter is a result from previous magnetic fields. Magnetic susceptibility is an additional important parameter, which measures the degree to which a material becomes magnetized when an external magnetic field is applied. Magnetic prospection depends on several parameters, such as the degree and orientation of induced and remanent magnetization of bodies, their magnetic susceptibility, their volume (or mass) and their distance from the sensor.

Magnetic measurements can be collected with single or multiple sensors configured to measure usually either the total magnetic field intensity or the total or vertical magnetic gradient (gradiometer mode). Recent equipment developments have introduced multiple sensor array carts to archaeology which, when combined with high-quality positioning and navigation systems, facilitate extremely high-sample density measurements over large areas extremely quickly (Fig. 14.10).

Contrary to its widespread application for archaeological prospection, field magnetometry plays a rather little role within geoarchaeological investigations. This is mainly because it is rather suited for the direct detection of ferrous or burnt materials than for understanding the stratigraphic matrix that contains archaeological material. One promising application, though, is mapping the extent of pedogenic soil minerals, which is more effective using single sensor equipment instead of gradiometry equipment.

In contrast, laboratory magnetic measurements are very useful for understanding the depositional conditions at archaeological sites (Dalan and Banerjee 1998). Magnetic susceptibility, particularly the frequency dependant component, is increased by archaeological occupation both through heating and the addition of organic material to the soil. The surface distribution of magnetic susceptibility values provides a proxy for the spatial distribution of occupation, and, in combination with other magnetic properties [e.g. anhysteretic remanent magnetization (ARM) and saturation isothermal remanent magnetization (SIRM)], it is possible to infer the intensity of habitation or to determine the provenance of archaeological materials (Thompson and Oldfield 1986; Dearing 1994).

14.2.5 Acoustic Procedures

Acoustic measurements are undertaken by creating repeatable acoustical waves that enter the ground and are refracted or reflected by changes in density before returning to a series of geophones (Fig. 14.11). The study of the arrival times, amplitude and form of the returned wavefronts provides information about the stratigraphy of the subsurface and the presence of archaeological material.

The most common acoustic techniques in archaeological research are seismic reflection and bathymetry, particularly for the reconstruction of underwater palaeolandscapes (i.e. Tizzard et al. 2015). The acoustic response of subsurface sediments measured by seismic reflection provides two- and three-dimensional representations of the geometry of sedimentary depositional units, their internal structures and their lithology. Seismic methods are particularly suited to aquatic surveys, as the acoustic properties of water allow the geophones to be used without direct coupling to the seafloor which greatly enhances survey speed compared to terrestrial surveys.

On land, seismic methods are useful for geoarchaeological investigations as their relatively deep depth of penetration is well suited to mapping large monumental structures and mounds or reconstructing the palaeotopography. Seismic refraction is particularly useful for determining the depth to bedrock within archaeological sites (Fig. 14.12). In contrast, seismic reflection is very valuable in terrestrial archaeology to map site stratigraphy, with surface wave techniques in particular showing great potential for wider use.

14.2.6 Other Geophysical and Geochemical Techniques

Microgravity measurements are carried out using sensitive devices that measure small variations in the Earth's gravitational field. These alterations are usually caused by air-filled voids in the subsurface (e.g. tunnels, cavities, caves and architectural features; see Fig. 14.13). They show a significant contrast between mass density and the surrounding loose sediments or the parent rock.

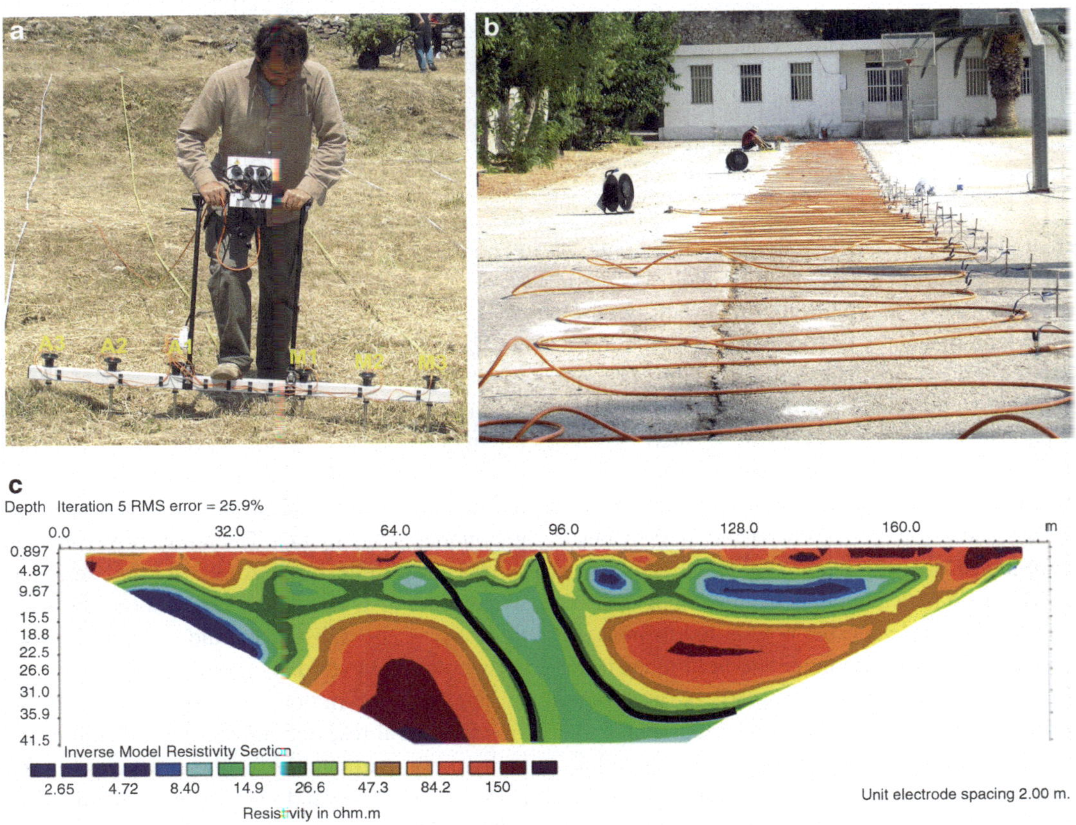

Fig. 14.9 (**a**) Geoscan Research Resistance meter (RM15) with multiplexer employing six electrodes (current electrodes As and potential electrodes Ms) at various configurations. (**b**) ERT extension of electrodes being used to map the area of the old Franciscan monastery of Rethymno (Mikrasiaton plaza) in Crete. (**c**) Stratigraphic results of an ERT transect measured above a fault (see the *black line*) within the suggested ancient port of Priniatikos Pyrgos at Istron, Crete (Sarris et al. 2014). (**d**) 2D representation of the 3D resistivity model shown as horizontal slices with increasing depth in the area of Mikrasiaton Square at the centre of the old town of Rethymno in Crete. The architectural relics of a Franciscan monastery become obvious towards the southern part of the surveyed area (Papadopoulos et al. 2008)

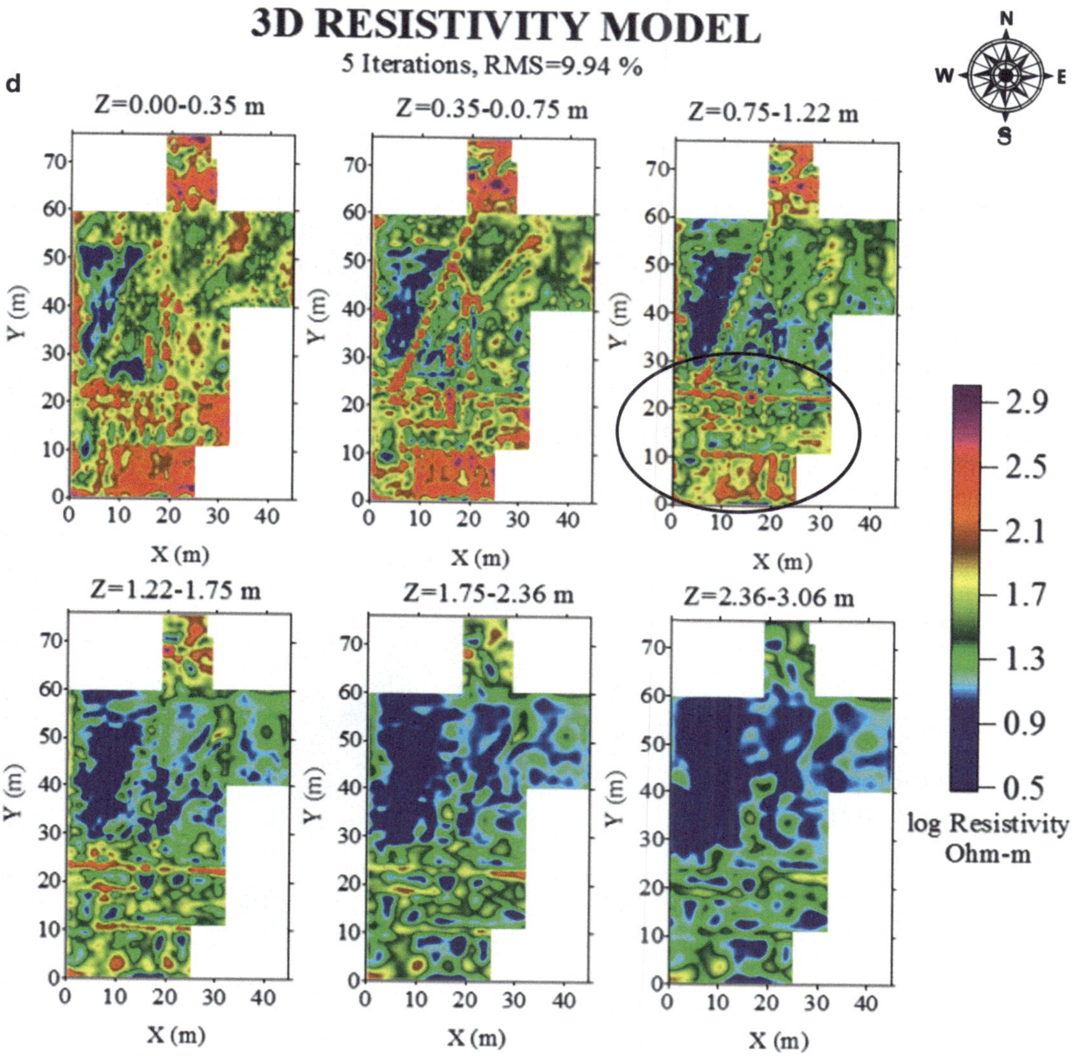

Fig. 14.9 (continued)

Gamma spectroscopy is another geophysical approach that measures the gamma rays emitted by radioactive materials. It has been widely used by soil scientists to map soil types through plotting ternary diagrams of K-Th-U concentration. Mapping soils with gamma spectroscopy is very effective, because the radiometric footprint of a soil reflects both the mineralogy of the parent rock and the effect of weathering. Even though this technique has not been widely used in archaeology, it qualifies well for validating soil mapping results obtained by complementary methods like electromagnetic induction.

Another approach is the mapping of novel and short-lived isotopes such as ^{137}Cs and ^{241}Am, which were deposited by the fallout from nuclear testing and accidents. The method provides an alternative for mapping disturbance on the basis that these isotopes are deposited on the surface and chemically immobile. Given that isotopes are distributed throughout the whole soil profile rather than just on the surface, a mixing of the soil layers can be presumed. Both Cs and Am approaches have the potential to assist geoarchaeological investigation by providing information about the sedimentary matrix of the archaeological site and determining if it has recently been disturbed by anthropogenic or natural processes.

Quite similar to isotope mapping is the chemical analysis of soils, which may indicate anthropogenic activity including agricultural and workshop activities, animal husbandry, construction of ditches, etc. Phosphate analysis, trace element analysis (Cu, Mn, Mo, Ca, Se, Sr, Zn, Cd, Cr, Cu, Ni, Pb) and assessment of chemical stability of organic chemical compounds are the most widely used approaches contributing to the recognition of former land-use practices and past human occupation in general. Calorimetric measurements, atomic absorption spectrometry, inductively coupled plasma mass spectrometry and gas chromatography-mass spectrometry are usually employed (Szostek et al. 2005; Price and Burton 2012; Manhita et al. 2014; Lauer et al. 2014)

14.3 Examples for Principal Applications

Geophysical methods have a wide range of application within archaeology in general, but their explicit use for solving geoarchaeological questions is principally related to three major aspects: the determination of sediment stratigraphy, the mapping of soil disturbance related to human occupation and the reconstruction of palaeolandscapes.

Sediment Stratigraphy Understanding the stratigraphy of an archaeological site is crucial for effective excavation planning, palaeoenvironmental reconstruction and putting excavation results into context, particularly with regard to geochronology. Yet geophysical methods are mostly currently used to simply locate and map archaeological sites and not to determine key stratigraphic features. The most important information that can be obtained using geophysics includes depth to bedrock, thickness and lateral extent of stratigraphic features and the location of archaeological material with respect to the stratigraphy.

Depth to bedrock and geomorphology of the surface provide important information about past depositional conditions and are key for correctly locating archaeological excavations so as to resolve the most complete record of occupation. Besides GPR and ERT, which are the most suitable techniques for investigating the vertical stratigraphy of archaeological sites, on-site or down-hole magnetic susceptibility measurements also provide promising results. When it comes to horizontal mapping, EMI, GPR, soil resistivity, ERT and gamma spectroscopy are particularly suitable.

a

Fig. 14.10 (**a**) A single fluxgate gradiometer sensor of Bartington G601. (**b**) An eight-sensor fluxgate gradiometer array of SENSYS surveying the Neolithic tell of Perdika 2 at Central Greece. (**c**) Indicative results from the single sensor magnetic survey of the Agora of Sikyon in Peloponnese (Sarris et al. 2008). (**d**) Multisensory survey of Almyriotiki Neolithic tell at Central Greece. Results have been superimposed on a WorldView-2-satellite image (2 June 2012) (Sarris et al. 2015a)

b

c

Fig. 14.10 (continued)

d

Fig. 14.10 (continued)

Stratigraphic Disturbance of Sediments and Soils Soil stratigraphy (i.e. the development of recognizable soil horizons) typically takes significant time to develop, and, as a result, its disturbance can provide crucial information about the historical record of an archaeological site. This perturbation can be mapped geophysically through a number of techniques. Among others, alteration of soil properties due to mixing of different components can be investigated by GPR, EMI, resistivity or gamma spectrometry, while changes of soil porosity are shown best by ERT and GPR. The latter approach is particularly appropriate for the detection of unmarked graves, as in the absence or following the decay of a coffin; soil perturbation caused by the dig is the most evident feature to be detected by geophysical methods (Moffat 2015; Conyers 2006).

Reconstruction of Palaeolandscapes Buried palaeolandscapes provide valuable information for the identification and the understanding of archaeological sites, particularly when surface features are absent. The reconstruction of palaeoenvironments allows an identification of areas with sediment accumulation (and, thus, prospective for preserving archaeological material) and areas characterized by landforms (such as rivers or shorelines) conducive to ancient land use. In addition, it can also provide information about past climatic conditions, which had an impact on human occupation history. Besides terrestrial palaeolandscape reconstructions, which are quite often based on multi-method approaches (e.g. by combining ERT, seismics, geochemical studies), aquatic palaeoenvironments that have been submerged due to

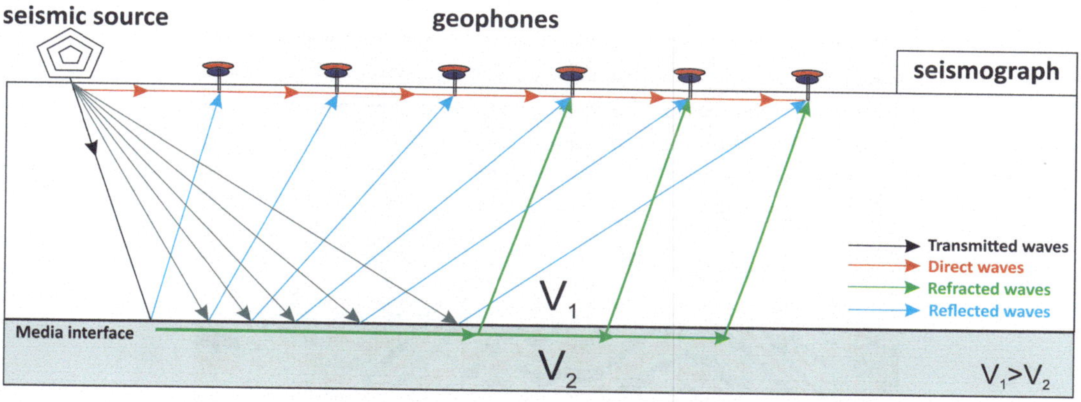

Fig. 14.11 Typical setup in a seismic survey. The seismic source is needed to generate controlled acoustical waves that are either reflected or refracted at the different media interfaces to be recorded by the geophones. Geophones convert the velocity of the registered waves into voltage

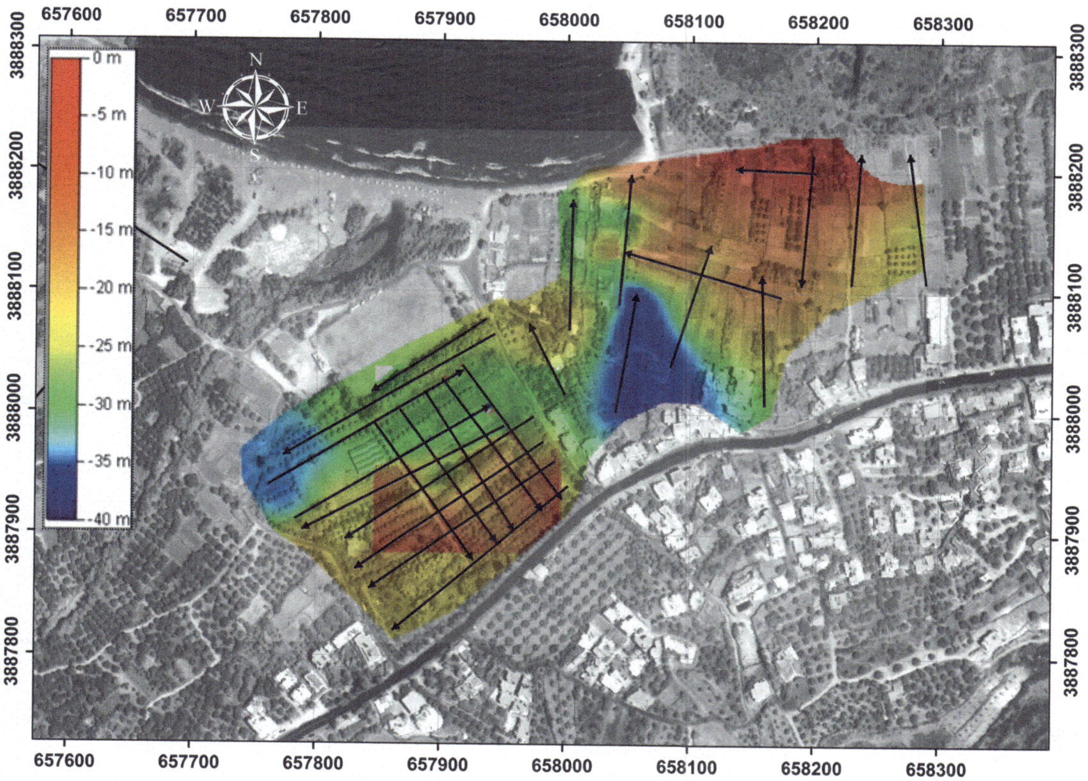

Fig. 14.12 Results of a seismic refraction survey at the area of Priniatikos Pyrgos (Istron, E. Crete, Greece), where an assumed ancient port has been hypothesized. The area has been completely covered by alluvial deposits and conglomerates as a result of past landslides and tectonic activity. The image represents the depth to the bedrock which reaches levels of about 40 m below surface, superimposed on a panchromatic Ikonos image (30 July 2001). The *arrows* represent the seismic transects along which the geophones were placed. The seismic results have also been verified through ERT measurements (Sarris et al. 2014)

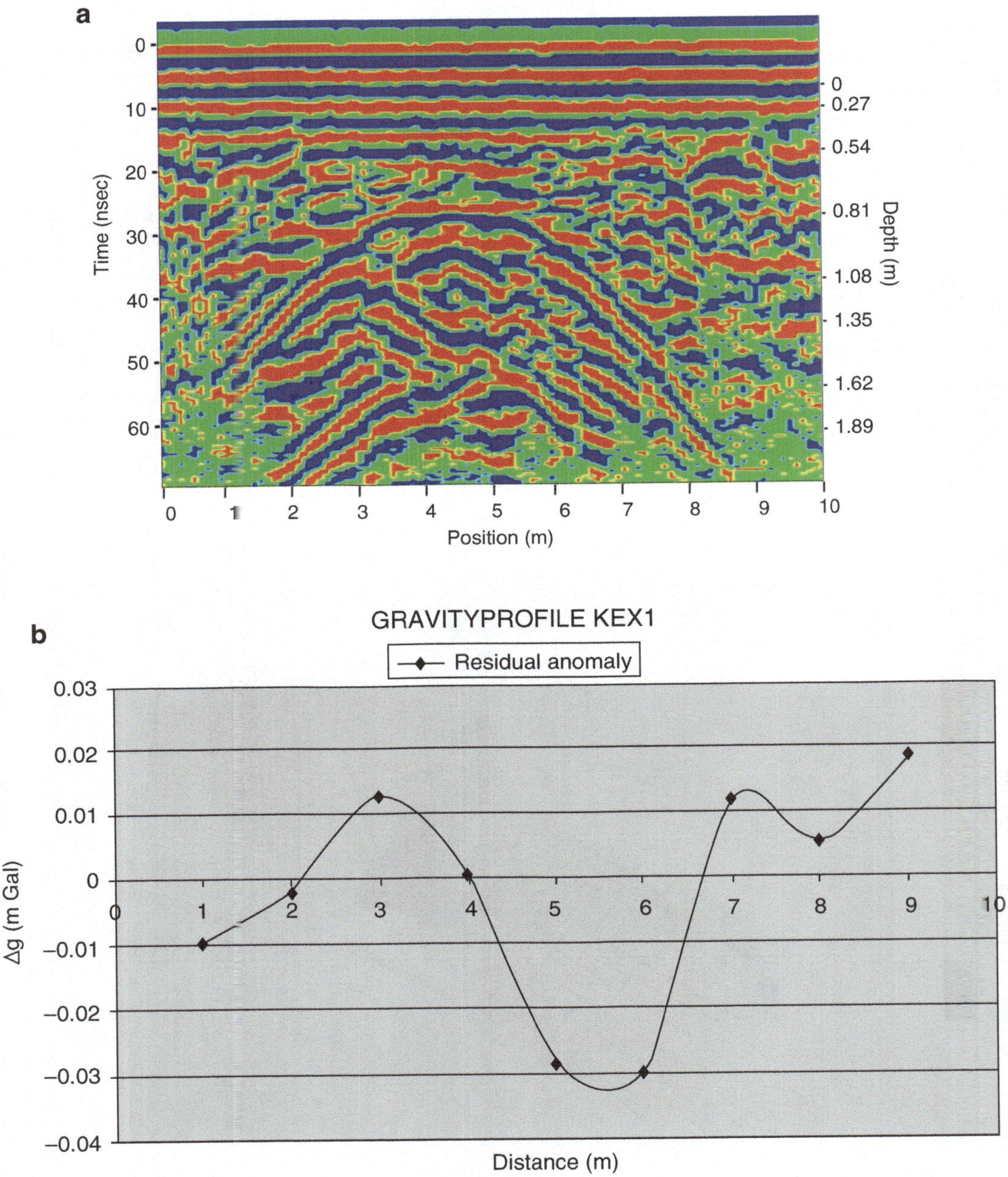

Fig. 14.13 (**a**) GPR radargram (taken with Noggin Plus GPR with 250 MHz antenna) above a Roman chamber tomb at the Kenchreai cemetery located at Koutsonglia Ridge, north of the harbour of Kenchreai in NW Peloponnese. The tomb has a diameter of 3.55 m and an interior height of about 2.5 m, located at a depth of 1.1 from the ground level. The amplitude of the GPR data is represented by *red* (positive reflection), *blue* (negative reflection) and *green* (minor dielectric constant differences between the media creating almost no reflective interfaces) colours. (**b**) Gravity residual anomalies recorded at the same transect using the Lacoste and Romberg model D land microgravitometer. The centre of the tomb chamber is located approximately at the middle of the transect (Sarris et al. 2007)

Table 14.1 Strengths and weaknesses of geophysical methods in geoarchaeological surveys

Method	Strengths	Weaknesses
Ground-penetrating radar	• Application in both shallow and medium depth investigations • Suitable for use in both rural and urban settings • Operation in limnic environments • Provision of high-resolution data • Fast data acquisition • Provision of stratigraphy and volumetric images of the subsurface • Ability to map geological and/or geomorphologic features • Determination of the depth to bedrock • Ability to detect voids, caves, tombs, wall structures	• Need of good contact with ground surface • Dependence on climatic conditions • Not possible to operate in marine investigations • Specialized way of data processing • Shielded antennas are required for surveying inside structures • Medium mobility and flexibility of instrumentation
Electromagnetic induction methods	• Application in both shallow and deep investigations • Suitable for use in rural settings • No need of contact with ground • Ideal for large-scale surveys • Large mobility of instrumentation • Provision of both soil conductivity and soil magnetic susceptibility • Provision of information about geology and lithology, stratigraphic changes and the depositional record of a landscape • Ability to detect ditches, pits, burnt architecture, palaeosoils, palaeochannels which may accumulate magnetic minerals through pedogenic or erosional processes (French 2003)	• Not suitable in urban environments • Easy to medium degree of data processing • Difficult to collect multi-sensor data as an array
Electrical resistance techniques	• Application in medium depth and deep investigations • Suitable for use in both rural and urban settings, limnic and sea investigations • Easy processing of the data • 2D and 3D modelling of the subsurface geological strata • Straightforward application in archaeological surveys • Ideal for the detection of deep depositional targets (e.g. ancient ports covered by alluvial deposits, ditches and palaeochannels)	• Need of contact with ground surface • Dependence on climatic conditions • ERT measurements need specialized data treatment • Expensive ERT instrumentation • ERT is relative bulky and slow for extensive mapping • Medium to hard mobility and flexibility of instrumentation
Magnetic methods	• Application in shallow depth investigations • Suitable for use in rural settings • No need of contact with ground • Ideal for extensive mapping surveys with high sampling resolution • Easy processing of the data • Large mobility of instrumentation • Detection of ferrous or burnt features, metal objects • Mapping the extent of pedogenic soil minerals	• Not suitable in urban environments • Dependence on terrain conditions • Very sensitive to metallic objects • Not appropriate for sediment stratigraphy studies
Magnetic susceptibility methods	• Application in shallow depth investigations • Suitable for use in rural settings • Ability to map the depositional conditions at archaeological sites and the occupation layers • Detection of extension and intensity of occupation • Provenance of archaeological materials • Proxy for the success of the magnetic survey	• Not suitable in urban environments • Need for Laboratory measurements • Not appropriate for high resolution or extensive surveys

(continued)

Table 14.1 (continued)

Method	Strengths	Weaknesses
Acoustic procedures	• Application in medium depth and deep investigations • Suitable for use in both rural and urban settings, limnic and sea investigations • Ability to map the sediment stratigraphy and reconstruct the palaeotopography and the depth to bedrock • Ability to map large monumental structures and mounds	• Need of good contact with ground surface • Specialized way of data processing • Slow method for mapping surveys • Medium to hard mobility and flexibility of instrumentation • Expensive instrumentation
Microgravity	• Application in shallow and medium depth investigations • Ideal for detection of voids, caves and monumental structures	• Very expensive instrumentation • Not suitable in urban environments • Tedious data processing • Not appropriate for extensive mapping
Chemical analysis	• Provision of information regarding agricultural and workshop activities, animal husbandry, construction of ditches • Reconstruction of past land-use practices	• Need for Laboratory measurements • Not appropriate for high resolution or extensive surveys
Gamma spectroscopy	• Fast survey • Ability to map soil types • Appropriate for sediment stratigraphy studies	• Sensitivity and reliability issues
Aerial and satellite remote sensing	• Provision of multispectral information • Provision of a digital terrain model (DEM) • Ideal for regional landscape studies	• Limitation to shallow depth investigations • Dependence of the time of imagery, crop growth, spatial resolution of the satellite platform • Specialized software for data processing

Fig. 14.14 (**a**) Differential GPS being used to survey an archaeological site in NE South Australia. (**b**) Robotic total station being used for mapping a cemetery in Queensland, Australia. (**c**) Camera attached to a kite for acquiring aerial photos from the Neolithic site of Perdika 2 at Thessaly, Central Greece. (**d**) Satellite image of the urban plan of the ancient city of Ferai (Central Greece); pansharpened IHS transformation of the 15 June 2009 Quickbird image showing linear features associated with roads to the northern fields. (**e**) Same image as in (**d**); modified simple ratio (MSR) transformation of the 4 May 2010 GeoEye-1 image indicating the projection of the above features to the southern field (Sarris et al. 2015b)

Fig. 14.14 (continued)

sea level rise are primarily investigated by acoustic methods (Tizzard et al. 2015).

14.4 Final Remarks

Geophysical techniques can make a significant contribution to geoarchaeological investigations. They allow non-invasive and rapid imaging of archaeological landscapes and help answer scientific questions by considering a site integrally along with its environmental surroundings. The methodological spectrum includes the investigation of the site stratigraphy, the mapping of remnants of past human occupation and the reconstruction of palaeolandscapes. Each method has its own merits and limitations, and it is through the combination of the various techniques that various aspects of the archaeolandscapes can be addressed (Sarris, 2013, 2015) (Table 14.1).

It has to be noted though that geophysical prospection techniques cannot and shall not independently address all questions arising from the archaeological research. Moreover, they are usually combined with other applications spanning from topographic mapping and geomorphometry of the landscape using LiDAR, DGPS units, robotic total stations and 3D laser scanners to aerial reconnaissance and satellite remote sensing (Fig. 14.14). Only the integration of the above-mentioned techniques with on-site excavations can provide a more holistic approach to the questions related to the interaction between human occupation and landscapes.

Acknowledgments This work was carried out under the project "ARCHERS: Advancing Young Researchers' Human Capital in Cutting Edge Technologies in the Preservation of Cultural Heritage and the Tackling of Societal Challenges" which is funded by an exclusive donation of the Stavros Niarchos Foundation ("Stavros Niarchos Foundation—FORTH Fellowships").

References

Annan AP (2009) Chapter - Electromagnetic principles of ground penetrating radar. In: Jol HM (ed) Ground penetrating radar theory and applications. Elsevier, Amsterdam, pp 1–40

Armstrong KL, Kalayci T (2015) Images of the past: magnetic prospection in archaeology. In: Sarris A (ed) Best practices of geoinformatic technologies for the mapping of archaeolandscapes. Archaeopress Publishing, Oxford, pp 1–12

Aspinall A, Gaffney C, Schmidt A (2008) Magnetometry for archaeologists. AltaMira Press, Lanham

Conyers LB (2006) Ground-penetrating radar techniques to discover and map historic graves. Hist Archaeol 40 (3):64–73

Conyers LB (2013) Ground-penetrating radar for archaeology. AltaMira Press, Lanham

Conyers LB (2015a) Analysis and interpretation of GPR datasets for integrated archaeological mapping. Near Surf Geophy 23(6):645–651

Conyers LB (2015b) Ground-penetrating radar for geoarchaeology. Wiley, London

Dalan RA (2007) A review of the role of magnetic susceptibility in archaeogeophysical studies in the USA: recent developments and prospects. Archaeol Prospect 15(1):1–31

Dalan RA, Banerjee SK (1998) Solving archaeological problems using techniques of soil magnetism. Geoarchaeology 13(1):3–36

Dearing J (1994) Environmental magnetic susceptibility. Using the Bartington MS2 system. Chi Publications, Kenilworth

French C (2003) Geoarchaeology in action. Studies in soil micromorphology and landscape evolution. Routledge, London

Jol H (1995) Ground penetrating radar antennae frequencies and transmitter powers compared for penetration depth, resolution and reflection continuity. Geophys Prospect 43(5):693–709

Hecht S, Fassbinder J (2006) Der Blick in den Untergrund: Magnetometrie und Geoelektrische Tomographie. Geoarchäologie. Geogr Rundsch 58(4): 38–45

Laigre L, Reynards E, Arnaud-Fassetta G, Baron L, Glenz D (2012) Characterisation of the Rhône river palaeodynamics in Central Valais (Switzerland) with the electrical resistivity tomography method. Géomorphol Relief Processus Environ 4:405–426

Lauer F, Prost K, Gerlach R, Pätzold S, Wolf M, Urmersbach S, Lehndorff E, Eckmeier E, Amelung W (2014) Organic fertilization and sufficient nutrient status in prehistoric agriculture? – Indications from multi-proxy analyses of archaeological topsoil relicts. Plos One. open access

Linford N, Linford P, Martin L, Payne A (2010) Stepped frequency ground-penetrating radar survey with a multi-element array antenna: results from field application on archaeological sites. Archaeol Prospect 17:187–198

Loke MH (2000) Electrical imaging surveys for environmental and engineering studies. A practical guide to 2-D and 3-D surveys. http://www.geo.mtu.edu/~ctyoung/LOKENOTE.PDF

Manataki M, Sarris A, Donati JM, Garcia CC, Kalayci T (2015) GPR: theory and practice in archaeological prospection. In: Sarris A (ed) Best practices of geoinformatic technologies for the mapping of archaeolandscapes. Archaeopress Publishing, Oxford, pp 13–24

Manhita A, Martins S, Costa J, Prazeres C, Rocha L, Dias C, Mirão J, Teixeira D (2014) A multi-analytical approach for the study of Neolithic pottery from the Great Dolmen of Zambujeiro (Évora, Portugal) – a preliminary study. e-Conserv J 2:66–78

Merz K, Green AG, Buchli T, Springman SM, Maurer H (2015) A new 3-D thin-skinned rock glacier model based on helicopter GPR results from the Swiss Alps. Geophys Res Lett 42:4464–4472

Moffat I (2015) Locating graves with geophysics. In: Sarris A (ed) Best practices of geoinformatic technologies for the mapping of archaeolandscapes. Archaeopress Publishing, Oxford, pp 45–53

Papadopoulos NG, Sarris A, Giapitsoglou C (2008) Mapping the buried archaeological remains in the area of the old Turkish school of Rethymno (Crete, Greece) through the application of surface ERT & GPR techniques. In: Zacharias N, Georgakopoulou M, Polykreti K, Fakorellis G, Vakouli T (eds) Proceedings of the 5th symposium of the Greek archaeometry society. Papazisis, Athens, pp 101–116

Papadopoulos N, Sarris A, Yi MJ, Kim JH (2009) Urban archaeological investigations using surface 3D ground penetrating and electrical resistivity tomography methods, joint issue of the Australian Society of Exploration Geophysicists, the Korean Society of Exploration Geophysicists and the Society of Exploration Geophysicists of Japan. Explor Geophys 40:56–68

Price D, Burton J (2012) An introduction to archaeological chemistry. Springer, New York

Sarris A (ed) (2015) Best practices of geoinformatic technologies for the mapping of archaeolandscapes. Archaeopress Archaeology: Archaeopress Publishing, Oxford

Sarris A (2013) Multi+ or manifold geophysical prospection?. In: Earl G, Sly T, Chrysanthi A, Murrieta-Flores P, Papadopoulos C, Romanowska I, Wheatley D (eds) Archaeology in the Digital Era – Volume II, e-Papers from the 40th Conference on computer applications and quantitative methods in archaeology (CAA2012), Amsterdam University Press, Amsterdam, p 761–770

Sarris A, Papadopoulos N (2012) Geophysical Surveying in Urban Centers of Greece. In: Boner W, Uhlirz S, Dollhofer L (eds) Proceedings of the 16th international congress on cultural heritage and new technologies, Museen der Stadt Wien - Stadtarchaologie, pp 96–114

Sarris A, Kalayci T, Simon FX, Donati J, Garcia CC, Manataki M, Cantoro G, Kalogiropoulou E, Karampatsou G, Argyriou N, Dederix S, Manzetti C, Nikas N, Vouzaxakis K, Rondiri V, Arachoviti P, Almatzi K, Efstathiou D, Stamelou E (2015b) Opening a new frontier in the Neolithic settlement patterns of Eastern Thessaly, Greece. International conference on communities, landscapes, and interaction in Neolithic Greece, Institute for Mediterranean Studies—FORTH, Rethymno. http://igean.ims.forth.gr/sites/default/files/publications/wSarris_et%20al_Neolithic_Greece_2015.pdf

Sarris A, Donati J, Kalayci T, Simon FX, Manataki M, Garcia C, Cantoro G (2015a) Amalgamation of satellite remote sensing and geophysical prospection for the investigation of ancient cities: two case studies from Demetrias and Pherai at the region of Magnesia, Thessaly, Central Greece, 8th Congress of Balkan Geophysical Society in Chania, Crete, Greece. http://politeia.ims.forth.gr/files/wmSarris_et_al_Balkan_2015.pdf

Sarris A, Papadopoulos N, Soupios S (2014) Contribution of geophysical approaches to the study of Priniatikos Pyrgos. In: Molloy BPC, Duckworth CN (eds) A cretan landscape through time: Priniatikos Pyrgos and environs, BAR International Series, vol 2634. Archaeopress, Oxford, pp 61–69

Sarris A, Papadopoulos N, Trigkas V, Kokkinou E, Alexakis D, Kakoulaki G, De Marco E, Seferou E, Shen G, Karaoulis M, Simirdanis K, Kondili F, Katifori M, Dogan M, Lolos Y, Kalpaxis T (2008) Recovering the urban network of ancient Sikyon through multi-component geophysical approaches. In: Posluschny A, Lambers K, Herzog I (eds) Layers of perception. Proceedings of the 35th international conference on computer applications and quantitative methods in archaeology (CAA), Berlin, Germany, 2–6 April, 2007. Kolloquien zur Vor- und Frühgeschichte 10:11–16

Sarris A, Dunn RK, Rife JL, Papadopoulos N, Kokkinou E, Mundigler C (2007) Geological and Geophysical Investigations in the Roman Cemetery at Kenchreai (Korinthia), Greece. J Archaeol Prospect 14:1–23

Scapozza C, Laigre L (2014) The contribution of electrical resistivity tomography (ERT) in Alpine dynamics geomorphology: case studies from the Swiss Alps. Géomorphol Relief Processus Environ 20(1):27–42

Schmidt A (2013) Earth resistance for archaeologists. AltaMira Press, Lanham

Simon FX, Moffat I (2015) Identification of shapes and uses of past landscapes through EMI survey. In: Sarris A (ed) Best practices of geoinformatic technologies for the mapping of archaeolandscapes. Archaeopress Publishing, Oxford, pp 25–34

Simon FX, Sarris A, Thiesson J, Tabbagh A (2015a) Mapping of quadrature magnetic susceptibility/magnetic viscosity of soils by using multi-frequency EMI. J Appl Geophys 120:36–47

Simon FX, Tabbach A, Sarris A (2015b) Practical assessment of a multi-frequency slingram EMI for archaeological prospection. In: Giligny F, Djindjian F, Costa L, Moscati P, Robert S (eds) Proceedings of the 42nd annual conference on CAA, CAA2014: 21st century archaeology- concepts, methods and tools, Archaeopress Archaeology, Oxford, p 43–49

Simyrdanis K, Papadopoulos N, Kim JH, Tsourlos P, Moffat I (2015) Archaeological investigations in the shallow seawater environment with electrical resistivity tomography. Near Surf Geophys 13:601–611

Szostek K, Głąb H, Lorkiewicz W, Grygiel R, Bogucki P (2005) The diet and social paleostratigraphy of Neolithic agricultural population of the Lengyel culture from Osłonki (Poland). Przegląd Antropologiczny – Anthropol Rev 68:29–41

Thompson R, Oldfield F (1986) Environmental magnetism. Springer, Netherlands

Tizzard L, Bicket A, De Loecker D (eds) (2015) Seabed prehistory: investigating the palaeogeography and early middle Palaeolithic archaeology in the Southern North Sea. Oxbow Books, Oxford

Tonkov N (2014) Geophysical survey at the early Neolithic site of Yabalkovo. In: Roodenberg J, Leshtakov K, Petrova V (eds) Yabalkovo vol 1, ATE- Ars et Technica Expicatus, Sofia University "St. Kliment Ohridski", Sofia, pp 73–78

Trinks I, Johansson B, Gustafsson J, Emilsson J, Friborg J, Gustafsson C, Nissen J, Hinterleitner A (2010) Efficient, large-scale archaeological prospection using a true three-dimensional ground-penetrating radar array system. Archaeol Prospect 17:175–186

Open Access This chapter is licensed under the terms of the Creative Commons Attribution 4.0 International License (http://creativecommons.org/licenses/by/4.0/), which permits use, sharing, adaptation, distribution and reproduction in any medium or format, as long as you give appropriate credit to the original author(s) and the source, provide a link to the Creative Commons license and indicate if changes were made.

The images or other third party material in this chapter are included in the chapter's Creative Commons license, unless indicated otherwise in a credit line to the material. If material is not included in the chapter's Creative Commons license and your intended use is not permitted by statutory regulation or exceeds the permitted use, you will need to obtain permission directly from the copyright holder.

A Geoarchaeological Approach for the Localization of the Prehistoric Harbor of Akrotiri, Thera

15

Katerina Theodorakopoulou, Yannis Bassiakos, Constantin Athanassas, Gerd Schukraft, Ingmar Holzhauer, Stefan Hecht, Bertil Mächtle, and Günther A. Wagner

Abstract

The disastrous volcanic eruption of Thera in the Aegean that happened in late seventeenth century BC (Late Bronce Age, LBA) buried under a thick mantle of volcanic ash the thriving city harbor of Akrotiri, situated at the southern edge of the island. The archaeological excavations at the site have witnessed the city's wealth and flourish; numerous, luxurious, and outlandish finds clearly indicate well-established maritime contacts between Akrotiri and eastern Mediterranean lands. Such maritime operations and overseas trade required, apparently, adequate harbor facilities. This paper deals with geoarchaeological and geophysical studies aiming at the localization of the buried harbor, a so far unrealized ambitious aim, in spite of the repeated intense attempts undertaken in the last decades. Preference for the relevant investigation was given to three small littoral valleys situated in small distances at both sides of the settlement, suggesting shallow bays before the Minoan eruption, hence probably having hosted the searched harbor(s). Prior to the fieldwork undertaken, all available geological and other related data were cartographically outlaid by means of GIS. Afterwards, in situ geomorphologic studies were conducted, followed by one littoral drilling and geophysical investigations including seismic refraction and electrical resistivity tomography (ERT). Both geophysics and drilling have shown that the hard pre-Minoan basement (consisting of dense andesitic lavas) is situated at depths of ca. 25 m in the Mavrorachidi valley.

K. Theodorakopoulou (✉)
Department of History, Archaeology and Social Anthropology, University of Thessaly, Volos, Greece
e-mail: ktheodorakopoulou@uth.gr

Y. Bassiakos
Laboratory of Archaeometry, N.C.S.R. "Demokritos", Athens, Greece

C. Athanassas
Centre de Recherche et d'Enseignement de Géosciences de l'Environnement (C.E.R.E.G.E.), Aix-en-Provence, France

G. Schukraft • I. Holzhauer • S. Hecht • B. Mächtle • G.A. Wagner
Institute of Geography, Heidelberg University, 69120 Heidelberg, Germany

© Springer International Publishing AG 2018
C. Siart et al. (eds.), *Digital Geoarchaeology*, Natural Science in Archaeology,
DOI 10.1007/978-3-319-25316-9_15

According to the seismic results, the recent shape of Potamos valley today seems to be similar to as during the Minoan times. Due to the relative minor infillings of the valley and the position of the valley floor above the recent sea level, there's no evidence of a former bay or harbor in that area. Moreover, the seismic data revealed that the Minoan surface (Cape Riva ignimbrite) seems to rise from the archaeological site of Akrotiri to the coastline in the southeast of the excavation up to 30 m a. s.l. Maybe there existed a former cliff that was covered by Minoan tephra but that also provided shelter for the village of Akrotiri. The investigated area south of the archaeological site offers a more appropriate site for the Minoan harbor. Future geoarchaeological investigations (drillings, geophysics) should concentrate both on the latter area and on Mavrorachidi valley (one of the three valleys mentioned before).

This study presents preliminary results from our ongoing research. It is anticipated that the synthesis of the available data along with the expected results from the remainder geophysical investigation, subsurface coring, chronometric dating, and submarine beach-rock studies will allow to determine the most probable harbor position(s) of Akrotiri.

Keywords
Akrotiri • Geophysical prospection • Harbor • Paleogeography • Thera

15.1 Introduction

Large-scale excavations conducted over the last 45—and more—years at Akrotiri on Thera, the southernmost island of the Cycladic archipelago in the Aegean, have brought to light one of the most important Bronze Age settlements in the region, buried under a thick mantle of volcanic ash. The settlement (over 10,000 m^2) was at the zenith of its development when it was completely buried by ash and pumice lapilli ejected from the volcano's crater. This eruption (ca. 1620 BC) has been estimated as four times as powerful as the Krakatoa eruption of 1883 AD (Friedrich et al. 2006).

Akrotiri was a well-organized city with a developed urban plan and an efficient drainage system under the paved streets. Connected with the sewage system were the sanitary facilities of the imposing two- and three-story houses, which were furnished with exquisite pieces of furniture and decorated with unique wall paintings. This high standard of living as revealed by both the architecture and the movable finds, indicates that the wealth of the city's inhabitants was largely the result of maritime trade with eastern Mediterranean lands, as it is reflected in the wall painting of the "Flotilla" from the accentuated building known as "West House" (Doumas 1983, 1991; Fig. 15.1).

The wealth and size of the city certainly equalled or surpassed the level of culture and wealth on Minoan Crete contemporaneous settlements. Akrotiri probably gained this high degree of wealth and culture through its foreign trade contacts and its position in the center of the Aegean waterways (Doumas 1991). Obviously, such maritime and overseas operations required adequate harbor facilities, which are difficult to locate because of the thick deposits of overlying volcanic material.

By using geoarchaeological techniques, we aim mainly at locating the city's harbor (s) and determine the LBA paleotopography. The geoarchaeological project started in 2011 by the Laboratory of Archaeometry, N.C.S.R "Demokritos," Athens, is strongly supported by the archaeologists of Akrotiri excavations.

Fig. 15.1 The "Flotilla" Fresco

In 2012 and 2014, the collaborating team from the Department of Geography of the University of Heidelberg conducted the geophysical prospection of the area. The first results of our studies, shortly commented, are presented below.

15.2 Pre-Minoan Paleotopography and Archaeological Significance

Early theories on the paleogeography of Thera prior to the Minoan eruption stressed that the small archipelago of Santorini enclosed a massif, known as the mythological island of "Stronghyle," which in Greek means *round* (e.g., Marinatos 1939; Pichler and Friedrich 1980; Aston and Hardy 1990), but were later disfavored by more sophisticated studies. Yet, the actual paleophysiography of pre-catastrophe Thera is still under debate; there are scholars advocating the existence of a volcanic massif to the north occupying most of the modern-day caldera limiting the extent of the pre-Minoan caldera to the south (e.g., Heiken and McCoy 1984, Heiken et al. 1990), while others delimited the boundaries of that depression further north (e.g., Druitt and Francaviglia 1992). Irrespective of the conclusions reached by those claims, they all stress the existence of a flooded pre-Minoan caldera, as suggested by the presence of erupted stromatolites (Friedrich et al. 1988; Eriksen et al. 1990), before the penultimate eruption of Santorini (~20 ka) but unspecified as to its volume, morphology, and position.

It is generally accepted that in prehistoric times, some small, sandy, and wind-protected bays could be the ideal places for harbor facilities, as there is rather no evidence (or very scant) of any relevant architectural remains. Anyhow, topography and geomorphology played an important role for the selection of the ideal place of prehistoric harbors. Additionally, the vicinity of a beach to the archaeological site, a wind-protected bay, the accessibility to roads for the easy transfer of products to the hinterland, etc. were essential factors for the prehistoric people to the location of a harbor. It is also noted that in prehistory, they usually used to have more than one harbor in a site, considering the weather conditions, seasons, magnitude/type of activities (harbors for small fishing boats or for trading activities), etc.

Akrotiri is considered as the "Pompeii" of the prehistory. Given the significance of the Minoan civilization, the searched localization of the prehistoric harbor will bear witness to the importance of hub function of prehistoric Thera in the trade routes of the eastern Mediterranean. As the presence of a well-defined harbor is missing from the up-to-date archaeological record, scholars have only hypothesized the influence of Thera on the cultures of the eastern Mediterranean. As a result, the discovery of the lost harbor of Thera would be greatly valued in the archaeology of the twenty-first century. The confirmation of a naval base on Akrotiri with the magnificence hinted in the fresco will allow the archaeologists to redraw the trading routes of the eastern Mediterranean during the Minoan times as well as to raise the importance of Thera in prehistoric seafaring (Doumas 1991, 1992a, b Friedrich and Sørensen 2010).

15.3 Methodology

A variety of approaches have been pursued for the indication of the likely position of the prehistoric harbor. Among them, the search for the LBA paleotopography at Akrotiri involved the "stripping off" the post-eruption deposits and considered the influence of the local submerging regime (Papageorgiou et al. 2010) in conjunction with the magnitude of the eustatic sea-level rise over the last four millennia. All available information about the geomorphology, archaeology, and geology of Thera, such as topographic and geological maps, papers, bathymetric data, and even unpublished reports, generated by former geological and archaeological surveys were collected and cartographically outlaid by means of GIS.

In the field, after the necessary geological and geomorphologic reconnaissance, we conducted an auger-type drilling at depths corresponding to the seventeenth-century BC paleo-strands, to identify the Minoan paleorelief. The drilling was performed in a valley known as Mavrorachidi, indicated as Site I in Fig. 15.2, approximately 650 m north of the modern strand. Loose materials as emerged from the drilling were collected for laboratory studies (sedimentological determination, physicochemical/mineralogical analysis, etc.); they also were examined by sieving for possible archaeological artifacts. Moreover, the littoral zone was geophysically investigated by employing the methods of seismic refraction and electrical resistivity tomography (ERT), aiming to determine the Minoan surface and the pre-Minoan bedrock.

The results from all the above individual approaches (sedimentology, archaeological identification, numeric dating, field data) have been merged to reconstruct the environmental features of the prehistoric landscape.

15.4 Results

15.4.1 Drilling

As mentioned, in Site I at Mavrorachidi valley, a drilling was performed ca. 700 m west of the excavated part of the settlement, for identifying the Minoan (pre-volcanic) paleorelief (D1, Fig. 15.2). The boring reached at a depth of ~25 m. As the drilling was of auger type (Fig. 15.3), it was not possible to obtain a continuous sediment core; thus, the complete and accurate tracking of the whole stratigraphic sequence was unattainable. The pre-Minoan ground surface, consisting of dense upper-Pleistocene andesitic lavas, was met at about 24.5–25.0 m depth. It is worthy to note that this hard rock constituting the basement was detected at the same depth also by the geophysical prospection (see below).

The sediments at the surface and up to 4 m below were mainly of fluvial origin, consisting of torrent-reworked pumice and gravel, hosting also voluminous hard andesitic boulders. Below them we met loose coarse volcanic sediments and lapilli tuffs until 14 m. Deeper, until 23 m, we had finer pumice (or partly—in the deeper levels—ignimbrite mechanically pulverized from the drilling head?), and until 23.5 m, there was a horizon of fine-grained sandy/gravelly sediment, of littoral or fluvial origin. Below this layer, there was a 0.5 m bed of andesitic coarse pebbles, and at approximately 24.0–24.5 m, we found the hard and compact andesite bodies, which also build up the nearby hill, known as Mesovouna. It is worth mentioning that one of the lower horizons, probably between the 14 and 23 m, hosted some archaeological findings like obsidian blades, shells, etc. that have been discovered. The material was given to the archaeologists for further study and keeping.

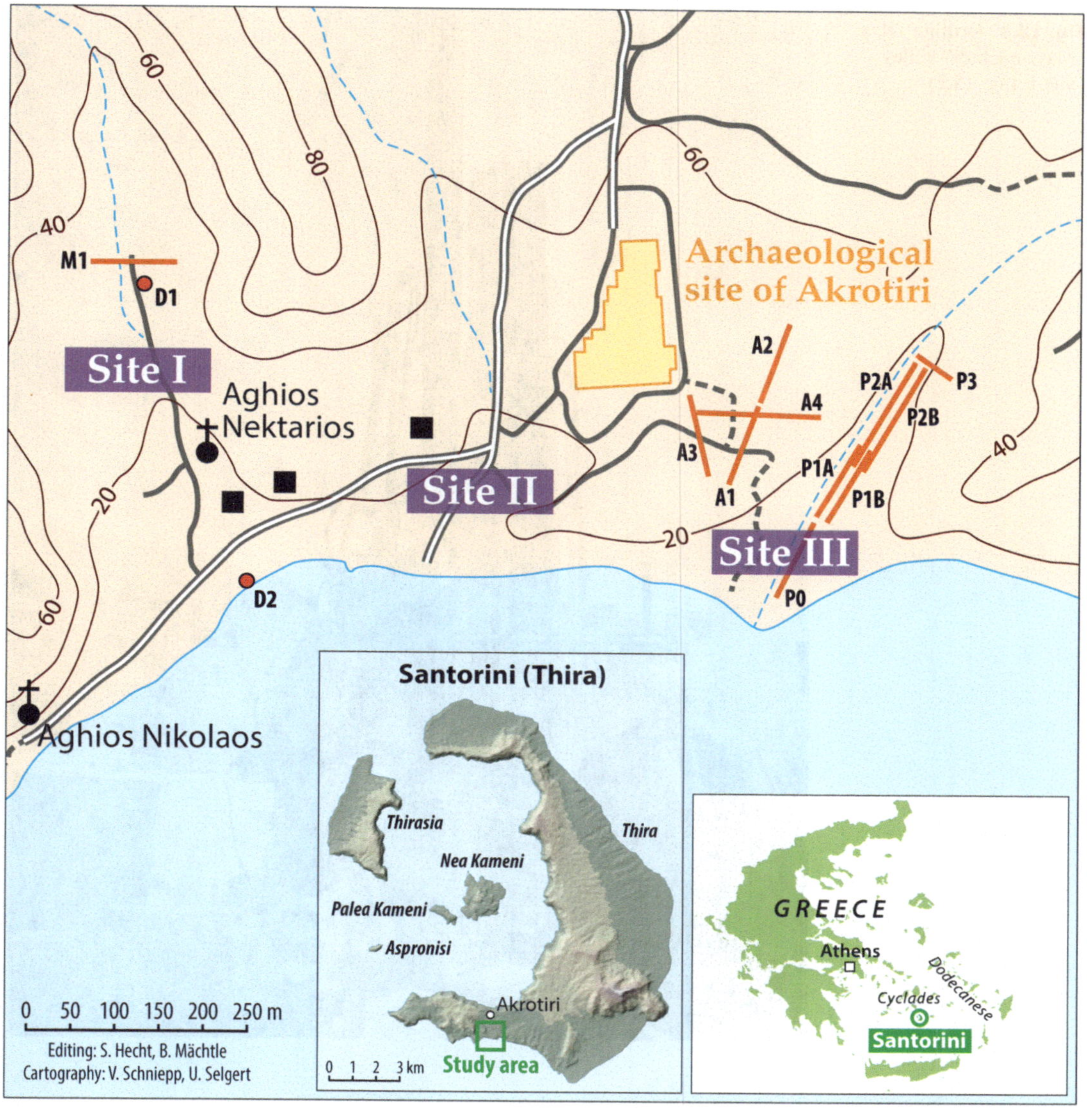

Fig. 15.2 The Akrotiri area showing the Sites I, II, and III of our research activities. D1, D2: drilling sites. A1–A4, P0–P3, M1: geophysical profiles

Additionally, drilling with a percussive "cobra-type" drill device was performed in front of the archaeological site close to the beach, but with no results because andesite boulders were too hard to go through. One other drilling performed at the beach, some meters SW of the previous drilling attempt, reached 7 m depth (D2, Fig. 15.2). It met sediments probably deposited in a former marine environment, consisting mainly of volcanic ash (Fig. 15.4).

In the near future, it is necessary to conduct drillings along the coastline by taking cores, in order to have a clearer picture of the stratigraphy of the area.

15.4.2 Geological Interpretations

One of the basic topics of our research had been also the distinction of pre-Minoan and post-

Fig. 15.3 Drilling D1 at "Mavrorachidi" valley (Site I, Fig. 15.2)

eruption deposits. We tried to determine the stratigraphy in the west and east of the archaeological site, the latter along with the "Potamos" or Marinatos ravine/valley. The sites are indicated as II and III in Figs 15.5 and 15.6, respectively.

Figures 15.5 and 15.6 demonstrate the volcanic stratigraphy near the archaeological site (Fig. 15.5) and in the surrounding area of Potamos little valley (Fig. 15.6). Volcanoclastic layers of the Minoan eruption dominate in the local stratigraphy, whereas limited outcrops of the pre-Minoan basement are exposed along the streambed of the nearby ravines. The upper parts of the stratigraphy indicate reworking of the volcanic material by syn-volcanic surface runoff. Local volcanic stratigraphy has maintained all four phases of the Minoan eruption, starting with a Plinian pumice phase that drapes the archaeological site itself, followed by thick phreatomagmatic deposits (~20 m) of the second and third phase, where the uppermost parts of the sequence indicate partial redeposition of the ejecta of the final phase. Interestingly, outcrops of the pre-Minoan paleosol and paleorelief are

Fig. 15.4 The stratigraphy of sediments by cobra-type drilling on the beach (D2, Fig. 15.2)

exposed occasionally along the streambed of Potamos valley, indicating that post-Minoan erosion has incised local streams down to more or less the original pre-Minoan surface. Unquestionable witness of the post-eruption age of the thick coarse sediments that overlay the Minoan tuffs is the pottery fragments discovered by some of us (K.T., K.A., Y.B., G.A.W.) during our surface surveys of 2012. Similar pottery fragments were also discovered in the overlying colluvia as indicated in Fig. 15.5, south and southwest of the excavated site. They all were given to the archaeologists for further study and keeping.

The shoreline evolution on the southern coast of Akrotiri peninsula is controlled by tectonism, by land-isostatic vertical movements, by sea eustatism, and by wave erosion. Recent kinematic studies on Santorini (Papageorgiou et al. 2007, 2010) demonstrate that southern Santorini is subsiding tectonically and isostatically at a rate of 2.8 ± 0.8 mm/a. This rate might indicate that the coastline of the 1620 BC lied at least 8–10 m below the modern sea surface taking also into the calculations an approximate related sea level of -4 m, owed to eustatism as deriving from estimations performed by Poulos et al. (2009).

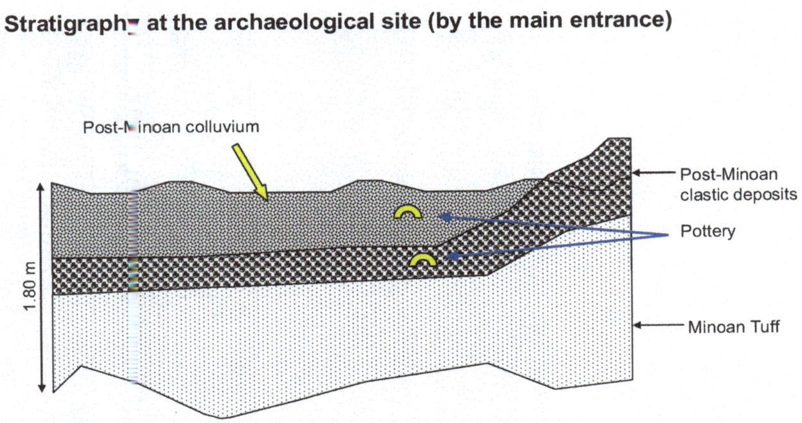

Fig. 15.5 The stratigraphy near the entrance of the archaeological site (Site II)

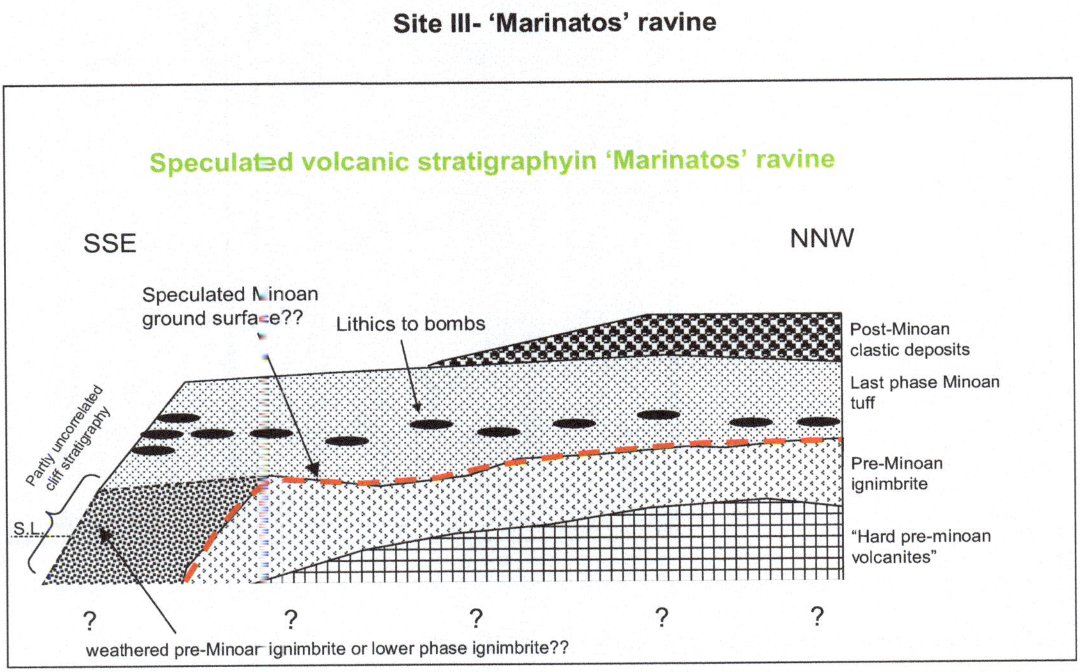

Fig. 15.6 The stratigraphy in "Potamos" valley, at Site III ("Marinatos" ravine)

15.4.3 Geochemical Analyses

Ten samples were sent for geochemical analyses by means of ICP-MS to Acme Analytical Laboratories (Vancouver) Ltd. The samples were taken from the Akrotiri strand corresponding to two horizons (pre-Minoan/Minoan) in order to distinguish the horizons from each other by any existing geochemical differentiation. Unfortunately, no important variation was observed on analyzed samples from the two horizons, making thus the differentiation very difficult. This might be owed to the loose and porous nature of the overlying horizons (mainly volcanic ash), in the sense that elements from the upper horizons have migrated to the lower horizons, ruling out the geochemical discrimination between the two stratigraphically different zones. However, advanced statistical analysis on the pertinent data has been planned, for investigating whether any "hidden" grouping of elements does still exist that might make possible any geochemical distinction between the two stratigraphically differing formations.

15.4.4 Geophysical Prospection

Geoarchaeological-geophysical surveys in 2012 and 2014 had been carried out in order to distinguish the Minoan surface from the overlying tephra and to reconstruct the land surface of the Late Bronze Age (LBA). For this purpose, seismic refraction measurements had been performed in the Mavrorachidi valley (Site I), in the Potamos valley (Site III), and in the southeastern part of the Akrotiri excavation site (Fig. 15.2). At Site II, it was not possible to conduct refraction seismics so far due to the ongoing traffic at the road nearby. Furthermore, geophysical investigations at the entrance of the museum are only feasible during the off-season.

For the determination of the seismic velocities, we used two geometric GEODE seismographs with 24 channels each, which were connected together. As a result, seismogram sections of 48 channels with geophone intervals of 2 m and 3 m could be recorded, respectively. The data interpretation was operated with two different software packages. REFLEX (Sandmeier 2007) was used for the application of wavefront inversion and subsequent network raytracing. The advantage of this method is the possibility to reconstruct sharp layer boundaries, especially between loose sediments and bedrock. Additionally, we used the software RAYFRACT (Rohdewald 2006) in order to obtain the spatial distribution of seismic velocities along the sections by means of seismic refraction tomography (SRT). The additional application of electrical resistivity tomography (ERT) was used to validate the seismic results and for the purpose of more detailed archaeological prospection. This contribution deals chiefly with the first results of the seismic measurements. A comprehensive introduction in both seismic and electrical investigations is given, e.g., by Reynolds (2011).

Figure 15.7 shows the seismic cross section in the upper part of Mavrorachidi valley in the close vicinity of the drilling site (Fig. 15.3). The transition from the volcanic tephra to the underlying andesitic basement is clearly visible due to a high contrast of seismic p-wave velocities between the tephra at the top (approx. 600 m/s) and the bedrock at the basis (2300 m/s). As already mentioned above, the reconstructed depth of the bedrock corresponds very well to the results of the auger-type drilling. Furthermore, the reconstruction of the buried Minoan surface allows an impression of the former valley shape. As the bedrock appears above the recent sea level, the existence of a bay in that area seems to be rather unlikely. Further investigations by means of drillings and geophysics more close to the seaside are necessary to examine the possibility of a Minoan harbor site in Mavrorachidi valley.

In the Potamos valley, some outcrops of the Cape Riva ignimbrite appear, which built the land surface of the Late Bronze Age in the surroundings of Akrotiri (cp. Russel and Stasiuk 2000). Starting at one of these outcrops, it was possible to trace the Cape Riva ignimbrite in the shallow subsurface. Since the p-wave velocities in the ignimbrite of about 500–600 m/s could be clearly measured directly at the outcrop, this seismic cross section at the eastern flank of the valley was also used to calibrate and to validate the acquired seismic data of all other profiles. This calibration was cross-checked using the seismic data at Akrotiri site (see below). The tomogram in Fig. 15.8 shows the spatial distribution of p-wave velocities for this section plus the reconstructed Minoan surface with the outcrop of the ignimbrite in the middle of the seismic line. Obviously, the cover of post-Minoan sediments in this area reaches not more than 4 m. A relatively minor post-Minoan infilling could also be recorded by means of several longitudinal sections in the valley floor (P0, P1, P2, Fig. 15.9). Hence, the shape of the Potamos valley today seems to be very similar to the valley shape in Minoan times. Furthermore, the existence of a large bay or a harbor site should be excluded for the Potamos valley, respectively.

The seismic sections A1–A4 were measured in the close vicinity of the archaeological site of Akrotiri (Fig. 15.2). First, the recorded data was processed by use of wavefront inversion and

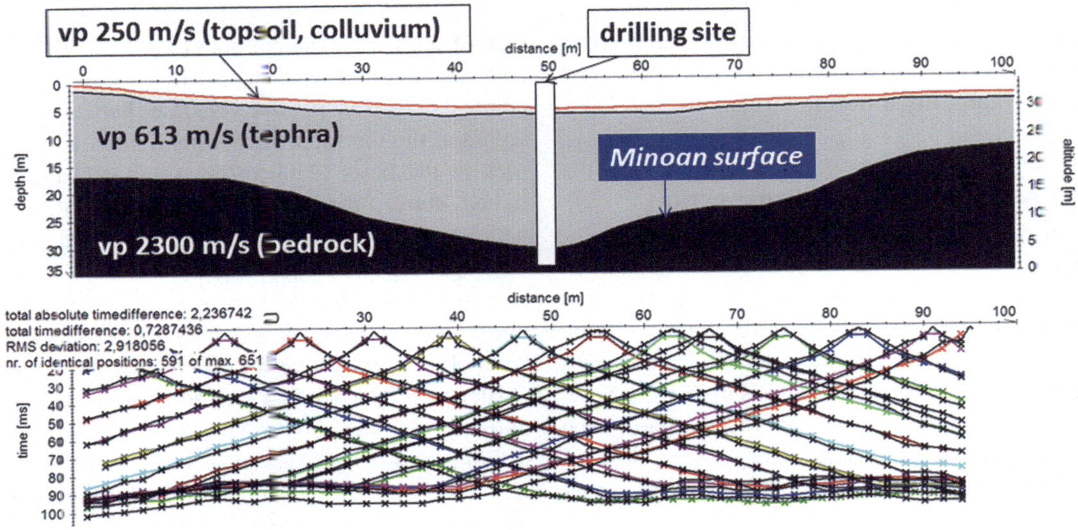

Fig. 15.7 Seismic profile M1 at Mavrorachidi valley (Fig. 15.2). Model of p-wave velocities derived from wavefront inversion and subsequent network raytracing (*above*). The observed travel times (*black*) at 14 shot point positions correspond very well with the computed synthetic travel times (*colored*) (*below*). A total absolute time difference of about 2.2 ms for 591 identical geophone positions is a very good value to support the proposed model (Sandmeier and Liebhardt 2005) (array of 48 geophones, spacing 2 m)

subsequent network raytracing. This routine helped to distinguish the boundary between the Minoan tephra (approx. 500 m/s vp) and the underlying Cape Riva ignimbrite (approx. 650 m/s vp). These relatively slight differences in p-wave velocities could be detected in all seismogram sections A1–A4. Second, seismic refraction tomography was applied to obtain the spatial distribution of p-wave velocities in order to trace the buried Minoan surface. The results of sections A1/A2 and A3 in Fig. 15.10 and section A4 in Fig. 15.11 show two different Minoan surface levels. The lower level appears in an altitude of about 20 m a.s.l. which corresponds very well with the outcrop of the Cape Riva ignimbrite at the western slope of the Potamos valley. In both cross sections A1/A2 and A3 (Fig. 15.10), the reconstructed Minoan surface rises up to the south and reaches approx. 30 m a.s.l., which means that the ignimbrite appears very close to the recent surface. Maybe a former cliff existed in that area, which is nowadays covered by the Minoan tephra. The position of Akrotiri behind a cliff could possibly help to protect the settlement against the seaside. Moreover, geomorphological features that imply excluding Potamos area as possible prehistoric harbor are the following: The Potamos drainage basin is substantially larger (and longer) in comparison with the similar ones existing in Site I (Mavrorachidi) and Site II (near the excavations' main entrance (see Fig. 15.2), hence diachronically transporting relatively higher amounts of torrential sediments from the catchment areas to the sea. At the same time, the pre-Minoan slopes of the Potamos valley (also today exposed) leave approximately 60 m width at distances ca. 250 m east—of southeast—of the settlement. Therefore, during flooding periods of the Upper Quaternary times, the increased amount of torrential debris should pass through this rather spacing valley, of which the depth was also shallow according to the aforementioned results of the geophysical investigation. Unfeigned witness of the voluminous torrential deposits transported over the millennia through this "narrow" and shallow valley is one terrestrial extrusion of detritus material, being created in front of the Potamos valley.

Obviously, the search for the missing harbor should be focused to the west of that cliff. Hence, the area southwest of Akrotiri village seems to be a more appropriate location for the prehistoric harbor.

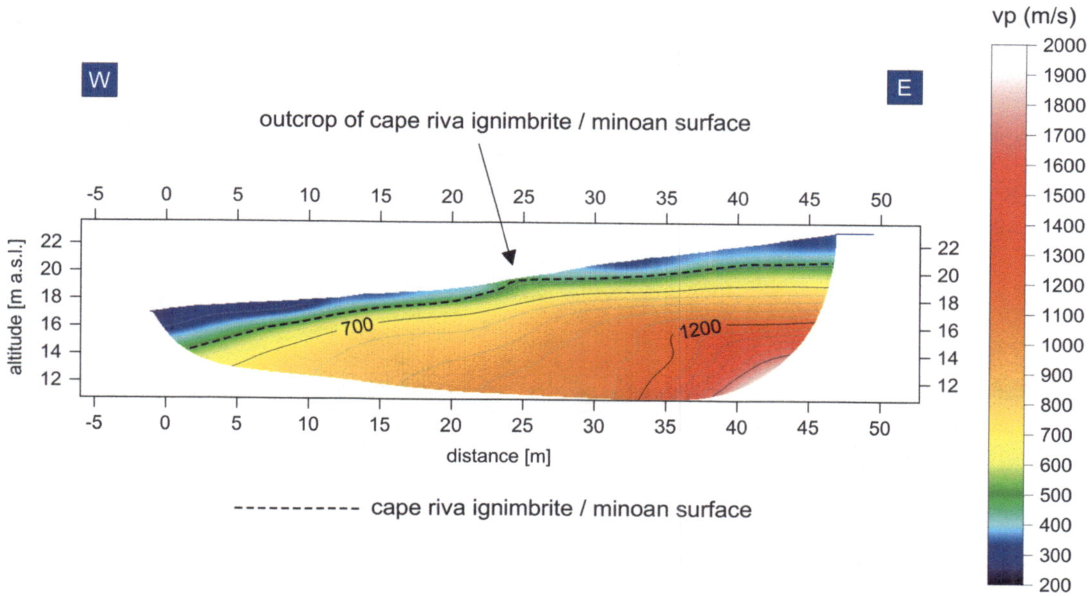

Fig. 15.8 Result of the seismic refraction tomography (SRT) P3 in the Potamos valley (Fig. 15.2). Starting at the outcrop of the Cape Riva ignimbrite, the Minoan surface could be traced in the subsurface (array of 24 geophones, 2 m equidistance)

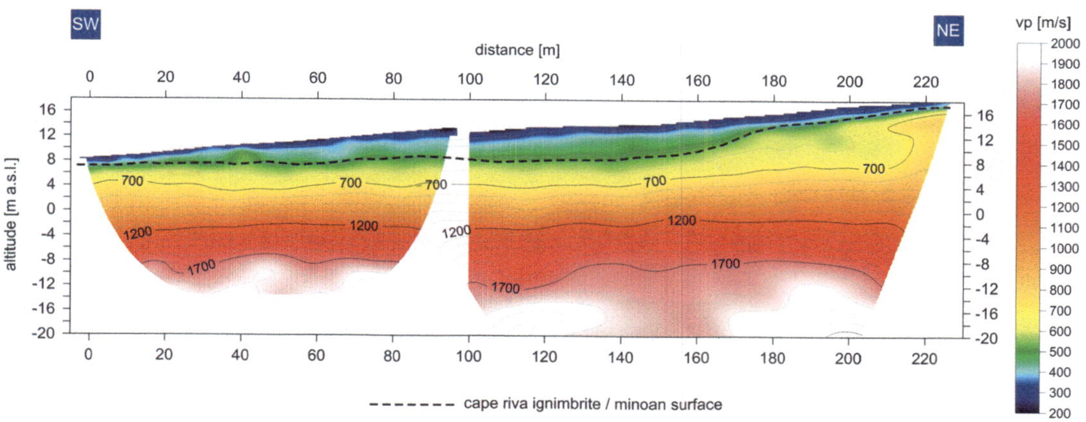

Fig. 15.9 Results of the seismic refraction tomography (SRT) in the Potamos valley. Longitudinal sections P1A (*left*, 48 geophones, 2 m equidistance) and P2A (*right*, 48 geophones, 3 m equidistance) along the valley floor (Fig. 15.2). Note that the thickness of the post-Minoan sediments reaches not more than 2–4 m

15.5 Discussion and Concluding Remarks

On the basis of the evaluated geological, geomorphological, and geophysical data, the valley known as "Potamos," indicated as Site III, in Fig. 15.3, seems to be excluded as an area that once hosted the "Minoan" harbor, although an exploratory drilling at depths beyond 12 m is required there. More suitable places for having fulfilled harbor necessities, according to our overall geomorphological view, seem to be the valley noted as Site II in Fig. 15.1 (in front of the archaeological site) and the one noted as Site I,

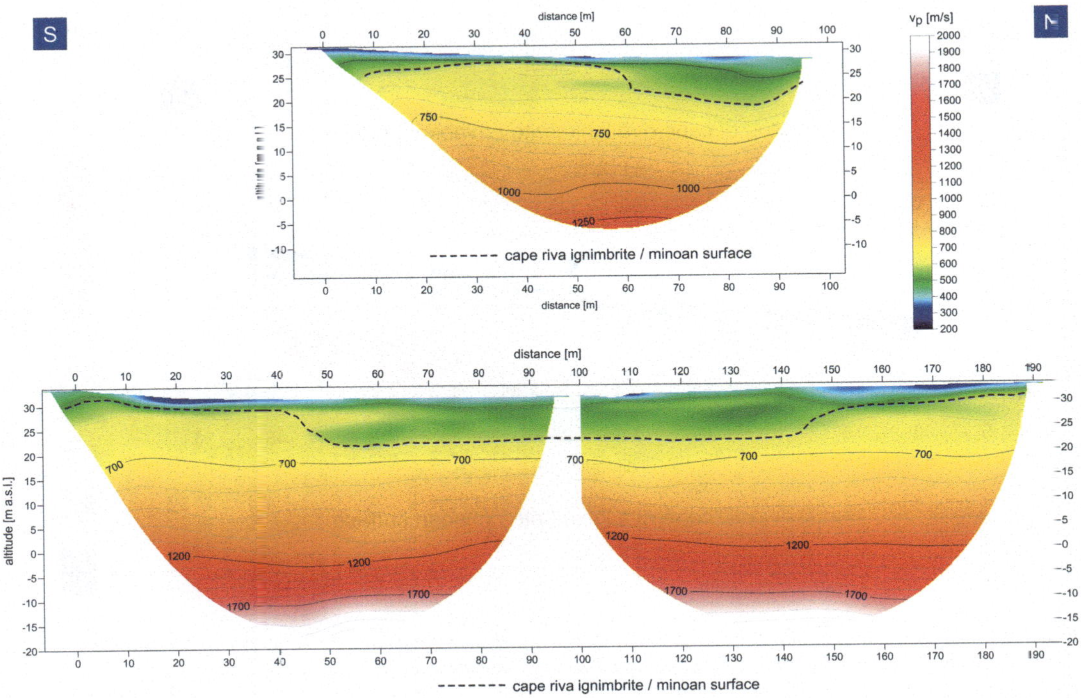

Fig. 15.10 Results of seismic refraction tomographies (SRT) A3 (*above*) and A1/A2 (*below*) southeast of Akrotiri archaeological site (Fig 15.2). Both sections show two main levels of the reconstructed Minoan surface. A lower one at an altitude of about 20 m a.s.l. and an upper level of about 30 m a.s.l (arrays of 48 geophones each, spacing 2 m)

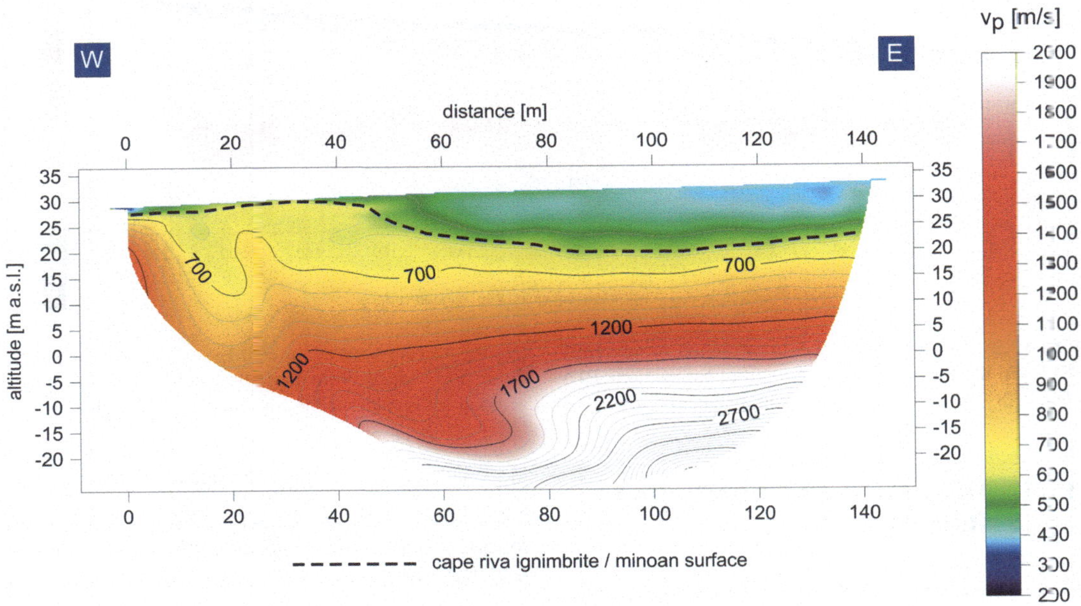

Fig. 15.11 Result of the seismic refraction tomography (SRT) A4 southeast of Akrotiri archaeological site (Fig. 15.2). In the western part of the profile (0–40 m), the Cape Riva ignimbrite seems to be very close to the surface (approx. 30 m a.s.l.) before dipping shallowly eastward to the second Minoan surface level of about 20 m a.s.l. (cp. Fig.15.10) (array of 48 geophones, 3 m equidistance)

the last lying approximately 700 m west of the excavated part of the Akrotiri settlement. One exploratory drilling performed approximately 650 m north of the strand in Site I at Mavrorachidi valley was of auger type, and it had not provided the required core; however, a tiny obsidian blade (as well as fragments of shells) discovered at the lower layers of the drilling comprises strong indication that a cultural layer, probably a coastland, once existed there, now buried under the pumices deposited from the action of the Minoan volcanic devastation. It is worth mentioning that the so far existing results of the field-based stratigraphic sequence as shown in Figs. 15.5 and 15.6, the geophysical investigation, and the drilling are in general consistent. In the case of Mavrorachidi valley, survey shows most coincidence as it concerns the estimated depth of the hard andesitic basement.

From a methodological point of view, the application of wavefront inversion and subsequent raytracing to invert the seismic data was very suitable to detect the depth of the andesitic basement in the Mavrorachidi valley precisely. Due to high contrasts of seismic p-wave velocities, this inversion routine provided the reconstruction of a sharp boundary between the tephra layers and the underlying bedrock, which could be successfully verified by the results of the auger drilling. In contrast, the use of the seismic refraction tomography (SRT) doesn't produce sharp layer boundaries, but it provides the distribution of seismic p-wave velocities across the whole seismic line at high spatial resolution. Therefore, this approach is appropriate to trace layer boundaries in the subsurface. As a result, it was possible to trace the Minoan surface (Cape Riva ignimbrite) at all sites, both in the Potamos valley and southeast of the Akrotiri excavation.

The outcomes of our studies indicate that the topography of the area, beyond the fluctuations of the sea level, sinking tectonic movements, and filling of the valleys under examination with Minoan volcanic ash (and subsequently disturbed by torrent reworking), has not significantly changed since the Minoan period. This is particularly true for the two main hills of the area that are situated west of the excavation, known as Kokkino Vouno (in Greek meaning red mountain) and Mesovouna ("intermediate" mountain). They have a pre-Minoan age; they consist of Pleistocene andesites and volcanic scoriae and are not covered by Minoan pumice. In conclusion, the presence of these two hills is determinant in the shaping of the today's landscape, in the same way as it was during the Akrotiri city thriving days. Therefore, it seems that the artistic representation of the "Flotilla" fresco very probably coincides with our observations and correlations of landscape details as implied in Fig. 15.12.

We consider that a key element in the search for the position of prehistoric port is an array of waterlogged to submerged beach rocks, which have been noted at the shallow seabed of the beach of Akrotiri. It is very likely that within this sequence of submerged ancient coast, to be inherent part of the ancient shoreline of the Minoan era which incorporates parts of the prehistoric port. Moreover, the need of extensive network of drillings, land, and underwater, followed by geotechnical coring in the region, is essential. It is necessary to take several cores in order to determine the stratigraphic and sedimentological sequence and have more accurate sampling for laboratory analyses and chronometric dating, by employing the luminescence technique (or any else more explicitly indicated method, depending on the nature of the sediments to be cored). Also the continuation of the geophysical investigation at the area in front of the entrance of the archaeological site is of great importance for the identification of possible architectural remains and the paleorelief.

As already mentioned, the localization of the buried harbor remains a so far unrealized ambitious aim, although a number of scholars have undertaken repeated attempts during the last decades. However, the expected utilization of the still unexploited geotechnical data coming from the boring works undertaken in the last years for the new shed of the excavation, the systematic and interdisciplinary character of the ongoing geoarchaeological project that has incorporated the new technologies emerged in the last years, and the undoubted progress in modern dating

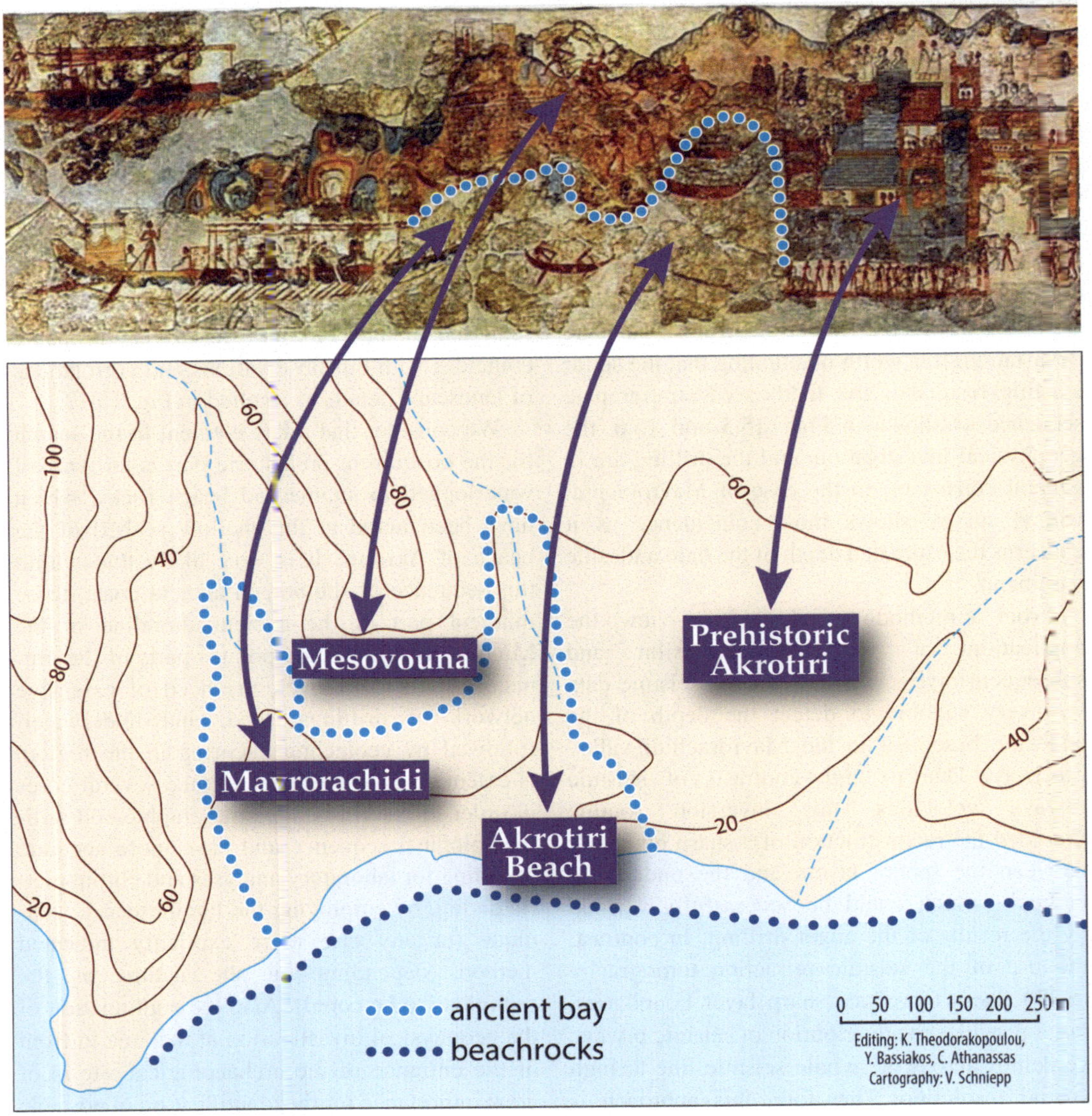

Fig. 15.12 Correlation of the geography of the prehistoric "Flotilla" fresco with the today's geomorphology

techniques might provide prospects of prosperity for success of the targeted purpose.

Acknowledgments We would like to thank the *Foundation of Education and European Culture (IPEP)*, the N.C.S.R. "Demokritos," as well as the painter Ute Grübnau-Grund, for their financial support and the director of Akrotiri Excavations Prof. Christos Doumas and his collaborators for licensing us to work at the archaeological sites and for unimpeded support of our activities.

References

Aston MA, Hardy PG (1990) The Pre-Minoan landscape of Thera: a preliminary statement. In Hardy A (ed) Thera and the Aegean world III, Earth sciences, vol 2. The Thera Foundation, London, pp 348–361

Doumas C (1983) Thera: the Pompeii of the ancient Aegean: excavations at Akrotiri. Thames and Hudson, London

Doumas C (1991) Thera's prehistoric harbor. The illustrated London news, Royal issue, 76–78
Doumas C (1992a) Akrotiri Theras (20 years of research - 1967–1987). Athens (in Greek)
Doumas C (1992b) The wall paintings of Thera. Thera Foundations, Athens. (in Greek)
Druitt TH, Francaviglia V (1992) Caldera formation on Santorini and the physiography of the islands in the late Bronze Age. Bull Volcanol 54:484–493
Eriksen U, Friedrich WL, Tauber H, Heinemeier J, Rud N, Thomsen MS, Buchardt B (1990) The Stronghyle caldera: geological, palaeontological and stable isotope evidence from radiocarbon dated stromatolites from Santorini. In: Hardy A et al (eds) Thera and the Aegean world III, vol 2. The Thera Foundation, London, pp 139–150
Friedrich W, Sørensen AH (2010) New light on the ship Fresco from Late Bronze Age Thera. Praehistorische Zeitschrift 85(2):243–257
Friedrich WL, Eriksen U, Tauber H, Heinemeier J, Rud N, Thomsen MS, Buchardt B (1988) Existence of a water-filled caldera prior to The Minoan eruption of Santorini, Greece. Naturwissenschaften 75:567–569
Friedrich WL, Kromer B, Friedrich M, Heinemeier J, Pfeiffer T, Talamo S (2006) Santorini eruption radiocarbon dated to 1627–1600 B.C. Science 312 (5773):548
Heiken G, McCoy F (1984) Caldera development during the Minoan Eruption, Thira, Cyclades, Greece. J Geophys Res 89:8441–8462
Heiken GH, McCoy FW, Sheridan M (1990) Paleotopographic and paleogeologic reconstruction of Minoan Thera. In: Hardy A (ed) Thera and the Aegean world III, Earth sciences, vol 2. The Thera Foundation, London, pp 370–378
Marinatos S (1939) The volcanic destruction of Minoan Crete. Antiquity 13(52):426
Papageorgiou E, Lagios E, Vassilopoulou S, Sakkas V (2007) Vertical and horizontal ground deformation of Santorini Island deduced by DGPS measurements. Proceedings of the 11th international congress, Greece. Bull Geol Soc Gr 40:1219–1255
Papageorgiou E, Tzanis A, Sotiropoulos P, Lagios E (2010) DGPS and magnetotelluric constraints on the contemporary tectonics of the Santorini volcanic complex, Greece. Proceedings of the 12th international congress, Geological Society of Greece. Bull Geol Soc Gr XLIII(1):344–356
Pichler H, Friedrich WL (1980) Mechanism of the Minoan eruption of Santorini. In: Doumas C (ed) Papers and proceedings of the second international scientific congress on Thera and the Aegean world II, vol II. Athens, pp 15–30
Poulos SE, Ghionis G, Maroukian H (2009) Sea-level rise trends in the Attico-Cycladic region (Aegean Sea) during the last 5000 years. Geomorphology 107:10–17
Reynolds JM (2011) An introduction to applied and environmental geophysics, 2nd edn. Wiley, Oxford
Rohdewald S (2006) RAYFRACT™ manual. Intelligent Resources Inc., Vancouver
Russel JK, Stasiuk MV (2000) Ground-penetrating radar mapping of Minoan volcanic deposits and the Late Bronze Age palaeotopography, Thera, Greece. In: WG MG, Griffiths DR, Hancock PL, Stewart IS (eds) The archaeology of geological catastrophes, vol 171. Geological Society, Special Publications, London, pp 105–121
Sandmeier KJ (2007) REFLEXW version 7.2. Sandmeier Geo, Karlsruhe
Sandmeier KJ, Liebhardt G (2005) Refraktionsseismik: iterative interpretationsmethoden. In: Knödel K, Krummel H, Lange G (eds) Handbuch zur Erkundung des Untergrundes von Deponien und Altlasten. Bd. 3: Geophysik, Springer, Berlin, pp 566–572

Merging the Views: Highlights on the Fusion of Surface and Subsurface Geodata and Their Potentials for Digital Geoarchaeology

16

Christoph Siart

Abstract

A multi-method research design based on geophysical prospecting (electrical resistivity tomography, seismic refraction), DEM generation (terrestrial LiDAR and total station) and GIS is applied for the first time to investigate geoarchaeologically relevant sites in an integrated way. Fusing multi-resolution surface and subsurface geodata provides profound insights into the formation, geometry and geomorphologic processes of karst depressions which serve as geoarchives in the Mediterranean area. Case studies from different locations on Crete are provided. In order to define crucial methodological requirements and guidelines for data fusion, both the impact of different elevation models and different geophysical methods and the influence of data resolution are assessed. Different approaches are presented along with their advantages and shortcomings, highlighting the various options offered by combining surface and subsurface geodata.

Keywords

Geophysical prospecting • Terrestrial laser scanning • Digital elevation models • GIS • Data fusion • Geomorphometry • Digital geoarchaeology

16.1 Introduction

Sediment-filled karst depressions like dolines are of great geoarchaeological interest, because they have been and still are of historico-cultural significance. Not only do they function as reservoirs that prevent sediments from erosion (Fig. 16.1), but they also serve as valuable zones for agriculture and settlement activities for thousands of years (Siart et al. 2008). Today, numerous archaeological remnants can be found inside and around such landforms. These findings insight into the economic exploitation and the

C. Siart (✉)
Institute of Geography, Physical Geography, Heidelberg University, 69120 Heidelberg, Germany
e-mail: qb120@uni-heidelberg.de

© Springer International Publishing AG 2018
C. Siart et al. (eds.), *Digital Geoarchaeology*, Natural Science in Archaeology,
DOI 10.1007/978-3-319-25316-9_16

cultural evolution of landscapes and moreover highlight the importance of human-environmental interactions when considering palaeoenvironmental history. Therefore, dolines can be regarded as so-called geoarchives. Prominent examples can be found in the mountains of Crete (Greece), where lots of karst landforms like dolines and poljes can be encountered due to area-wide existence of carbonate rocks, which are subject to chemical dissolution. For a long time, Cretan dolines have just rudimentarily been studied, mostly with regard to some general geomorphologic properties (e.g. Fabre and Maire 1983; Bartels 1991; Egli 1993), but not in terms of their geoarchaeological significance, their subsurface geometry and their sediment fill. Hence, no detailed prediction about the actual archive function and the depth structure was possible.

In order to gain comprehensive insight into these karst geoarchives and to better understand local human-environmental relationships, an integral, multi-method research approach is key. In this context, it is important to take a look at the theoretical structure of sediment-filled dolines first. Seen from a geomorphologic point of view, they represent two-component systems with a surface and a subsurface part (Waltham et al. 2005; Ford and Williams 2007), which is why data on both entities is required if comprehensive investigations are to be conducted (Fig. 16.1). As for their ground level, useful topographic information can be obtained from digital elevation models (DEM). They are often free of charge and ready to use, but their poor spatial resolution is mostly insufficient for analysing meso- to microscale relief features (Siart et al. 2009). The precision of DEM can be enhanced by total station measurements and DGPS tracking, but both techniques involve enormous efforts and carry the risk of operator bias (selective sampling in the field, Armesto et al. 2009). In fact, the most detailed topographic geodata for landform investigations stems from LiDAR applications. Above all, terrestrial laser scanning (TLS) allows for a complete 3D capture of objects with a high point density that outperforms the resolution of airborne datasets and therefore even permits studying very small and subtle surface features as well as vertical objects (Hämmerle and Höfle 2018).

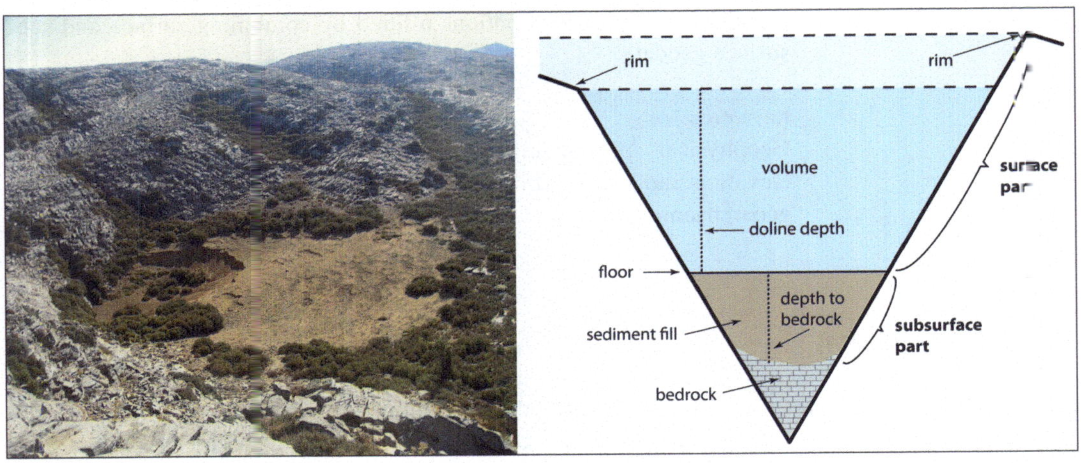

Fig. 16.1 *Left*: Sediment-filled solution doline in Crete. The harsh contrast between eroded limestone slopes and the sediment bottom of the landform is highlighted by shrub vegetation at the margins. *Right*: Theoretical two-component model of a typical solution doline consisting of a surface and a subsurface art, displayed as a cross section

Regarding the subsurface of karst landforms, the best and fastest way to obtain reliable data is geophysical prospecting (Stepisnik and Mihevc 2008; Ravbar and Kovacic 2010). However, as described by Sarris et al. (2018), each geophysical technique is prone to specific shortcomings, which is why multi-method approaches are indispensable to yield reliable results and to avoid ambiguities within the context of interpretation. Electrical resistivity tomography (ERT) and seismic refraction tomography (SRT) are increasingly used in this context, because of their easy-to-use and non-destructive properties (Hoover 2003; Sheehan et al. 2005; Bechtel et al. 2007). Besides producing valuable information on geomorphodynamics, they also allow detecting buried remains and deducing former land use techniques by providing an insight into the underground and the thickness of sediment deposits (Gaffney 2008; Hecht 2009). However, ERT and SRT generally only generate 2D data, and on top of that, the use of tomographic data is very complex in karst terrains, as dolines often show heterogeneous geometrical characteristics and a high level of geophysical noise (Van Schoor 2002; Terzic et al. 2007).

Despite this variety of methodologies, Cretan karst landforms have only been investigated as to either surface or subsurface morphology for a long time. The paper at hand addresses this shortcoming by presenting a novel approach that is based on testing different combinations of selected techniques, which collaboratively allow studying sediment-filled karst depressions in their entirety. It highlights different ways of defining the morpho-structural properties of dolines; of analysing the geometrical, geophysical and geoarchaeological characteristics of their infills (sediment thickness, underground topography, archaeological artefacts); and of developing 3D models that illustrate the subsurface of buried karst reliefs. Hence, insights into the formational processes of karst landforms and their historico-cultural significance are provided. In methodological terms, the potentials and benefits of fusing different types of geodata for investigations at the human-environmental interface are to be demonstrated. The three case studies presented below also exemplify the concept of digital geoarchaeology (Siart et al. 2018), as valuable synergies between geoscientific, archaeological and computer scientific expertise are generated and used. Each one is based on different applications and datasets to determine the most suitable setup for investigating karst geoarchives, but taken as a whole, they are all part of an iterative, integral research design and strongly interrelate.

16.2 Case Study 1: Zominthos (Central Crete, Greece)

The study area of Zominthos is located in the Ida Mountains of Central Crete (1180 m a.s.l.). Intense karstification led to the formation of numerous sediment karst-filled depressions such as dolines. With respect to the local human impact, one of the most outstanding features is the existence of several remnants of the Minoan culture (villa of Zominthos; Neopalatial period, ~1650 BC) that used to colonize and exploit the area during the Bronze Age. At around 1200 BC, the area was completely deserted due to interplay of climate deterioration and societal upheavals (Siart 2010). Up until today, it was only periodically used for stock breeding, while agriculture or settlement activities are absent.

16.2.1 Methods

A total of 16 ERT profiles based on a 100-electrode system (Geotom, Schlumberger and dipole-dipole configurations) was measured in the doline of Zominthos to evaluate the depth of the sediment fills and to identify potential buried artefacts nearby the Minoan settlement. Resultant data was post-processed with RES2DINV software. SRT transects (Geometrics, 48-channel system) were measured at the exact same locations for comparison with ERT studies, since both methods show different sensitivities and are prone to individual advantages and limitations (Schrott and Sass 2008). REFLEX and RAYFRACT software packages were used to process the 2D outcomes. As documented by Siart et al. (2011), the methodological implementation of surface and subsurface data fusion can be achieved by GIS-based processing of digital elevation models according to geophysical datasets. Hence, a 3D model of the subsurface topography was deduced from the interpretation of geophysical findings in order to investigate the underground structure. Therefore, relief data was collected by total station mapping (Leica TPS 700) and GPS measurements before conversion into a high-resolution digital terrain model (DTM; spacing, .5 m) with Surfer 8 and ArcGIS 9.3. Depth levels of doline fills were delineated by vectorization, interpolated and transformed into the third dimension. Specific seismic velocity values (v_p > 2000 m/s) and earth resistivities ($R >$ 200 Ωm) served as crucial limits to identify the maximum depth of the sediments (Siart et al. 2011). The DTM was then modified by reducing the surface Z-values by the depth of fills, which led to the lowering of the superficial topography and to the exposure of the approximate bedrock of the doline. In areas devoid of geophysical data coverage, resistivities and velocities measured in the surroundings were interpolated to create the continuous subsurface topography of the Zominthos karst depression. The newly generated digital terrain and subsurface model (DTSM) was converted into TIF format and draped by high-resolution satellite images (QuickBird; pixel size, 0.61 m).

16.2.2 Results and Interpretation

As indicated by the geophysical outcomes, the doline of Zominthos is filled with huge amounts of loose sediments (colluvium) that reach depths of up to 30 m (Fig. 16.2). This finding is consistent on almost all transects and can be verified by ERT and SRT (Siart 2010). Both

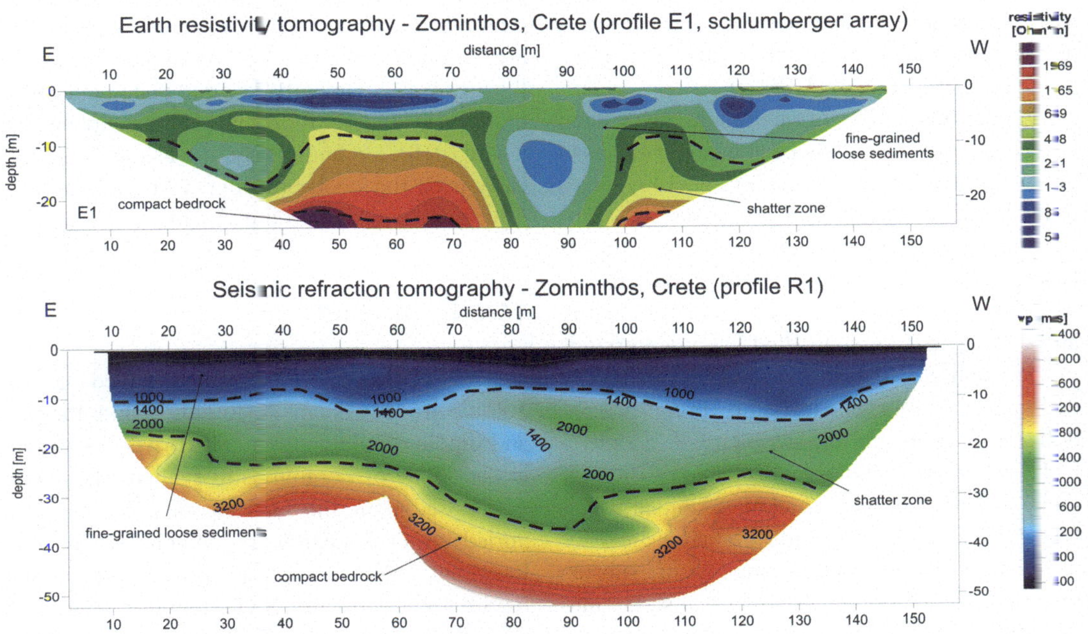

Fig. 16.2 Geophysical tomographies of the Zominthos doline. *Top*: ERT E1 indicating loose sediments atop with thicknesses of up to 25 m (low resistivities), an intermediate shatter zone comprised of fine-grained material mixed with coarse sediments and the basal bedrock (high resistivities). *Bottom*: SRT profile R1 measured in parallel to E1, confirming the three-part structure of the subsurface. As shown by the high values at the bottom, the depth to bedrock is sometimes even higher than indicated by ERT (>30 m)

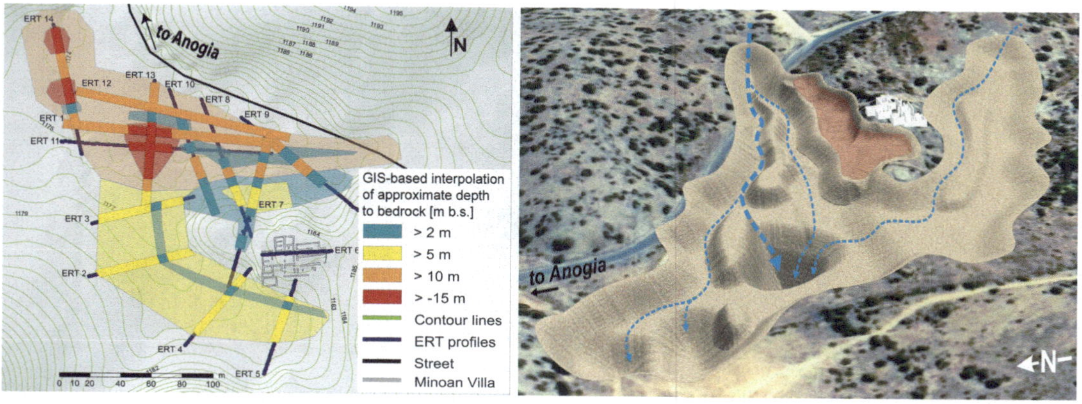

Fig. 16.3 *Left*: GIS-based vectorization and interpolation of geophysical outcomes. *Right*: DTSM of Zominthos karst depression showing the undulating bedrock and the presumed drainage pattern that might have been subaerial during Minoan occupation

geophysical techniques helped detect the actual limestone bedrock at the bottom of the karst landform.

After fusion and transfer of all 2D geodata into the third dimension, several buried channels can be observed in the DTSM, which flank the settlement hill of the Minoan villa (Fig. 16.3). These trenches were used for drainage in Minoan times quite likely (Siart 2010). A hypothetical subsurface runoff pattern can be presumed, extending from the south-eastern part of the catchment area to the centre of the doline (Fig. 16.3). It illustrates the dominant underground water flow and highlights areas of high susceptibility to percolation.

To conclude, the doline of Zominthos is characterized by a subsurface relief that appears totally different from today's local surface morphology. Our results reveal an extremely heterogeneous underground topography characterized by strong geophysical variations and remarkable differences in depth to bedrock, which must be considered typical of karst terrains (Deceuster et al. 2006).

16.3 Case Study 2: Kritsa-Latô (East Crete, Greece)

Kritsa-Latô, the second area of investigation, is located in the Dikti mountains in east Crete. Analogous to Zominthos described in Case Study 1, the geological setting largely consists of limestone in which numerous karst landforms developed. The study site itself is situated on the margins of a deeply incised doline (approx. 70 m deep) at 278 m a.s.l. Here, too, one can find a close spatial, socioeconomic relationship between a karstic geoarchive and past human activity. In the near vicinity, archaeologists excavated the ancient settlement of Latô, which was occupied from Mycenaean to Hellenistic times, and experienced its cultural bloom under the Dorian rule (Demargne 1901). However, no evidence of subsequent occupation during mediaeval times was found so far. The sites' geoarchaeological importance is stressed by numerous surface findings (e.g. pottery shards from Minoan to modern times), as well as a tapped spring and partly buried wall remains on the doline floor (slopes of a water-filled pond, Fig. 16.4), which provide evidence for ancient land use in and around the doline. Although the area was colonized and exploited during parts of the mid and late Holocene, only little is known about the relationships between man and his environment. As in Zominthos, the topics to be investigated relate to both the surface and the subsurface topography, the depth of sediment fills as well as the palaeoenvironmental reconstruction by means of a multimethod approach.

Fig. 16.4 Aerial view of the doline floor (*left*) with location of ERT measurements (*white lines* and *rectangle*), TLS positions (*black dots*) as well as the dried-out spring (*centre*) and the water-filled pond, in which massive buried wall remains appear (*right*)

16.3.1 Methods

Three ERT profiles were measured to assess the subsurface topography of the doline and the depth of sediment fills (Fig. 16.4). A 100-electrode system (Geotom) was used with different profile lengths (25–200 m) and both Schlumberger and dipole-dipole arrays. With regard to several geophysical anomalies that showed on the 2D transects (areas of geoarchaeological interest), a 3D electrode configuration was applied (pole-pole array; grid, 10 × 10 m; electrode spacing, 1 m). Post-processing was based on RES2DINV and RES3DINV, while calibration of resistivity values was carried out through comparison with ERT results from other karst geomorphologic studies (e.g. Siart 2010). Terrestrial laser scanning (TLS; Hämmerle and Höfle 2018) was applied to provide highly accurate topographic information of the doline surface. Data acquisition was conducted by using a time-of-flight scanner (Riegl VZ-400) with a narrow infrared laser beam and a fast scanning mechanism. A Nikon D300(s) digital camera with a fisheye lens on top of the scanner provided additional RGB information. In total, eight different scan positions were collected for complete coverage of the study site (Fig. 16.4). Different scanning modes (range, 600 m; acquisition rate, 40,000 measurements/s vs. range, 350 m; acquisition rate, 122,000 measurements/s) were used for area-wide high-resolution data. Post-processing included coarse registration or reflector-based matching of single scan positions and subsequent fine processing by means of multi-station adjustment (iterative closest point (ICP) algorithm; Hoffmeister 2018). As the geomorphologic interpretation required unambiguous datasets, redundant details such as vegetation were deleted in a multistep approach using the RiSCAN Pro software (Siart et al. 2012). Two DTM (pixel sizes, 0.25 m and 2 m) were derived by meshing the improved point clouds and used for additional morphometric analysis in ArcGIS 9.3. In order to assess local geomorphodynamic processes and former land use practices, ERT were georeferenced and superimposed on the DTM to pinpoint the exact position of subsurface structures and to correlate them to microrelief features on the surface.

16.3.2 Results and Interpretation

The TLS outcomes allow investigating the studied landforms in high detail, in particular as to geomorphometric information derived by DTM analysis with RiSCAN Pro and ArcGIS 9.3 (e.g. slope, depth, diameter, gradient, aspect; Siart et al. 2012). Assuming that the subsurface geometry, which borders the sediment-filled part of the doline, resembles an inverted cone (Fig. 16.1)—a fact deduced from the ERT profiles in analogy to the findings of Sustersic (2006)—the subaerial volume (2.37M m^3) and the volume of the sediment fill (0.245M m^3) can be approximated for the very first time. As documented by ERT, the Kritsa-Latô doline exhibits huge sedimentary fills with depths of at least 25 m (Fig. 16.5). Since the techniques' vertical penetration only reached down to about 35 m b.s., even thicker accumulations must be presumed, which clearly highlight demonstrating the sediment trap function of karst depressions. This fact also highlights the crucial socioeconomic value of karst depressions in terms of suitability for agriculture and land use. Considering the archaeological findings, which prove human impact ever since the first millennium BC, the Kritsa-Latô doline represents a diachronic phenomenon.

Special attention must be paid to a zone of high geoarchaeological interest in the centre of the doline. Here, the surface drainage pattern derived by GIS-based hydrologic analysis indicates a superficial gully with a depth of less than half a metre. It represents the most prominent mesoscale geomorphic feature in the entire landform. Starting from the doline margin, where a well-structured accumulation of limestone blocks has been erected to capture an ancient spring that has dried up nowadays (Fig. 16.4), it slopes down to the centre of the doline floor. However, this ditch does not correspond to the subsurface findings that were indicated by 3D ERT. Instead, the high resistivity zone consists of limestone boulders in the near subsurface that are embedded in a thick accumulation of fine-grained sediments. The blocks must be of anthropogenic origin and quite likely form part of an ancient water reservoir (Siart et al. 2012). In order to unravel this issue, TLS data and geophysical outcomes must be put into a spatial context that provides insights into the whole extent of the water harvesting system: The near-surface high resistivity values are exactly in line with both the partly unearthed wall remains in the centre of the doline floor and the captured spring at its margin (Fig. 16.6). One can therefore presume a prehistoric drainage channel that was constructed to direct surface

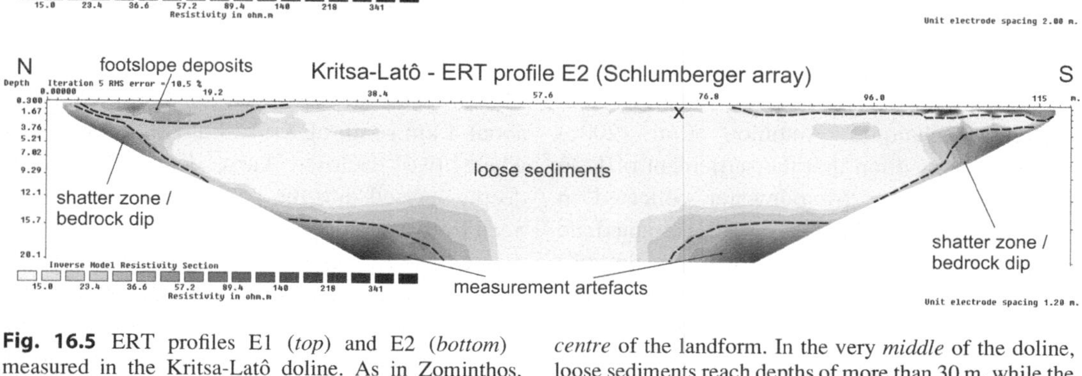

Fig. 16.5 ERT profiles E1 (*top*) and E2 (*bottom*) measured in the Kritsa-Latô doline. As in Zominthos, the subsurface shows a threefold setup with thick accumulations of loose sediments atop, an intermediate shatter zone and the basal bedrock that dips towards the *centre* of the landform. In the very *middle* of the doline, loose sediments reach depths of more than 30 m, while the limestone bedrock wasn't detectable due to insufficient penetration depth of ERT

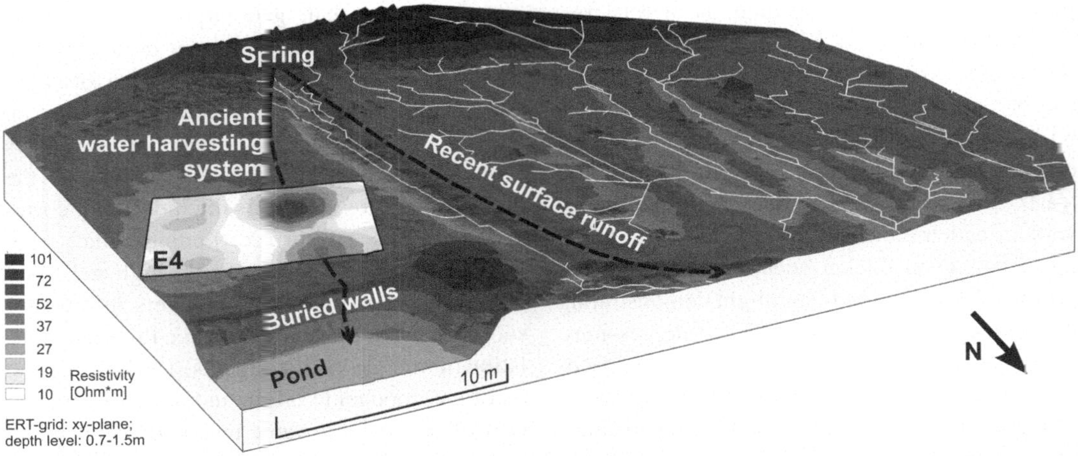

Fig. 16.6 Block section of the ancient water harvesting system in the Kritsa-Latô doline. A depth slice of the 3D ERT E4 was georeferenced and draped over the TLS-based DTM to allow for an integral interpretation of surface and subsurface geodata. High resistivities in E4 (*dark colours*) can be interpreted as limestone blocks in a depth of 0.7–1.5 m b.s., which are in clear alignment with the tapped spring on the southern margin and the unearthed wall remains in the water-filled pond. In contrast, the recent surface runoff pattern derived by GIS is not consistent with its ancient counterpart. Today, the most prominent mesoscale feature is a gully that drains to the northwest of the doline floor

runoff into a walled cistern. According to the GIS-based hydrologic surface analysis, the recent surface gully shown in the DTM is completely disconnected from the ancient system, though, and shows no spatial correlation with the subsurface findings. Even though dating the water harvesting system still remains vague due to the absence of chronometric data, the findings could possibly be attributed to the Dorian era. Several cisterns unearthed in the proximate site of Latô substantiate these findings (Siart et al. 2012). Among others, Angelakis et al. (2007) refer to the sophisticated achievements of hydrotechnical engineering during Minoan and Mycenaean times, e.g. tapping of springs and irrigation techniques. Antoniou et al. (2006) additionally mention that the settlement of Latô was supplied only by rainwater collected in cisterns since no springs existed in its immediate vicinity. Apart from that, the subsurface findings identified in the sediment fill could also be part of a drainage system constructed to prevent arable land from flooding. Since the flat bottoms of sediment-filled dolines were easy to use, extensive draining was common in Bronze Age Greece (Showleh 2007; Koutsoyiannis et al. 2008).

The nearby settlement of Latô and the lack of findings from post-Dorian to mediaeval times suggest that the local occupation of the Kritsa-Latô doline took place during the first millennium BC. Evidentially, the availability of water was essential for subsistence in the karstified and, thus, arid mountains of Crete as early as the Bronze Age (Angelakis and Koutsoyiannis 2003; Panagiotopoulos 2007).

16.4 Case Study 3: Kroustas (East Crete, Greece)

The third case study was carried out in Kroustas, about 4 km south of Kritsa-Latô at 355 m a.s.l., where two enclosed karst depressions were deeply incised into the parent carbonate rocks. Seen from a geomorphologic point of view, these dolines stand out due to their high depth-to-width ratio, their steep walls and their flat, sediment-covered floor. As in the previous examples from Zominthos and Kritsa-Latô, the major aim was to elicit human-environmental interactions by focusing on karst archives using a multi-method research design.

16.4.1 Methods

Field work included TLS, differential GPS measurements (DGPS), geophysical prospecting as well as post-processing and fusion of digital datasets with GIS (for details on equipment and work steps, see Case Study 2). First, TLS was used to capture the entire topography of the doline complex at a total of 12 different scan positions (Fig. 16.7). After co-registration of single scan positions with DGPS coordinates, multi-station adjustment was accomplished with RiSCAN Pro. As the dataset included surface objects such as dense bushes and shrubs, which generally hamper geomorphologic analyses, the actual ground surface was modelled by the removal of vegetation using OPALS software (Höfle and Rutzinger 2011; Siart et al. 2013). After meshing the point clouds, a DTM with 0.5 m spacing was generated. Geophysical prospecting included acquisition of three SRT profiles, which were post-processed with REFLEX and RAYFRACT software, as well as three 2D ERT transects that were treated with RES2DINV. Subsequently, surface and subsurface geodata were processed and merged with GIS. The fusion workflow allowed generating a DTSM by combining highly detailed topographic surface data (i.e. DTM) and subsurface depth profiles derived by jointly interpreting SRT and ERT. In a semi-automatic procedure with GRASS GIS, a manually defined area of the DTM doline bottom was replaced by subsurface data (lowering of Z-values; comp. Case Study 1). This allowed deriving an elevation raster model from the geophysical depth profiles, which was ultimately inserted into the DTM by modelling the transition zone with a defined slope (Siart et al. 2013).

16.4.2 Results and Interpretation

As shown by the SRT profiles that reach maximum depth levels between 50 and 60 m, the subsurface can be subdivided into three distinct main layers (Fig. 16.7): an uppermost zone (low values, $P < 1000$ m/s) corresponding to

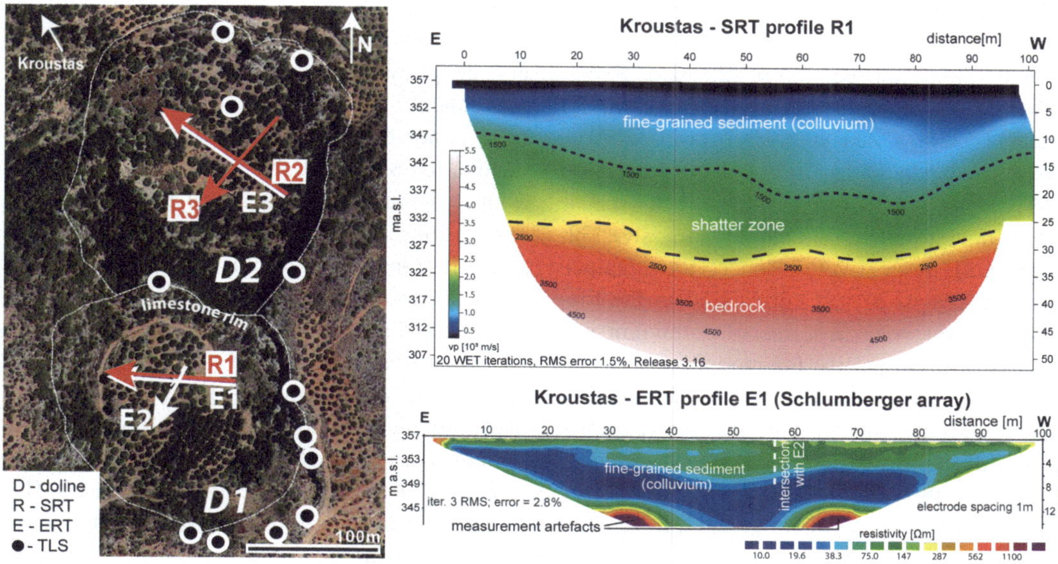

Fig. 16.7 *Left*: Aerial view of the Kroustas dolines with location of ERT and SRT transects (*white* and *pink lines*) and TLS positions (*black dots*). *Right*: SRT R1 and ERT E1 were measured in the southern doline D1 and intersect in about the *middle*. As in Zominthos and Kritsa-Latô, the subsurface consists of three different zones (overlying colluvium, intermediate shatter zone and basal limestone). The bedrock extends in high depth levels of about 30 m b.s. and was only detectable by SRT

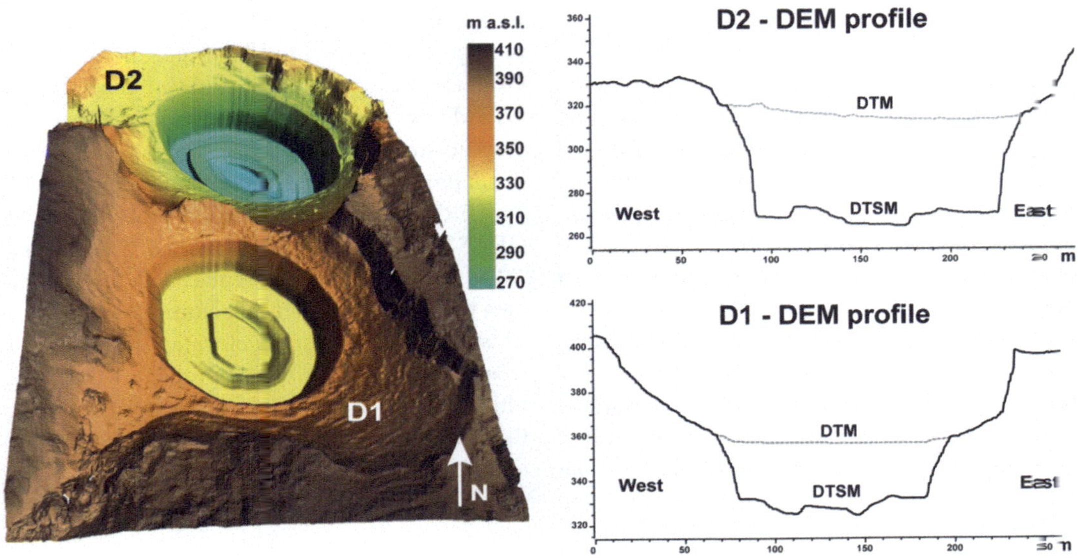

Fig. 16.8 The integrated DTSM of Kroustas (*left*) displays the actual course of the parent rock both in the superficial part of the doline and the area covered by sediments. The latter was masked out to provide an image of the empty landform in its entirety. Subsequently, geomorphometric calculations such as generation of topographic cross sections (*right*) were feasible

fine-grained sediments with thicknesses up to 20 m, an underlying shatter zone in which fractured limestone intermingles with unconsolidated material and the solid bedrock of the doline. The boundary between the latter two oscillates between depths of 25 and 45 m, showing P-wave values higher than 2000 m/s (Hecht 2003). Measured ERT profiles reach maximum depths of 15–20 m but in comparison to SRT do not reach down to the bedrock because of insufficient penetration and an excessive colluvial infill, respectively. Hence, the obtained values (0–250 Ωm) just correspond to the topmost refractor zone in terms of colluvium and therefore do not qualify for modelling the subsurface topography. However, both the low electrical resistivities and the low P-wave velocities can be allocated to loose sediments—a finding which is in absolute accordance with data from Kritsa-Latô and Zominthos. As depth to bedrock generally proves to be the sole parameter that allows for establishing a reliable and detailed DTSM, only SRT could be used for defining the subsurface shape of the Kroustas dolines. According to Siart (2010), P-wave velocities higher than 2000 m/s were chosen for GIS-based processing of the DTM (lowering of the doline bottom while masking out near-surface sediments). The integrated DTSM, which resulted from the fusion of laser scanning and geophysical data (Fig. 16.8), was used for subsequent geomorphometric investigations (e.g. generation of cross sections; for further details, see Siart et al. 2013). It allowed studying a karst depression as an actual two-component system for the very first time.

Seen from a geomorphic point of view, the Kroustas dolines were formed by carbonate dissolution and neotectonics that led to large funnel-shaped hollows during the quaternary. Geoelectrical tomographies prove the complex and unpredictable properties of the buried karst system, particularly the irregular depth to bedrock and the heterogeneous subsurface. The findings point to a so-called collapse doline as opposed to solution dolines investigated in Case Studies 1 and 2. The absence of sediment stratification deduced from geophysical outcomes indicates permanent low energetic influx under rather stable geomorphodynamic conditions.

Due to the fine-grained sediments and their good water capacity, the location was predestinated for human land use in former times, as is the case with Zominthos and Kritsa-Latô. Future studies will have to focus on potential archaeological remains, which have not been investigated so far.

16.5 Conclusions

To sum up, the paper at hand tries to assess the concept of data fusion in terms of answering geomorphologic and geoarchaeological research questions on the one hand and with regard to the methodology on the other hand. While the former explicitly pertains a better understanding of past human-nature interactions and the reconstruction of palaeoenvironments, the latter focuses on the technological approach itself, which can be described best as digital geoarchaeology—a concept that bridges the gap between archaeology, geosciences and informatics.

16.5.1 Geomorphologic and Geoarchaeological Implications of Data Fusion

The case studies from Crete reveal completely new insights into the geometrical, geophysical and geoarchaeological properties of enclosed karst depressions. As shown by the results, comprehensive datasets that complement each other are indispensable to understand the functional principle of dolines. One of the most common features of all investigated landforms is the fact that they are filled with thick deposits of loose sediments (up to 50 m), which point to the great age of corresponding landforms (Siart et al. 2011, 2013). Dolines must therefore be considered as sediment reservoirs and important geoarchaeological archives, which have been subject to permanent influx or throughput of material. Apart from their significance for geomorphologic investigations, both the surface and the subsurface geodata provide valuable information on the geoarchaeological evolution of studied regions. The process of filling was triggered by extensive land use, which caused a severe degradation of vegetation and ultimately led to massive erosion in the mountains of Crete. In chronological terms, this palaeoenvironmental transformation was closely related to the rise and fall of the Minoan civilization (Siart 2010). In addition, several geophysical anomalies suggest the existence of water harvesting and drainage systems that can be put into a chronological context with the help of archaeological expertise: Pre-Christian land use of dolines becomes quite evident since the flat topography, the thick sedimentary fill and the hydrologic favour of many Cretan dolines were of major importance for economic exploitation ever since the Bronze Age (Siart et al. 2008). Back then, just like today, sufficient availability of soil and water was crucial for agriculture, stock breeding and settlement activities in karst terrains. Hence, the dolines of Zominthos, Kritsa-Latô and Kroustas served as favourable locations as they represented fertile oases in literally dry regions.

As highlighted by all three examples from Crete, the novel approach of data fusion helps to better understand local geomorphodynamics, landform evolution processes and human impact. The results offer first-time insights into the human-environmental interactions in the mountains of the island and prove that combining surface and subsurface geodata offers promising prospects for future studies. In turn, further investigations will help to obtain a higher chronological resolution of the geomorphologic and geoarchaeological findings.

16.5.2 Methodological Synopsis

The presented case studies clearly demonstrate that fusion of 3D geodata enables to establish an integral image of karst landforms for the very first time. The multi-method research design allows quantity assessment of loose sediments accumulated within dolines and makes geomorphometric analyses feasible (Siart et al. 2013). However, the outcomes also prove that studying geoarchaeological archives is carried out best using integrated research designs, while monodisciplinary approaches possibly might not be sufficient to fully capture landforms and human-environmental interactions as a whole. Several different workflows can be applied for that purpose, but it is important to consider that techniques, which proved beneficial at one specific site, may not unconditionally be used elsewhere in the same way. If we take a look at the surface topography of landforms, TLS turns out as most qualified for non-invasively acquiring precise topographic data. It absolutely outperforms DTM or DSM derived by other remote-sensing techniques (e.g. satellite imagery, total station measurements), since microrelief structures become visible and outcomes are devoid of operator's bias. When implemented in GIS, TLS-based data can be used for morphometric analyses that better the understanding of geomorphologic features or processes as well as for a combination with geophysical results. With regard to the subsurface of karst depressions, both SRT and ERT are of great value, but each method is prone to specific limitations. While SRT allows identifying the actual bedrock and offers higher depth penetration, ERT is more sensitive to the internal differentiation of the underground and it is especially suitable, if buried archaeological remains or local geomorphodynamics are to be investigated. However, the most decisive question concerns the quantity and quality of geophysical data required for thoroughly exploring sediment-filled dolines. Typically, the precision of outcomes derived from data fusion can be improved by increasing the number of geophysical transects (Siart et al. 2011). Nevertheless, the individual morphology of landforms governs all variables for geophysical prospecting and TLS. In case of funnel-like solution depressions (e.g. Zominthos, Kritsa-Latô), which often show very heterogeneous subsurface structures, a higher quantity of transects may be required (Siart 2010). ERT turns out as the most qualified technique here, as lesser penetration depth is needed and microscale structures appear more distinctly. In contrast, collapse dolines as in Kroustas can be investigated with fewer tomographies and lower DTM resolutions due to their less complex, well-defined appearance (cauldron shape). Usually, they are characterized by distinctive vertical slopes and thick sediment fills that exceed those of solution dolines, which both facilitate the acquisition of data in the field. The sole application of ERT is not recommendable, though, as it has an insufficient penetration depth. SRT is the first choice if the course of the bedrock is to be identified.

It is also important to note that proper fusion of surface and subsurface geodata must consider the diverse characteristics of primary single-method outcomes. In particular, this holds true for scale and resolution. While TLS captures highly detailed topographic data, initial geophysical results require a priori interpretation and additional calibration (e.g. by other geophysical methods; Schrott and Sass 2008). Apart from that, DTSM are not capable of revealing the actual subsurface in detail, but provide examples that help approximate the underground structure. Such digital models do not display certain stages of landscape development at certain points in time, since they only give an impression of specific geomorphologic structures.

Even though surface and subsurface geodata have prima facie different properties and must be processed differently, they prove to be highly valuable once applied in combination. As illustrated by the examples from Crete, ERT, SRT and TLS provide the best and most detailed information for precise geomorphologic and geoarchaeological studies. Their fusion allows accessing an absolutely new level of interpretation, as they consider the third dimension for the very first time and enable a synopsis of results from different methodologies, which have not been used in combination before. This is exactly where the concept of digital geoarchaeology reveals itself, as multiple geodata, geomorphologic know-how, computerized methods, archaeological findings and humanity expertise overlap and synergize within contextualization.

References

Angelakis A, Koutsoyiannis D (2003) Urban water engineering and management in ancient Greece. In: Stewart B, Howell T (eds) The encyclopedia of water science. Marcel Dekker, New York, pp 999–1008

Angelakis A, Savvakis Y, Charalampakis G (2007) Aqueducts during the Minoan era. In: Angelakis A, Koutsoyiannis D (eds) Insights into water management: lessons from water and wastewater technologies in ancient civilisations. International Water Association Publishing, London, pp 95–102

Antoniou G, Xarchakou R, Angelakis AN (2006) Water cistern systems in Greece from Minoan to Hellenistic period. In: International Water Association (ed) Proceedings of 1st IWA international symposium on water and wastewater technologies in ancient civilizations, 28–30 October 2006, International Water Association Publishing, Iraklio, pp 457–462

Armesto J, Ordóñez C, Alejano L, Arias P (2009) Terrestrial laser scanning used to determine the geometry of granite boulder for stability analysis purposes. Geomorphology 106:271–277

Bartels G (1991) Karstmorphologische Untersuchungen auf Kreta. Erdkunde 45:27–37

Bechtel T, Bosch F, Gurk M (2007) Geophysical methods. In: Goldscheider N, Drew D (eds) Methods in karst hydrogeology. Taylor and Francis, London, pp 171–200

Deceuster J, Delgranche J, Kaufmann O (2006) 2D crossborehole resistivity tomographies below foundations as a tool to design proper remedial actions in covered karst. J Appl Geophys 60(1):68–86

Demargne P (1901) Les ruines de Goulas ou l'ancienne ville de Latô en Crète. Bull Corresp Hell 25:282–307

Egli BR (1993) Ökologie der Dolinen im Gebirge Kretas (Griechenland). Ph.-D thesis, University of Zürich

Fabre G, Maire R (1983) Néotectonique et morphogenèse insulaire en Grèce: Massif du Mont Ida (Crète). Méditerranée, 2, troisième série, tome 48, pp 39–49

Ford D, Williams P (2007) Karst hydrogeology and geomorphology. Wiley, Chichester

Gaffney C (2008) Detecting trends in the prediction of the buried past: a review of geophysical techniques in archaeology. Archaeometry 50:313–336

Hämmerle M, Höfle B (2018) Introduction to LiDAR in geoarchaeology from a technological perspective. In: Siart C, Forbriger M, Bubenzer O (eds) Digital geoarchaeology – new techniques for interdisciplinary human-environmental research. Springer, Heidelberg, pp 167–182

Hecht S (2003) Differentiation of loose sediments with seismic refraction methods - potentials and limitations derived from case studies. Z Geomorphol Suppl 132:89–102

Hecht S (2009) Viewing the subsurface in 3D: Sediment tomography for (Geo-)Archaeological prospection in Palpa, Southern Peru. In: Reindel M, Wagner GA (eds) New technologies for archaeology. Springer, Heidelberg, pp 87–102

Hoffmeister D (2018) 3D laser scanning for geoarchaeological documentation and analysis. In: Siart C, Forbriger M, Bubenzer O (eds) Digital geoarchaeology – new techniques for interdisciplinary human-environmental research. Springer, Heidelberg, pp 183–200

Höfle B, Rutzinger M (2011) Topographic airborne LiDAR in geomorphology: a technological perspective. Z Geomorphol 55(2):1–29

Hoover R (2003) Geophysical choices for karst investigations. In: Beck B (ed) Sinkholes and the engineering and environmental impacts of karst. American Society of Civil Engineers, Reston, pp 529–538

Koutsoyiannis D, Zarkadoulas N, Angelakis A, Tchobanoglous G (2008) Urban water management in ancient Greece: legacies and lessons. J Water Res Plan Manage-ASCE 134(1):45–54

Panagiotopoulos D (2007) Minoische Villen in den Wolken Kretas. Antike Welt 38:17–24

Ravbar N, Kovacic G (2010) Characterisation of karst areas using multiple geo-science techniques, a case study from SW Slovenia. Acta Carsol 39(1):51–60

Sarris A, Kalayci T, Moffat I, Manataki M (2018) An introduction to geophysical and geochemical methods in digital geoarchaeology. In: Siart C, Forbriger M, Bubenzer O (eds) Digital geoarchaeology – new techniques for interdisciplinary human-environmental research. Springer, Heidelberg, pp 215–236

Schrott L, Sass O (2008) Application of field geophysics in geomorphology – advances and limitations exemplified by case studies. Geomorphology 93:55–73

Sheehan J, Doll W, Watson D, Mandell W (2005) Application of seismic refraction tomography to karst cavities. In: Kuniansky E (ed) US Geological Survey Scientific Investigations Report 2005–5160. USGS, Reston, pp 29–38

Showleh T (2007) Water management in the bronze age: Greece and Anatolia. In: Angelakis A, Koutsoyiannis D (eds) Insights into water management: lessons from water and wastewater technologies in ancient civilisations. International Water Association Publishing, London, pp 77–84

Siart C (2010) Geomorphologisch-geoarchäologische Untersuchungen im Umfeld der minoischen Siedlung von Zominthos. Ein Beitrag zur Erforschung der holozänen Landschaftsgeschichte Zentralkretas. Heidelberger Geographische Arbeiten 130, Heidelberg

Siart C, Eitel B, Panagiotopoulos D (2008) Investigation of past archaeological landscapes using remote sensing and GIS: a multi-method case study from Mount Ida, Crete. J Archaeol Sci 35:2918–2926

Siart C, Bubenzer O, Eitel B (2009) Combining digital elevation data (SRTM/ASTER), high resolution satellite imagery (Quickbird) and GIS for geomorphologic mapping: a multi-component case study on Mediterranean karst in Central Crete. Geomorphology 112:106–121

Siart C, Hecht S, Brilmayer Bakti B, Holzhauer I (2011) Analysis and 3D visualisation of Mediterranean subsurface karst features based on tomographic mapping (Zominthos, Central Crete). Z Geomorph NF Suppl 55 (3):315–335

Siart C, Ghilardi M, Forbriger M, Theodorakopoulou K (2012) Terrestrial laser scanning and electrical resistivity tomography as combined tools for the geoarchaeological study of the Kritsa-Latô dolines (Mirabello, Crete, Greece). Géomorphol Relief Processus Environ 2012(1):59–74

Siart C, Forbriger M, Nowaczinski E, Hecht S, Höfle B (2013) Fusion of multi-resolution surface (terrestrial laser scanning) and subsurface geodata (ERT, SRT) for karst landform investigation and geomorphometric quantification. Earth Surf Proc 38 (10):1153–1147

Siart C, Forbriger M, Bubenzer O (2018) Digital geoarchaeology - bridging the gap between archaeology, geoscience and computer science. In: Siart C, Forbriger M, Bubenzer O (eds) Digital geoarchaeology – new techniques for interdisciplinary human-environmental research. Springer, Heidelberg, pp 1–10

Stepisnik U, Mihevc A (2008) Investigation of structure of various karst formations in limestone and dolomite bedrock with application of electrical resistivity imaging. Acta Carsol 37(1):133–140

Sustersic F (2006) A power function model for the basic geometry of solution dolines: considerations from the classical karst of south-central Slovenia. Earth Surf Proc 31:293–302

Terzic J, Sumanovac F, Buljan R (2007) An assessment of hydrogeological parameters on the karstic island of Dugi Otok, Croatia. J Hydrol 343:29–42

Van Schoor M (2002) Detection of sinkholes using 2D electrical resistivity imaging. J Appl Geophys 50:393–399

Waltham T, Bell F, Culshaw M (2005) Sinkholes and subsidence. Karst and Cavernous rocks in engineering and construction. Springer, Heidelberg

Index

© Springer International Publishing AG 2018
C. Siart et al. (eds.), *Digital Geoarchaeology*, Natural Science in Archaeology,
DOI 10.1007/978-3-319-25316-9

The manufacturer's authorised representative in the EU is Springer Nature Customer Service Centre GmbH, Europaplatz 3, 69115 Heidelberg, Germany. If you have any concerns regarding our products, please contact ProductSafety@springernature.com

Printed and bound by CPI Group (UK) Ltd, Croydon, CR0 4YY
25/09/2026
02222859-0020